Island of Reil (Insula) in the Human Brain

Mehmet Turgut • Canan Yurttaş
R. Shane Tubbs
Editors

Island of Reil (Insula) in the Human Brain

Anatomical, Functional, Clinical and Surgical Aspects

Editors
Mehmet Turgut
Adnan Menderes University
School of Medicine
Aydın
Turkey

Canan Yurttaş
Ege University
Department of Anatomy
Izmir
Turkey

R. Shane Tubbs
Seattle Science Foundation
Seattle
Washington
USA

ISBN 978-3-319-75467-3   ISBN 978-3-319-75468-0   (eBook)
https://doi.org/10.1007/978-3-319-75468-0

Library of Congress Control Number: 2018939859

© Springer International Publishing AG, part of Springer Nature 2018
This work is subject to copyright. All rights are reserved by the Publisher, whether the whole or part of the material is concerned, specifically the rights of translation, reprinting, reuse of illustrations, recitation, broadcasting, reproduction on microfilms or in any other physical way, and transmission or information storage and retrieval, electronic adaptation, computer software, or by similar or dissimilar methodology now known or hereafter developed.
The use of general descriptive names, registered names, trademarks, service marks, etc. in this publication does not imply, even in the absence of a specific statement, that such names are exempt from the relevant protective laws and regulations and therefore free for general use.
The publisher, the authors and the editors are safe to assume that the advice and information in this book are believed to be true and accurate at the date of publication. Neither the publisher nor the authors or the editors give a warranty, express or implied, with respect to the material contained herein or for any errors or omissions that may have been made. The publisher remains neutral with regard to jurisdictional claims in published maps and institutional affiliations.

Printed on acid-free paper

This Springer imprint is published by the registered company Springer International Publishing AG part of Springer Nature
The registered company address is: Gewerbestrasse 11, 6330 Cham, Switzerland

To my niece, Kassidy Raines who is resilient

R. Shane Tubbs

Preface

The anatomist, physician and surgeon Johann Christian Reil (1759–1813) (Fig. 1) was born in Rhaude (Ostfriesland) in East Frisia, Germany, and studied the nervous system in detail. He began his medical studies at Göttingen and finished at Halle. Reil is most remembered eponymously via the isle or island of Reil. However, his name is also associated with many other anatomical structures of the human body (e.g. Reil's ansa, Reil's covered band, Reil's sulcus, Reil's triangle). He was one of the first to use alcohol for tissue preservation and perfected this method by adding alkali so that tissues such as the brain would be hardened. Such techniques allowed Reil to use blunt dissection and trace cerebral pathways to detail the insula in his 1796 publication *Exercitationum anatomicarum fasciculus primus*. He also identified and studied the cerebellum, lenticular nucleus, locus coeruleus, arcuate fasciculus, corona radiata, internal and external capsules, claustrum and the lemniscal system. Reil followed the teachings of Unzer (1727–1799) who believed that the reflexes were due to traceable anatomical tracts in the brain. Reil was head of the Vitalist school and coined the term 'psychiatry'. Reil and other Vitalists championed the establishment of hospitals for the humane treatment of the insane [1]. Reil also focused on the problematic anatomical bridges between the cerebrospinal and ganglionic systems provided by the rami with which his predecessors Bichat and Johnstone had struggled. Reil's solution was comparing the rami to semi-conductors he termed *Apparat der Halbleitung*, representing in modern terms an electrical 'make-and-break' key between the two systems [2]. He believed physiologically that these

Fig. 1 Drawing of Johann Christian Reil

structures prevented conduction of improper signals, but in a pathological state it became a conductor of the signals, connecting the organic and animal lives.

Hidden in the depths of the Sylvian fissure, the insula, as studied by Reil, has in general received less attention than its cerebral lobe counterparts. This 'silent' region of the brain is associated with a host of diseases and functions but only recently has many of these been elucidated. Therefore, a tome dedicated to this part of the brain is both timely and necessary. It is the hope of the editors that this text will serve as a nidus to propel and expedite further research on this underrepresented part of the brain.

Aydın, Turkey Mehmet Turgut
Izmir, Turkey Canan Yurttaş
Seattle, Washington, USA R. Shane Tubbs

References

1. Isler H. Neurology and the neurological sciences in the German-speaking countries. In: Finger S, Boller F, Tyler KL, editors. Handbook of clinical neurology, History of neurology, vol. 95. Philadelphia: Elsevier; 2010.
2. Clarke E, Jacyna LS. Nineteenth-century origins of neuroscientific concepts. Berkeley: University of California Press; 1987.

Contents

Part I Anatomy of the Human Insula

1. **The Insular Cortex: Histological and Embryological Evaluation** 3
   Yigit Uyanikgil, Turker Cavusoglu, Servet Celik, Kubilay Dogan Kilic, and Mehmet Turgut

2. **Gross Anatomy of the Human Insula** 15
   Igor Lima Maldonado, Ilyess Zemmoura, and Christophe Destrieux

3. **Surgical Anatomy of the Insula** 23
   Christophe Destrieux, Igor Lima Maldonado, Louis-Marie Terrier, and Ilyess Zemmoura

4. **Anatomy of the Insular Arteries** 39
   Servet Celik, Okan Bilge, Canan Yurttaş, and Mehmet Turgut

5. **Anatomy of the Insular Veins** 55
   Servet Celik, Okan Bilge, Mehmet Turgut, and Canan Yurttaş

6. **Middle Longitudinal Fasciculus in the Human Brain from Fiber Dissection** 71
   Igor Lima Maldonado, Ilyess Zemmoura, and Christophe Destrieux

7. **Structural Connectivity of the Insula** 77
   Jimmy Ghaziri and Dang Khoa Nguyen

8. **Insular Pharmacology** 85
   Hasan Emre Aydın and İsmail Kaya

9. **Neuroimaging Techniques for Investigation of the Insula** 91
   Ersen Ertekin, Özüm Tunçyürek, Mehmet Turgut, and Yelda Özsunar

10. **Measurements of the Insula Volume Using MRI** 101
    Niyazi Acer and Mehmet Turgut

11. **Neurocognitive Mechanisms of Social Anxiety and Interoception** 113
    Yuri Terasawa and Satoshi Umeda

Part II Functions of the Human Insula

12 Participation of the Insula in Language 123
Alfredo Ardila

13 Lateralization of the Insular Cortex 129
Michael J. Montalbano and R. Shane Tubbs

14 Gustatory Areas Within the Insular Cortex 133
Richard J. Stevenson, Heather M. Francis, and
Cameron J. Ragg

15 Role of the Insula in Human Cognition and Motivation 147
Oreste de Divitiis, Teresa Somma, D'Urso Giordano,
Mehmet Turgut, and Paolo Cappabianca

16 Role of the Insula in Visual and Auditory Perception 151
Matthew Protas

17 The Anterior Insula and Its Relationship to Autism 157
Seong-Jin Moon, Lara Tkachenko, Erick Garcia-Gorbea,
R. Shane Tubbs, and Marc D. Moisi

18 Role of the Insular Cortex in Emotional Awareness 161
Fareed Jumah

19 Alterations of Reil's Insula in Alzheimer's Disease 169
Paul Choi, Emily Simonds, Marc Vetter, Charlotte Wilson,
and R. Shane Tubbs

20 Contributions of the Insula to Speech Production 175
Christoph J. Griessenauer and Raghav Gupta

**21 Processing Internal and External Stimuli in the Insula:
A Very Rough Simplification** 179
Alfonso Barrós-Loscertales

**22 The Role of the Insula in the Non-motor Symptoms
of Parkinson's Disease** 191
Braden Gardner

Part III Clinical Disorders Related with Insula

23 Insular Cortex Epilepsy 197
Manish Jaiswal

24 Insular Ischemic Stroke 203
Bing Yu Chen, Olivier Boucher, Christian Dugas, Dang Khoa
Nguyen, and Laura Gioia

**25 Attention, Salience, and Self-Awareness:
The Role of Insula in Meditation** 213
Jordi Manuello, Andrea Nani, and Franco Cauda

26 **Neuropsychological Deficits Due to Insular Damage**........ 223
    Olivier Boucher, Daphné Citherlet,
    Benjamin Hébert-Seropian, and Dang Khoa Nguyen

27 **The Role of the Insula in Schizophrenia**................... 239
    Cameron Schmidt

Part IV Surgery of the Insular Cortex

28 **Surgery of Insular Diffuse Low-Grade Gliomas**............ 255
    Karine Michaud and Hugues Duffau

29 **Surgical Strategy for Insular Cavernomas**................ 263
    Mehmet Turgut, Paulo Roberto Lacerda Leal, and
    Evelyne Emery

30 **Role of the Insula in Temporal Lobe Epilepsy
    Surgery Failure**....................................... 271
    Vamsi Krishna Yerramneni, Alain Bouthillier, and
    Dang Khoa Nguyen

31 **Neuropsychology in Insular Lesions Prior-During
    and After Brain Surgery**.............................. 281
    Barbara Tomasino, Dario Marin, Tamara Ius, and Miran Skrap

Index.. 293

Part I
Anatomy of the Human Insula

The Insular Cortex: Histological and Embryological Evaluation

Yigit Uyanikgil, Turker Cavusoglu, Servet Celik, Kubilay Dogan Kilic, and Mehmet Turgut

Abbreviations

ACC   Anterior cingulate cortex
CB   Calbindin
CR   Calretinin
FIC   Frontoinsular cortex
PV   Parvalbumin
VEN   von Economo neurons

1.1 Definition and Etymology

The insula was first identified by Johann Christian Reil, but the rising of interest on this subject started at the beginning of the twentieth century with Brodmann and von Economo [1].

The insula (Latin island) is also called the Island of Reil, the insular cortex, or the insular lobe. This part of the brain is the triangular neocortex area just below the lateral (Sylvian) fissure. The insular cortex is not visible on an exterior view of the brain, as it is fully covered laterally by opercula of the parietal, frontal, and temporal lobes [2].

The insular cortex is the "forgotten" or understudied lobe of the mammalian brain. It is located in the lateral sulci in each cerebral hemisphere. This part of the cerebrum is part of the cerebral cortex which folds deep within the lateral. It cannot be seen from the surface because of the three opercula (parietal, temporal, frontal) which cover it, and is positioned between the piriform, orbital, motor, sensory and auditory cortices of higher order [3].

Over the lateral surface of the insula, the frontal, temporal and parietal opercula are found (Fig. 1.1) [4].

It is thought that the insular cortex is involved in various functions related to the regulation of the homeostasis of the body in connection with consciousness and emotions. These functions are self-awareness, motor control, cognitive functioning, interpersonal experience, and perception. Psychopathological processes take place in relation to these [5].

Y. Uyanikgil, Ph.D. · T. Cavusoglu, M.D.
K. D. Kilic, M.Sc. (✉)
Department of Histology and Embryology,
Ege University School of Medicine, Izmir, Turkey
e-mail: yigit.uyanikgil@ege.edu.tr;
kubilay.dogan.kilic@ege.edu.tr

S. Celik, M.D.
Department of Anatomy, Ege University School of Medicine, Izmir, Turkey
e-mail: servet.celik@ege.edu.tr

M. Turgut, M.D., Ph.D.
Department of Neurosurgery, Adnan Menderes University School of Medicine, Aydın, Turkey

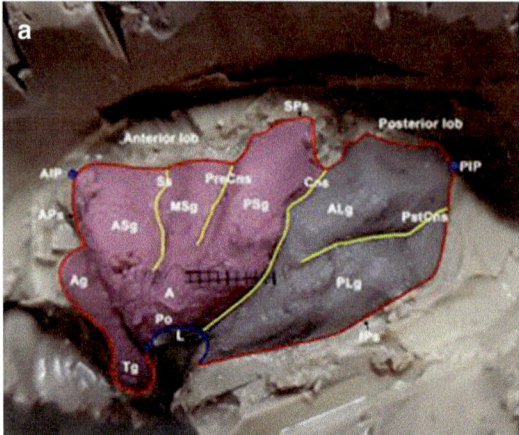
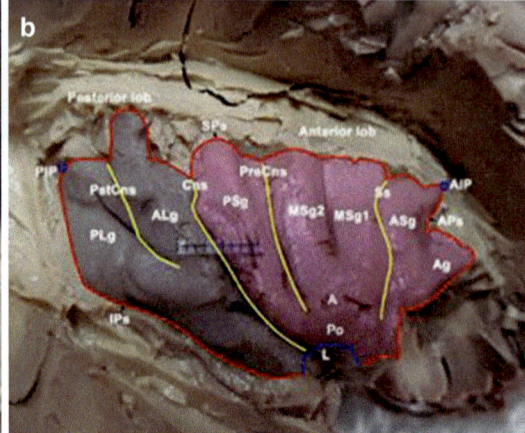

Fig. 1.1 (a) Left insula. (b) Right insula. *AIP* anterior insular point, *PIP* posterior insular point, *Cns* central sulcus, *PreCns* precentral sulcus, *PstCns* postcentral sulcus, *Ss* short insular sulcus, *SLs* superior limiting sulcus, *ILs* inferior limiting sulcus, *ALs* anterior limiting sulcus, *ASg* anterior short insular gyrus, *MSg* middle short insular gyrus, *MSg1* middle short insular gyrus 1, *MSg2* middle short insular gyrus 2, *PSg* posterior short insular gyrus, *ALg* anterior long insular gyrus, *PLg* posterior long insular gyrus, *Ag* accessory gyrus, *Tg* transverse gyrus, *L* limen of the insula, *A* apex of the insula, *Po* pole of the insula

1.2 Embryology of the Insula

Embryologically, the insular cortex differs from the telencephalon. Deep limbic structures may form a group in the limbic lobe [6].

The primordial insula is initially located on the free lateral surface of the cerebral hemisphere, but during further development, it lags behind and becomes gradually overgrown by adjacent regions of the hemispheres [7].

During development, the insular cortex develops from an allocortex to an isocortex. The first developing cortical structure during fetal brain development is the insula. Inferior cortical regions form the insular limen; in this region, brain cortex development begins at the sixth week of gestation. From the anterior prosencephalon, the telencephalic vesicles develop [8]. The foramen of Monro is formed at the junction between the diencephalon and telencephalon. The latter is also formed from the alar and basal plates [9].

The neocortex of the telencephalon develops disproportionately and causes the cerebral hemisphere to move in the posteroinferior and outward direction. The temporal lobe is the tipping point of this expansion. This structure's central window opens into the insula. Residuals of the paleocortex and archicortex lie within nondeveloped paralimbic and limbic structures. The windows themselves give Sylvian fissure later [10].

The posterior telencephalon expands between the 14th and 16th week of gestation. Between the third and fourth month, the appearance of the Sylvian fissure accompanies this occurrence. Between week 16 and 17, the posteroinferior periinsular sulcus, which originates from the earlier linear furrow surface of the lateral hemispheral region at the 13th week, is shaped. Periinsular sulci and the Sylvian fossa appear simultaneously; together they form a margined indentation at week 18. The rotation and compaction of neural tissues undergo the insula beneath the Sylvian fissure; these structures gradually cover it with temporoparietal cortex and then the frontal lobes when further stages of cortical proceed [11]. The ganglionic cumulus of the paleocortical mantle lies deep within the insula before week 16. This cell cluster located in the lateral thalamus undergoes lateral division and gives rise to the caudate nucleus and putamen latter. The medial part of this cell cluster forms the amygdaloid body. Septal nuclei are formed from the anterior ventral aspect of the paleocortex. The hippocampus and associated fornix are formed by the archicortex. Archicortex is located on the medial-dorsal surface of the telencephalic vesicle. The hippocampus is displaced toward the temporal lobe and leaves behind its fiber tract, the fornix, in the medial wall of the lateral ventricle while the neocortex is expanding and the hemisphere is rotating. The fornix is part of the paleocortex. It forms the limbic system with the mammillary body. Mesocortex is a transition region between the archicortex and paleocortex. The lateral wall of the developing ventricle can be observed. The cortical folding process preserves the connections that develop during the first stage of neural tube development. The latest part of cortical folding is buried in the insula to maintain connections. The continued presence of the connecting fibers explains the reason that the insula is richly connected to nearby and remote centers within the fully developed brain (Table 1.1) [12].

Cortical folding separates insula into two parts; these parts are not equal. We classify them depending on their size. Larger parts are located at the front and are called the anterior insula, while the smaller part is located at the back and is called the posterior insula. The insular central sulcus extends from the superior periinsular sulcus to the limen insula. Anteriorly, there are three short gyri, and posteriorly, there are two long gyri and accessory gyri. Accessory gyri have fiber connections to the orbitofrontal and lateral olfactory areas in its anteroinferior portion.

Three areas have been identified in the insular cortex. These cellular layers are defined as:

1. Granular area (Posterior)
2. Anterior agranular area (Trilaminar)
3. Nongranulated area (Pericentral)

Around the 15th–16th week, the cortical plate becomes obvious at the insular region. Cell apices go through the ventricles, a layer of pseudostratified neuroepithelium forms the telencephalic vesicle except for the external vascular connective tissue, the basal membranes of these cells are adjacent to each other. These neuroepithelial cells migrate toward the ventricles to form the proliferative zone during cell division. Cell fate, differentiation, maturation, and developing into glia and neuroblast occur here. The intermediate and the marginal zones enlarge in proportion with the postmitotic neuroblast population. The cortical plate forms from the marginal zone. In that zone, we can find mature neurons and migrating neuroblasts. Efferent and afferent projections of the insula reach adjacent and distant cortical regions. The posterodorsal part of the insula has a connection with the primary and secondary somatosensory cortices, the supplementary motor area, the temporal cortex, and the retro insular area. It specializes in motor and auditory movements. The anteroventral part has a connection with the cingulate, periamygdaloid, and entorhinal cortices and specializes in viscero-autonomic olfactory and gustatory functions. Anatomical limits between cytoarchitectural divisions of the insula were not defined clearly. In the chart below, we summarized the developmental process of the insula (Table 1.1).

In a 50-day-old human embryo, the meningeal compartment and essential vascular components could be recognized [13]. Before the beginning of the intracerebral microvascularization of the brain, the meningeal compartment establishes [14].

Table 1.1 Milestones in the development of the insula according to the weeks

Week	Developmental event
13	The posteroinferior periinsular sulcus originates from the earlier linear furrow surface of the lateral hemispheric region
14	The posterior telencephalon expands posterior, inferior, and lateral ways
15	The cortical plate becomes obvious at the insular region
16	The posteroinferior periinsular sulcus is shaped
18	The periinsular sulci and the Sylvian fossa appear simultaneously, together they form a margined indentation

1.3 Cell Movements and Radial Migration Pathway

In the early stages of development, cells move in the neuroepithelium before the non-mature neurons separate from the ventricular zone. Proliferative precursor cells pass from one rhombomere to another; cells are limited to a particular rhombomere when they become postmitotic in the hindbrain [15–17].

Studies on developing cortical regions described radial pathway of development [18, 19]. This pathway follows the radial disposition of the germinative zones of the neural tube. The structure of these zones are pseudostratified epithelium.

The alignment of postmitotic neurons with a radial glial fiber system during the cortical formation period led to the general hypothesis that it provides a scaffold for the directed migrations at the radial cells [20–23]. In vitro and in vivo supports have come from scientific researches [24–27]. Almost all of the billions of neuronal precursors in the human cortex migrate along glial fibers.

During development, glial cells take into account the properties and functions of developing neurons. In the early period, glial cells elaborate the processes that run along the wall of the developing nerve tube [19, 28, 29]. These cells have a role in the primary pathway for directed migrations (Fig. 1.2) [21].

1.4 Histology of the Insula

The insula is divided into three subareas depending on the neuron granularity in the cortical layer IV. The insula's cytoarchitecture composition is not the same in all layers; it changes from three to six [30].

According to discrimination, the ventral and anterior part of the insula is part of the agranular neocortex, the posterior and dorsal insula show distinct granularity, and the middle insula is disorganized. It should also be noted that almost all of the ventral prefrontal area of humans exhibit a subdivision of the agranular insular cortex. This lower part of the insular cortex is located immediately adjacent to the orbital cortex and is therefore referred to as the frontoinsular cortex [31].

When neuron clusters are found in the frontoinsular cortices, termed von Economo neurons (VENs). It is thought that VENs in the frontoinsular cortex together with the agranular insular cortex of the anterior and ventral insula have a role in social awareness. Along with that, granular and dysgranular insular regions are important for receiving input about the body's inner homeostatic state and somatosensory input from the periphery. Anterior insula has a key role in processing emotional, motivational, sensory, and cognitive stimuli and gustation.

Three concentrically separated regions are identified. They are rostroventral agranular zone, caudodorsal granular zone, and dysgranular (Ig) zone [32, 33]. Agranular and granular terms refer to the absence or presence of internal granular layer IV. Granular cells are rare in layer IV and do not exhibit complex laminar differentiation, called dysgranular, in the intermediate zone. A special feature of the anteriorial cortex is that the fifth layer contains a large number of large spindle cells in addition to the pyramidal neurons. In the anterior cingulate cortex, similar cells can be observed [34]. Studies showed these cells can be found in humans and great apes but have not been observed in other mammalians since 2007 [35, 36]. Hof and van der Gucht reported them in whales, and then Hakeem et al. reported them in elephants [37, 38]. VENs are hypothesized to have a part in the networks which undergo complex social cognition, self-awareness, and decision-making [35, 39, 40]. VENs often present a corkscrew-like morphology, and a vertical alignment resembling migrating neurons, which was observed in the infant brain [39, 41].

The granular insular cortex is located in visceral area. In humans, it is classified as the neocortex and is distinguished from adjacent allocortex by the presence of granular layers. These layers are the external granular layer (II) and internal granular layer (IV) and by differentiation of the external pyramidal layer (III) into sublayers. It is located in the posterior part of the insula [42, 43].

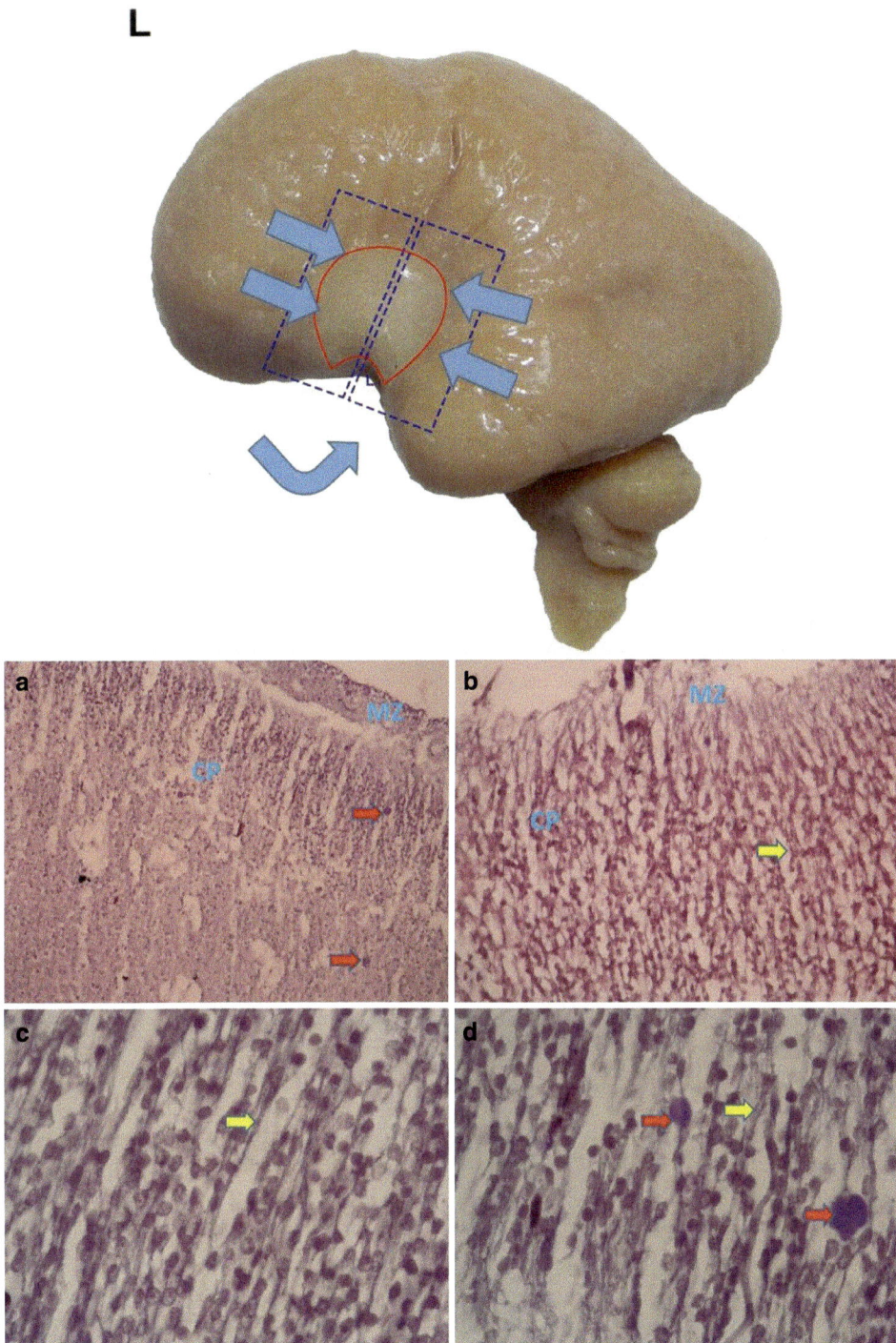

Fig. 1.2 In the development of the fetal brain, cell migration happens from inside to outside, and the structure folds laterally. In H&E stained slides, cell migrations through radial glial cell process can be observed during the further stages of development. (**a**) Migrating neurons in the fetal brain (×10 magnification), (**b**) radial glial cells of developing brain (×10 magnification), (**c**) migrating neurons in the fetal brain (×40 magnification), (**d**) radial glial cells of developing brain (×40 magnification) (*L* left, *MZ* marginal zone, *CP* cortical plate, red arrow: migrating neuron, yellow arrow: radial glial cell)

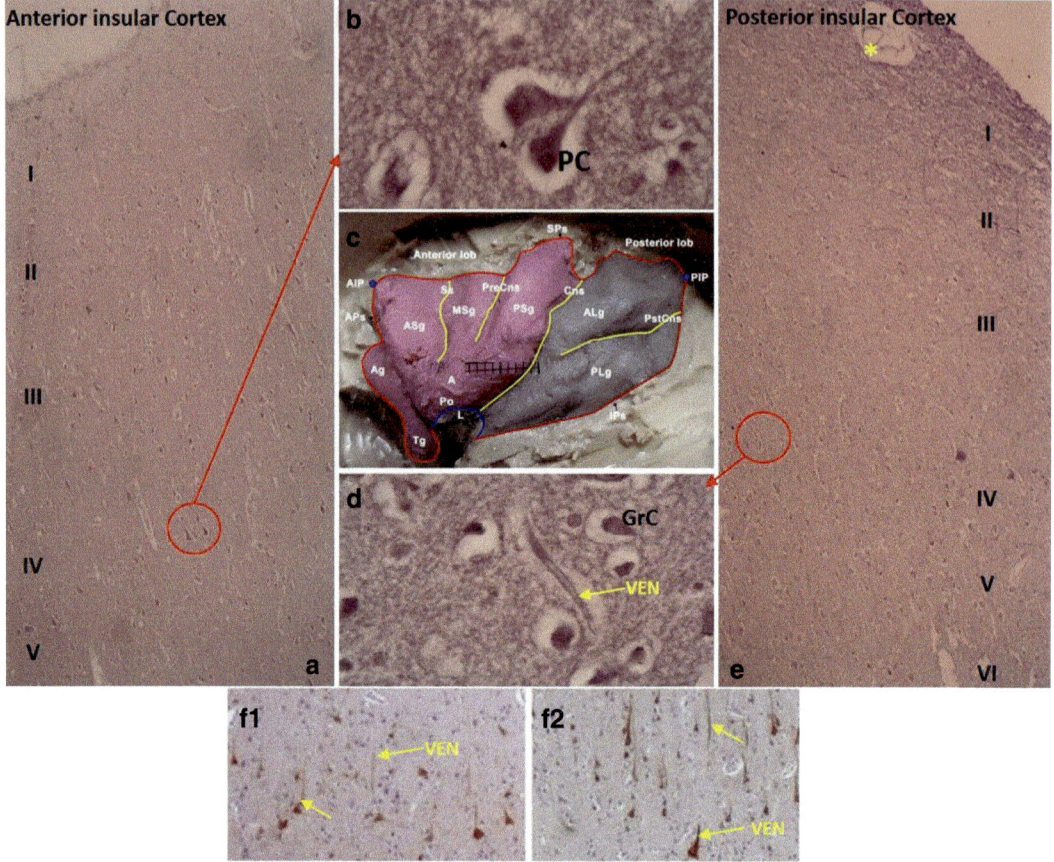

Fig. 1.3 Comparison of the anterior and posterior insular cortex. (**a**) Anterior insular cortex (×10 magnification), (**b**) pyramidal neurons located in the second and fourth layer of the anterior insular cortex. Granular cells located in the second and fourth layer of the posterior insular cortex (×40 magnification), (**c**) lobes of the insula, (**d**) von Economo neurons between the third and fourth layers, (**e**) posterior insular cortex (×10 magnification) (luxol fast staining). F1, F2. SMI32 immunohistochemical staining of von Economo neurons (×20 magnification). *PC* pyramidal cell, *GrC* granular cell, yellow asterisk perforating artery, *VEN* von Economo neuron. *F1–F2 are modified from Chance SA, Sawyer EK, Clover LM, Wicinski B, Hof PR. Crow TJ (2013) hemispheric asymmetry in the fusiform gyrus distinguishes Homo sapiens from chimpanzees. Brain Struct Funct 218(6):1391–405*

The agranular insula is a part of the cerebral cortex. In humans, it is classified as the allocortex and is distinguished from the adjacent neocortex by absence of the external granular layer (II) and internal granular layer (IV). Located in the anterior part of the insula, the posterior portion of the orbital gyri, and the medial part of the temporal lobe (Fig. 1.3) [42, 43].

The part of the insula in the intermediate region is broadly related to all other parts of the insula, motor, somatosensory, and pain processing, while the fragment at the back of the insula is a representation of the body that constantly learns about autonomic functions. In newborns, an asymmetry develop in the right hemisphere with VENs predominating the right hemisphere, and with similar VEN counting in the left hemisphere, right hemisphere develops in the fronto-insular cortex (FIC). In contrast, the number of VENs in the right and left anterior cingulate cortex (ACC) are comparable throughout adulthood. Specifically, VENs are the first in prenatal development at 35 weeks of gestation, whereas they are rare in ACC, with a high number in FIC (37,700 FIC in right and 25,620 in left) [39]. The postnatal increase in the number of VENs reached eight times in the right FIC (about 322,910) and four times in the left SA (about

74,000), compared with numbers at birth, which reached a peak at eight months. This is opposed to having no previous asymmetry in the ACC (VEN numbers are around 200,000 in both hemispheres). When the age of 4 is reached, the number of VENs on the right is reduced by half, while the number of VENs on the ACC and left FIC are comparable to eight months of age [39]. In contrast to the large number of VENs reported in the ACC of people with VENs, only VENs are reported to have a complete term fatal chimpanzee (approximately 5.3% of the total neuronal cells of the Vb layer), which is similar to the percentage at eight months postnatal life in humans (total neuron, which represents 4.3% of the total number) [44]. Thus, 1–2% of the total neurons in the V layer are reduced in order to obtain adult percentages [39]. Studies have shown that VENs strongly express non-phosphorylated neurofilament protein epitopes similar to large pyramidal cells, although they are not labeled with calcium-binding proteins parvalbumin (PV), calbindin (CB), or calretinin (CR) [45–47]. Colocalization of PV, CB, and CR with the main inhibitor neurotransmitter g-aminobutyric acid does not consider VENs as an inhibitor neuron class [45, 48, 49]. There are studies showing that VENs are involved in schizophrenia (SZ)-1 [39, 50]. Quantitative analyses of the phylogenetic variation in the biochemical profile of VENs are immunoreactive for ATF3, IL4Ra, and NMB in higher frequencies than other neuronal classes [51].

The ontogeny of VENs reveals that this neuronal line shows up in prenatal development in ACC and FIC, increases in late neonates, and reaches adult numbers at four years of age [39]. It is unclear whether VENs evolved after birth or migrated to specific cortical regions during postnatal development.

Layer V contains numerous ordinary pyramids and aberrant pyramidal elements of two types, which are triangular, bifid, or trifid cells and fusiform cells [19].

The triangular, bifid, or trifid cells have two, three, or more ascending dendritic branches. These branches extend into the first (plexiform) layer of the cortex, and a single-basal dendrite forms from the soma into a tuft of diverging dendrites. The axon arises from the lower end of the basal dendritic stem.

The fusiform cells are characterized by a radially oriented dendritic shaft, in which the soma forms a simple, spindle-shaped thickening. The apical dendrite ascends to the plexiform layer. The basal dendritic shaft is generally devoid of branches and forms, at a varying distance from the soma, a tuft of descending dendrites. The soma and the initial portion of the apical dendrite give rise to delicate horizontal dendrites. The axon rises from the end of the basal dendritic stem and descends to the white matter. Before leaving the cortex, it gives rise to collateral branches.

Typically, the agranular frontoinsular area forms a transitional zone between the frontal operculum and the insula proper, and contains numerous slender, elongated, fusiform cells. These elements are not only confined to the insula but also occur in the ACC. They are abundant in the layer V of the cortical areas, but scattered elements of the same type can also be found in layer III [52].

The insula is an isocortex (neocortex) comprised of six distinct layers. Brodmann divided the insula into two regions, separated by the central sulcus: an anterior agranular region (containing pyramidal neurons in layers II and IV) and a posterior granular region (containing granular cells in layers II and IV). Modern belief depends on the development of the concentric model in which the agranular region is located ventral-anteriorly and the granular region is located dorsal-posteriorly, separated by a large band of the dysgranular cortex [53].

VENs are much larger than nearby granular and pyramidal cells and have symmetric long narrow apical and basal dendritic arbors [54]. On the basis of their size, it is suggested that VENs project to other areas of the brain as a fast relay transmission system. Postmortem studies in humans indicate that VENs primarily express transcription factors consistent with involvement in interoceptive functions (e.g., pain, immune, visceral) and subcortical projection targets (e.g., striatum, superior colliculus, basal pons, spinal cord).

1.5 Histology of the Insular Vascularization

Like all arteries, these ones comprise three different tunicas, from inside to out; they are intima, media, and adventitia. In tunica intima we can see endothelium and subendothelial connective tissue. Between intima and media, an internal elastic lamina can be seen as a separator. This lamina provides mechanical strength. In tunica media there are packed layers of smooth muscle cells. Collagen and elastin fibers surround these cells [55]. Intracranial arteries do not have external elastic lamina between tunica media and tunica adventitia; which is different from the extracranial arteries. Tunica adventitia is often observed as quite thinner compared to the body. The wall-lumen ratio of the intracranial arteries is lower than the extracranial arteries. The intracranial extradural space arteries show similar structure to large arteries, while the cerebrospinal fluid surrounds the intradural arteries. In the brain parenchyma, they keep these occurrences. Enlarged perivascular spaces, termed the Virchow–Robin spaces, enable the presence of this process. Later astrocytes layer covers the intradural arteries (glia limitans) [55, 56].

Three distinct and interrelated components take place in the human brain. They are the extracerebral or meningeal compartment, the intracerebral dual extrinsic compartment, and the intrinsic microvascular compartment. Sequential embryonic development begins in the early embryo with the establishment of the extracerebral meningeal compartment that covers the surface of the growing brain followed by that of the intracerebral extrinsic microvascular compartment, from which the intrinsic microvascular compartment evolves subsequently [14, 55].

The extracerebral or meningeal compartment contains three essential constituents: dural, arachnoidal, and pial. The dural lamella contains the main venous sinuses, the arachnoidal lamella contains the brain main arteries and veins, and the pial lamella contains the pial capillary anastomotic plexus [14].

1.6 Anatomical Connections

The anterior part of the insula is divided into three or four short gyri with shallow sulci. The anterior insula takes a direct projection from the basal part of the ventral medial nucleus of the thalamus and receives a particularly large input from the central nucleus of the amygdala. In addition, the anterior insula itself comes out of the amygdala. The anterior insula, the temporal and occipital lobes, and the operative and orbitofrontal cortex are connected to the triangular and operative segments of the inferior frontal gyrus [57].

The posterior insula is predominantly directed toward the dorsal face of the lateral and central amygdaloid nuclei. The posterior insula is formed by long gyri. The posterior insula communicates with the secondary somatosensory cortex and receives input from the posterior part. This region receives highly specialized inputs to give homeostatic information such as pain, temperature, itchiness, local oxygen status, and sensual touch from the ventromedial nucleus of the thalamus (posterior segment) [58].

The sulcus of Reil is triangular. It separates the insula from the neighboring gyri of the operculum in the above, behind, and front [32].

1.7 Functions of the Insular Cortex

There are two main functional parts of the insula: anterior and posterior. The anterior insula is responsible for decoding information pertaining to its body representation and subjective emotional experiences, processing olfactory and gustatory inputs. The posterior insula receives input from the ventromedial nucleus of the thalamus and processes information pertaining to touch, temperature, and pain. All known functions can be summarized like: (1) salience [59], (2) conduct disorder [60], (3) somatosensory functions [61], (4) pain and temperature perception [62], (5) viscerosensation [63], (6) drug addiction [64], (7) olfaction [65], (8) visceromotor control [66], (9) somatomotor control [67], (10) motor plasticity [68], (11) cognitive control [69], (12) speech production [70], (13) bodily awareness [71], (14) self-recognition [72], (15) auditory functions [73], (16) social emotions [74], (17) homeostasis [72], (18) vestibular functions [75], (19) taste [76], (20) schizophrenia [77], and (21) individual emotions [78].

As regards the putative functional significance of the insular cortex as a whole, recent attempts at synthesizing the enormous amount of experimental neuroanatomical, physiological, clinical, and neuroimaging data available on the insula create a coherent picture of human insular functioning [72]. His concept may be summarized as follows:

1. The posterior, middle, and anterior sectors of the insula represent three different stages or levels of integration in a posteriorly-to-anteriorly directed processing stream.
2. Afferents from the solitary nucleus, from the phylogenetically new pathway from lamina I spinal neurons, converge upon the posterior insula, where they provide a primary interoceptive representation of the physiological condition of the body.
3. The mid-insula is to be considered as a polymodal integrative zone, where the interoceptive information from the posterior insula is represented and associated with inputs from multiple other sources. Prominent among these are higher-order sensory cortices, providing emotionally salient information from the external world, and the cingulate cortex and amygdala, providing homeostatic information related to the current motivational state.
4. The information processing for self-awareness is essential.

This chapter—which was written for elaborating morphological concepts of the insula and contributing to the various pathologies and surgical procedures to be performed in this area—is important for both clinical medical and basic medical researchers. Current findings in brain imaging and direct cortical stimulation make safer surgical procedures in the insular region possible.

References

1. Triarhou LC. The cytoarchitectonic map of Constantin von Economo and Georg N. Koskinas. In:Microstructural Parcellation of the Human Cerebral Cortex. Berlin Heidelberg: Springer; 2013. p. 33–53.
2. Vogt BA, Pandya DN. Cingulate cortex of the rhesus monkey: II. Cortical afferents. J Comp Neurol. 1987;262(2):271–89.
3. Türe U, Yaşargil DC, Al-Mefty O, Yaşargil MG. Topographic anatomy of the insular region. J Neurosurg. 1999;90(4):720–33.
4. Dronkers NF. A new brain region for coordinating speech articulation. Nature. 1996;384(6605):159.
5. Herbert BM, Pollatos O. The body in the mind: on the relationship between interoception and embodiment. Top Cogn Sci. 2012;4(4):692–704.
6. Hui KK, Liu J, Makris N, Gollub RL, Chen AJ, Moore C, Kennedy D, Rosen BR, Kwong KK. Acupuncture modulates the limbic system and subcortical gray structures of the human brain: evidence from fMRI studies in normal subjects. Hum Brain Mapp. 2000;9(1):13–25.
7. Toi A, Lister WS, Fong KW. How early are fetal cerebral sulci visible at prenatal ultrasound and what is the normal pattern of early fetal sulcal development? Ultrasound Obstet Gynecol. 2004;24(7):706–15.
8. Kalani MYS, Kalani MA, Gwinn R, Keogh B, Tse VC. Embryological development of the human insula and its implications for the spread and resection of insular gliomas. Neurosurg Focus. 2009;27(2):E2.
9. Benet A, Hervey-Jumper SL, Sánchez JJG, Lawton MT, Berger MS. Surgical assessment of the insula. Part 1: surgical anatomy and morphometric analysis of the transsylvian and transcortical approaches to the insula. J Neurosurg. 2016;124(2):469–81.
10. Villemure JG, Daniel RT. Peri-insular hemispherotomy in paediatric epilepsy. Childs Nerv Syst. 2006;22(8):967–81.
11. Chi JG, Dooling EC, Gilles FH. Gyral development of the human brain. Ann Neurol. 1977;1(1):86–93.
12. Mettler FA. Corticifugal fiber connections of the cortex of macaca mullatta. The frontal region. J Comp Neurol. 1935;61(3):509–42.
13. Streeter GL (1918) The developmental alterations in the vascular system of the brain of the human embryo. Carnegie Institution of Washington.
14. Marín-Padilla M. The human brain intracerebral microvascular system: development and structure. Front Neuroanat. 2012;2012:6.
15. Jessell TM, Lumsden A. Inductive signals and the assignment of cell fate in the spinal cord. In: Molecular and cellular approaches to. Neural Dev. 1997, 1997;290
16. Lumsden A. The cellular basis of segmentation in the developing hindbrain. Trends In Neurosci. 1990;13(8):329–35.
17. Lumsden A, Keynes R. Segmental patterns of neuronal development in the chick hindbrain. Nature. 1989;337(6206):424–8.

18. y Cajal SR (1955) Studies on the cerebral cortex (limbic structures). Year Book Publishers.
19. y Cajal SR (1995) Histology of the nervous system of man and vertebrates (Vol. 1). Oxford University Press, USA.
20. Rakic P. Neuron-glia relationship during granule cell migration in developing cerebellar cortex. A Golgi and electonmicroscopic study in Macacus rhesus. J Comp Neurol. 1971;141(3):283–312.
21. Rakic P. Mode of cell migration to the superficial layers of fetal monkey neocortex. J Comp Neurol. 1972;145(1):61–83.
22. Rakic P. Neuronal migration and contact guidance in the primate telencephalon. Postgrad Med J. 1978;54:25–40.
23. Sidman RL, Rakic P. Neuronal migration, with special reference to developing human brain: a review. Brain Res. 1973;62(1):1–35.
24. Edmondson JC, Hatten ME. Glial-guided granule neuron migration in vitro: a high-resolution time-lapse video microscopic study. J Neurosci. 1987;7(6):1928–34.
25. Fishell GORD, Hatten ME. Astrotactin provides a receptor system for CNS neuronal migration. Development. 1991;113(3):755–65.
26. Hatten ME. The role of migration in central nervous system neuronal development. Curr Opin Neurobiol. 1993;3(1):38–44.
27. Gao WQ, Hatten ME. Immortalizing oncogenes subvert the establishment of granule cell identity in developing cerebellum. Development. 1994;120(5):1059–70.
28. Kölliker A (1890) Zur feineren anatomie des zentralen nerven-systems. Engelmann.
29. Retzius G (1894) Die neuroglia des Gehirns beim Menschen und bei Saeugethieren. von Gustav Fischer.
30. Bauernfeind AL, de Sousa AA, Avasthi T, Dobson SD, Raghanti MA, Lewandowski AH, Zilles K, Semendeferi K, Allman JM, Craig AD, Hof PR, Sherwood CC. A volumetric comparison of the insular cortex and its subregions in primates. J Hum Evol. 2013;64(4):263–79.
31. Mesulam M, Mufson EJ. Insula of the old world monkey. III: efferent cortical output and comments on function. J Comp Neurol. 1982;212(1):38–52.
32. Mesulam MM, Mufson EJ. The insula of Reil in man and monkey. In:Association and auditory cortices. US: Springer; 1985. p. 179–226.
33. Bonthius DJ, Bonthius NE, Napper R, Astley SJ, Clarren SK, West JR. Purkinje cell deficits in nonhuman primates following weekly exposure to ethanol during gestation. Teratology. 1996;53(4):230–6.
34. von Economo C. Eine neue Art Spezialzellen des Lobus cinguli and Lobus insulae. Zschr ges Neurol Psychiatr. 1926;100:706–12.
35. Allman JM, Watson KK, Tetreault NA, Hakeem AY. Intuition and autism: a possible role for Von Economo neurons. Trends Cogn Sci. 2005;9(8):367–73.
36. Nimchinsky EA, Gilissen E, Allman JM, Perl DP, Erwin JM, Hof PR. A neuronal morphologic type unique to humans and great apes. Proc Natl Acad Sci. 1999;96(9):5268–73.
37. Hof PR, Van Der Gucht E. Structure of the cerebral cortex of the humpback whale, Megaptera novaeangliae (Cetacea, Mysticeti, Balaenopteridae). Anat Rec. 2007;290(1):1–31.
38. Hakeem AY, Sherwood CC, Bonar CJ, Butti C, Hof PR, Allman JM. von Economo neurons in the elephant brain. Anat Rec. 2009;292(2):242–8.
39. Allman JM, Tetreault NA, Hakeem AY, Manaye KF, Semendeferi K, Erwin JM, Park S, Goubert V, Hof PR. The von Economo neurons in frontoinsular and anterior cingulate cortex in great apes and humans. Brain Struct Funct. 2010;214(5–6):495–517.
40. Allman JM, Tetreault NA, Hakeem AY, Manaye KF, Semendeferi K, Erwin JM, Park S, Goubert V, Hof PR. The von Economo neurons in the frontoinsular and anterior cingulate cortex. Ann N Y Acad Sci. 2011;1225(1):59–71.
41. Allman JM, Hakeem A, Watson KK. Book review: two phylogenetic specializations in the human brain. Neuroscientist. 2002;8(4):335–46.
42. Zilles K. 27: architecture of the human cerebral cortex. In: Paxinos G, Mai JK, editors. The human nervous system. 2nd ed. Amsterdam: Elsevier; 2004. OCLC 54767534.
43. Chance SA, Sawyer EK, Clover LM, Wicinski B, Hof PR. Crow TJ (2013) Hemispheric asymmetry in the fusiform gyrus distinguishes *Homo sapiens* from chimpanzees. Brain Struct Funct 218(6):1391–1405.
44. Hayashi K, Morishita R, Nakagami H, Yoshimura S, Hara A, Matsumoto K, Nakamura T, Ogihara T, Kaneda Y, Sakai N. Gene therapy for preventing neuronal death using hepatocyte growth factor: in vivo gene transfer of HGF to subarachnoid space prevents delayed neuronal death in gerbil hippocampal CA1 neurons. Gene Ther. 2001;8(15):1167.
45. Nimchinsky EA, Vogt BA, Morrison JH, Hof PR. Spindle neurons of the human anterior cingul. Ate cortex. J Comp Neurol. 1995;355(1):27–37.
46. Campbell MJ, Morrison JH. Monoclonal antibody to neurofilament protein (SMI-32) labels a subpopulation of pyramidal neurons in the human and monkey neocortex. J Comp Neurol. 1989;282(2):191–205.
47. Hof PR, Morrison JH. Neurofilament protein defines regional patterns of cortical organization in the macaque monkey visual system: a quantitative immunohistochemical analysis. J Comp Neurol. 1995;352(2):161–86.
48. DeFelipe J. Types of neurons, synaptic connections and chemical characteristics of cells immunoreactive for calbindin-D28K, parvalbumin and calretinin in the neocortex. J Chem Neuroanat. 1997;14(1):1–19.
49. Hendry SHC, Jones EG. GABA neuronal subpopulations in cat primary auditory cortex: co-localization with calcium binding proteins. Brain Res. 1991;543(1):45–55.

50. Schumacher J, Laje G, Jamra RA, Becker T, Mühleisen TW, Vasilescu C, Mattheisen M, Herms S, Hoffmann P, Hillmer AM, Georgi A, Herold C, Schulze TG, Propping P, Rietschel M, McMahon FJ, Nöthen MM, Cichon S. The DISC locus and schizophrenia: evidence from an association study in a central European sample and from a meta-analysis across different European populations. Hum Mol Genet. 2009;18(14):2719–27.
51. Stimpson CD, Tetreault NA, Allman JM, Jacobs B, Butti C, Hof PR, Sherwood CC. Biochemical specificity of von Economo neurons in hominoids. Am J Hum Biol. 2011;23(1):22–8.
52. Hofman MA, Falk D. Cerebral cortical development in rodents and primates. Evolution of the primate brain: from neuron to behavior. Prog Brain Res. 2012;195:45.
53. Shura RD, Hurley RA, Taber KH. Insular cortex: structural and functional neuroanatomy. J Neuropsychiatry Clin Neurosci. 2014;26(4):iv–282.
54. Farjardo C, Escobar MI, Buritica E, Arteaga G, Umbarila J, Casanova MF, Pimienta H. Von economo neurons are present in the dorsolateral (dysgranular) prefrontal cortex of humans. Neurosci Lett. 2008;435:215–8.
55. Krings T, Mandell DM, Kiehl TR, Geibprasert S, Tymianski M, Alvarez H, terBrugge KG, Hans FJ. Intracranial aneurysms: from vessel wall pathology to therapeutic approach. Nat Rev Neurol. 2011;7(10):547–59.
56. Alpers BJ, Berry RG, Paddison RM. Anatomical studies of the circle of Willis in normal brain. AMA Arch Neurol Psychiatry. 1959;81:409–18.
57. Liegeois-Chauvel C, Musolino A, Chauvel P. Localization of the primary auditory area in man. Brain. 1991;114(1):139–53.
58. Shi CJ, Cassell MD. Cortical, thalamic, and amygdaloid connections of the anterior and posterior insular cortices. J Comp Neurol. 1998;399(4):440–68.
59. Wiech K, Lin CS, Brodersen KH, Bingel U, Ploner M, Tracey I. Anterior insula integrates information about salience into perceptual decisions about pain. J Neurosci. 2010;30(48):16324–31.
60. American Psychiatric Association. DSM-IV® sourcebook, vol. 1: American Psychiatric Pub.; 1994.
61. Robinson CJ, Burton H. Organization of somatosensory receptive fields in cortical areas 7b, retroinsula, postauditory and granular insula of M. fascicularis. J Comp Neurol. 1980;192(1):69–92.
62. Kang Y, Williams LE, Clark MS, Gray JR, Bargh JA. Physical temperature effects on trust behavior: the role of insula. Soc Cogn Affect Neurosci. 2010;6(4):507–15.
63. Stephani C, Vaca GFB, Maciunas R, Koubeissi M, Lüders HO. Functional neuroanatomy of the insular lobe. Brain Struct Funct. 2011;216(2):137–49.
64. Naqvi NH, Bechara A. The insula and drug addiction: an interoceptive view of pleasure, urges, and decision-making. Brain Struct Funct. 2010;214(5–6):435–50.
65. Kurth F, Zilles K, Fox PT, Laird AR, Eickhoff SB. A link between the systems: functional differentiation and integration within the human insula revealed by meta-analysis. Brain Struct Funct. 2010;214(5–6):519–34.
66. Tatschl C, Stöllberger C, Matz K, Yilmaz N, Eckhardt R, Nowotny M, Dachenhausen A, Brainin M. Insular involvement is associated with QT prolongation: ECG abnormalities in patients with acute stroke. Cerebrovasc Dis. 2006;21(1–2):47–53.
67. Mutschler I, Wieckhorst B, Kowalevski S, Derix J, Wentlandt J, Schulze-Bonhage A, Ball T. Functional organization of the human anterior insular cortex. Neurosci Lett. 2009;457(2):66–70.
68. Weiller C, Chollet F, Friston KJ, Wise RJ, Frackowiak RS. Functional reorganization of the brain in recovery from striatocapsular infarction in man. Ann Neurol. 1992;31(5):463–72.
69. Cole MW, Schneider W. The cognitive control network: integrated cortical regions with dissociable functions. Neuroimage. 2007;37(1):343–60.
70. Ackermann H, Riecker A. The contribution (s) of the insula to speech production: a review of the clinical and functional imaging literature. Brain Struct Funct. 2010;214(5–6):419–33.
71. Karnath HO, Baier B. Right insula for our sense of limb ownership and self-awareness of actions. Brain Struct Funct. 2010;214(5–6):411–7.
72. Craig AD, Craig AD. How do you feel—now? The anterior insula and human awareness. Nat Rev Neurosci. 2009;10(1)
73. Bieser A. Processing of twitter-call fundamental frequencies in insula and auditory cortex of squirrel monkeys. Exp Brain Res. 1998;122(2):139–48.
74. Cacioppo JT, Decety J. Social neuroscience: challenges and opportunities in the study of complex behavior. Ann N Y Acad Sci. 2011;1224(1):162–73.
75. Grüsser OJ, Pause M, Schreiter U. Vestibular neurones in the parieto-insular cortex of monkeys (Macaca fascicularis): visual and neck receptor responses. J Physiol. 1990;430(1):559–83.
76. Verhagen JV, Kadohisa M, Rolls ET. Primate insular/opercular taste cortex: neuronal representations of the viscosity, fat texture, grittiness, temperature, and taste of foods. J Neurophysiol. 2004;92(3):1685–99.
77. Crespo-Facorro B, Kim JJ, Andreasen NC, O'Leary DS, Bockholt HJ, Magnotta V. Insular cortex abnormalities in schizophrenia: a structural magnetic resonance imaging study of first-episode patients. Schizophr Res. 2000;46(1):35–43.
78. Zaki J, Davis JI, Ochsner KN. Overlapping activity in anterior insula during interoception and emotional experience. Neuroimage. 2012;62(1):493–9.

Gross Anatomy of the Human Insula

Igor Lima Maldonado, Ilyess Zemmoura, and Christophe Destrieux

2.1 Introduction

When the lips of the lateral fissure are separated from each other, a new group of sulci and gyri appear. They are arrayed together in the form of an island, which is the reason why the German anatomist, Johann Christian Reil, designated them as "the insular lobe".

Vicq d'Azyr and Monro also paid attention to this region in the eighteenth century, but Reil was the first to publish a description with some degree of detail in 1809 [1–4]. Since the publication of the fourth edition of the Paris *Nomina Anatomica* in 1975, it has been systematically considered a cerebral lobe [5, 6].

The insula is partially composed of the so-called mesocortex. Its topography and physiology are intermediate to the isocortex, which covers most of the cerebral hemispheres, and the allocortex, which is present in the (phylogenetically older) amygdala and hippocampus [4, 5].

2.2 Topography

Embedded in and covered by the adjacent brain lobes, the insula forms the medial limit of the lateral (Sylvian) fissure in each cerebral hemisphere. However, it does not occupy the full extent of that fissure. Anteroinferiorly, the initial portion of the fissure has a preinsular topography. Similarly, it extends posterior to the last gyri of the insula, forming the retroinsular segment.

The two lips of the lateral sulcus correspond to portions of the cerebral hemisphere that have developed laterally to the insula [7]. They completely cover it and assume the role of true opercula. We distinguish two opercula: one is superior or frontoparietal and the other is inferior or temporal. The lower border of the frontal and parietal lobes form the superior operculum, while the inferior operculum is formed entirely by the superior temporal gyrus (Fig. 2.1a).

From a topographic point of view, the insula is considered the outer shield of a well-defined cerebral central core [5, 8, 9]. Attached to the upper portion of each half of the midbrain, the central core corresponds in each hemisphere to the insula, basal nuclei, thalamus, and internal capsule.

I. L. Maldonado, M.D., Ph.D. (✉)
Departamento de Biomorfologia, Instituto de Ciências da Saúde, Universidade Federal da Bahia, Salvador, Brazil

Serviço de Neurocirurgia, Unidade Neuro-Músculo-Esquelética, C.H.U. Prof. Edgard Santos, Salvador, Brazil

UMR 1253, iBrain, Université de Tours, Inserm, Tours, France

LE STUDIUM Loire Valley Institute for Advanced Studies, Orléans, France

I. Zemmoura, M.D., Ph.D. · C. Destrieux, M.D., Ph.D.
CHRU de Tours, Service de Neurochirurgie, Tours, France

UMR 1253, iBrain, Université de Tours, Inserm, Tours, France

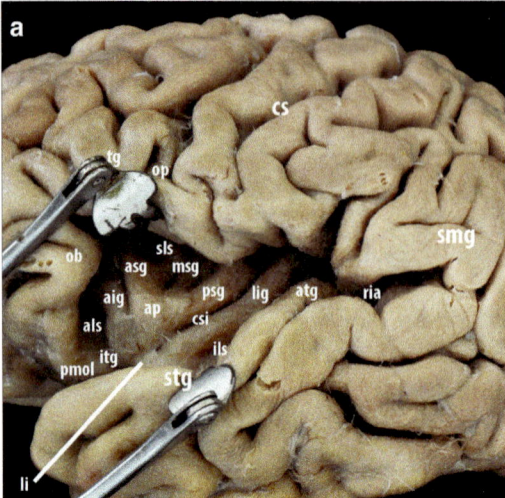

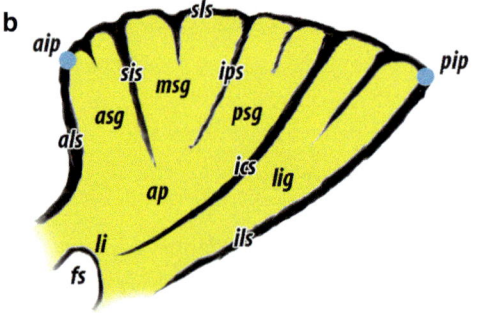

Fig. 2.1 General anatomy of the human insula. (**a**) Sulci and gyri of a left insular lobe; as seen when the superior and inferior opercula are separated from each other. (**b**) Schematic representation showing the main anatomical features of the insula. *Aip* Anterior insular point, *Als* anterior limiting sulcus, *ap* insular apex, *asg* accessory short insular gyrus, *fs* falciform sulcus, *ics* insular central sulcus, *ls* inferior limiting sulcus, *ips* insular precentral sulcus, *li* limen of the insula, *lig* long insular gyrus, *msg* middle short gyrus, *pip* posterior insular point, *psg* posterior short gyrus, *sis* short insular sulcus, *sls* superior limiting sulcus

2.3 Limits

The contour of the insular lobe is delimited by the circular or peri-insular sulcus, which is marked, but interrupted by the transverse gyrus of the insula. Considering the roughly triangular shape of the insular base, this sulcus is generally divided into three parts: the anterior, posterior, and inferior (or posteroinferior) limiting sulci (Fig. 2.1b).

The deep anterior limiting sulcus separates the anterior border of the insula from the orbital portion of the inferior frontal gyrus and from the posterior orbital gyrus. In the form of a true fissure, it is vertically oriented or has a slightly oblique direction, pointing inferiorly and posteriorly. The depth of this sulcus is in relation to the anterior limb of the internal capsule. The upper extremity of the anterior limiting sulcus is at the level of the anterior horn of the lateral ventricle, where the ventricle surrounds the head of the caudate nucleus [5].

The superior limiting sulcus is the longest of the three. It separates the superior surface of the insula from the superior operculum. It is related successively from anterior to posterior to the medial aspect of the triangular portion of the inferior frontal gyrus, the opercular portion of the same gyrus, the pre-central gyrus, and the sub-central gyrus, and a cortical fold that closes inferiorly to the central sulcus bridging the pre- and postcentral gyri.

The inferior limiting sulcus has a strong obliquity, pointing inferiorly and anteriorly. While the posterior segment of this sulcus separates the insula from the retroinsular part of the lateral fissure, the anterior segment separates it from the superior temporal gyrus.

Since the pole of the insula is connected to the inferior frontal gyrus and the superior temporal gyrus at the level of the limen insulae, the aforementioned sulci do not surround the insula completely. This conformation prevents the inferior limiting sulcus from being continuous with the anterior limiting sulcus. As a consequence, this lobe is not a true island, as it remains connected to superficial gyri of the neighboring lobes; it may thus be compared to a quasi-island or peninsula [10].

2.4 General Form

The insula presents a general form that resembles that of an oblique pyramid with a triangular base and a low height. Its base is large and is medially orientated in continuity with the rest of the cerebral hemisphere. One of the vertices of the base is directed anteroinferiorly and corresponds to the insular pole.

The apex, situated posterior and lateral to the pole, is the lateralmost portion of the insular cortex. It is directed laterally, anteriorly, and inferiorly; it therefore has a location eccentric to the base. It is also directed toward the external opening of the Sylvian fissure, without reaching or exceeding this aperture. Under surgical conditions, it can be seen or directly accessed through the so-called anterior Sylvian point, the site of the lateral sulcus in which the triangular portion of the inferior frontal gyrus separates from the superior temporal gyrus widening the sulcus. From this point, the lateral sulcus sends out its anterior, ascending, and posterior rami.

Anteriorly and inferiorly to the apex, the limen insulae connect the frontal, insular, and temporal lobes. The middle cerebellar artery turns approximately 90° just anterior or immediately distal to the limen, as it changes direction to enter the posterior ramus of the lateral fissure [11].

Two aspects of the insula are more clearly distinct: the anterior and the lateral. The anterior aspect is smaller in extension. It is covered by the fronto-orbital portion of the superior operculum, consisting of the posterior portion of the posterior orbital gyrus and the orbital portion of the inferior frontal gyrus.

The lateral aspect is rich in insular sulci and gyri. Thus, it has a more complex shape than the anterior one. Its upper third is covered by parts of the frontal-parietal operculum, namely the triangular and opercular portions of the inferior frontal gyrus, the subcentral gyrus, and the anterior basal portion of the supramarginal gyrus. The lower two-thirds are covered by the superior temporal gyrus.

2.5 Sulci and Gyri

In relation to its internal anatomical organization, the insula is formed by a set of gyri that are arranged centrifugally and extend from the apex to the base. These gyri have significant individual variations, so it is not possible to determine a pattern that corresponds to all cases, only to most of them.

At its lateral surface, a sulcus that is much longer than the others leaves the superior limiting sulcus and describes a strongly oblique trajectory anteriorly and inferiorly until the limen insulae. This sulcus, namely, the central sulcus of the insula, is easily recognizable since no other sulcus in the lateral aspect of the insula descends as inferiorly as it does. Its direction is roughly the same as that of the central sulcus of the superolateral surface of the cerebral hemisphere. It divides the insula into two clearly distinct parts: one is the anterior lobule of the insula, and the other is the posterior lobule of the insula. In a minority of cerebral hemispheres, estimated to be less than 10%, the central sulcus of the insula may not be well defined or may be interrupted by connecting gyri [4].

The anterior lobule of the insula consists of three short gyri that have a common origin in the irregularly rounded apex. They are distinguished according to their position as anterior, middle, and posterior.

The anterior short gyrus extends along the anterior limiting sulcus. It is directed obliquely, superiorly, and anteriorly toward the inferior frontal gyrus. There may be an anatomical variation with bifurcation of its upper extremity. The point at which the anterior border of this gyrus joins the anterior and superior limiting sulci has been named the "anterior insular point" [4].

The middle short gyrus is the smallest of the three. It follows a roughly vertical trajectory, starting from the apex toward the opercular portion of the inferior frontal gyrus. It is separated from the anterior insular gyrus by the short insular sulcus, and from the posterior gyrus by the precentral sulcus of the insula.

The posterior gyrus also arises from the insular apex. Its inferior extremity is tip-shaped. It extends obliquely, superiorly, and posteriorly along the central sulcus of the insula, forming the anterior bank of that sulcus and terminating superiorly with a generally bifurcated (or even trifurcated) extremity. In a lateral view, this extremity projects approximately at the level of the subcentral gyrus of the cerebral hemisphere.

Regardless of the three main gyri mentioned above, one or two accessory gyri are also often present. They are variable, short, and deeply situated anterior to the apex and to the anterior gyrus of the insula. They connect the anterior lobe of the insula to the anterior portions of the inferior frontal gyrus.

At a lower location, crossing the middle cerebral artery, a small transverse gyrus connects the apex area to the orbital region in the inferior aspect of the frontal lobe. It reaches the posteromedial lobule of the frontal lobe, which is composed of the posterior portion of the medial orbital gyrus and the medial portion of the posterior orbital gyrus, extending along and lateral to the lateral olfactory stria [4, 5]. The transverse gyrus of the insula may be hypoplastic in up to 14% of cerebral hemispheres [4]. The transverse and accessory gyri form the insular pole, located in the inferiormost and anteriormost portion of this lobe. [4].

The posterior lobule of the insula is smaller than the anterior one. It is delimited anteriorly by the central insular sulcus and posteriorly by the inferior limiting sulcus. It is comprised of two long, strongly oblique gyri: one is anterior, along the central sulcus of the insula; and the other is posterior, adjacent to the inferior limiting sulcus.

These two gyri are sometimes only slightly separated from each other because of the shallow, undeveloped character of the postcentral insular sulcus in approximately one-fourth of cerebral hemispheres [4]. They originate from an inferior common tip that is continuous with the superior temporal gyrus. In their superior portions, they may bifurcate and thus form two to four secondary gyri that meet the frontoparietal operculum at the level of the postcentral gyrus. The term "posterior insular point" has been used to describe the posterior-superior limit of the posterior insular lobule, where the superior and inferior limiting sulci meet. From this point, the posterior ramus of the lateral fissure continues as a retroinsular space.

2.6 Deep Relationships

Deep within the insular gyri, the subcortical white matter of the extreme capsule roughly accompanies the contours and accidents of the insular sulci and gyri. In horizontal sections, small conical extensions of this capsule are seen in the form of spines that are directed toward the insular cortex (Fig. 2.3a).

Through it, the insular gyri are indirectly related to the claustrum, the most superficial nucleus of that region. The claustrum consists of a thin sheet of gray matter with a thickness of 1–2 mm, extending along the deep face of the insula, interposed between the extreme capsule and the external capsule, a portion of white matter that covers the lenticular nucleus laterally.

In horizontal and coronal sections, the periphery of the claustrum curves laterally, following the form of the cerebral cortex that covers the Sylvian fissure. In its lower portion, it widens considerably, except for the anteroinferior part in proximity to the limen of the insula. At this site, the thickness of the white matter bundles (uncinate and inferior fronto-occipital) compete for space with the gray substance of the claustrum [12–14].

The lateral aspect of the lentiform nucleus corresponds to the lateral aspect of the putamen. Since it has a surface area that is smaller than that of the insula itself, and considering the fact that the internal capsule laterally inclines around this nucleus, the depths of the anterior, superior, and inferior limiting sulci get significantly close to the fibers of the internal capsule [5].

In relation to the other, deeper basal nuclei, the anterior half of the lateral face of the insula corresponds internally to the head of the caudate nucleus. Its posterior half corresponds to the thalamus and other portions of the caudate nucleus. The distance from one of the insular points (anterior or posterior) to the ventricular ependyma is little more than 1 cm. Beyond the limiting sulci, the insula is surrounded by the arcuate fasciculus, which is part of the superior longitudinal fasciculus complex (Figs. 2.2 and 2.3).

The superior limiting sulcus follows the course of the frontal horn of the lateral ventricle, as well as the body and part of the atrium. The inferior limiting sulcus accompanies the temporal horn and part of the atrium for about 80% of its length [4]. About 1 cm of the anterior portion of that sulcus corresponds to the temporal stem.

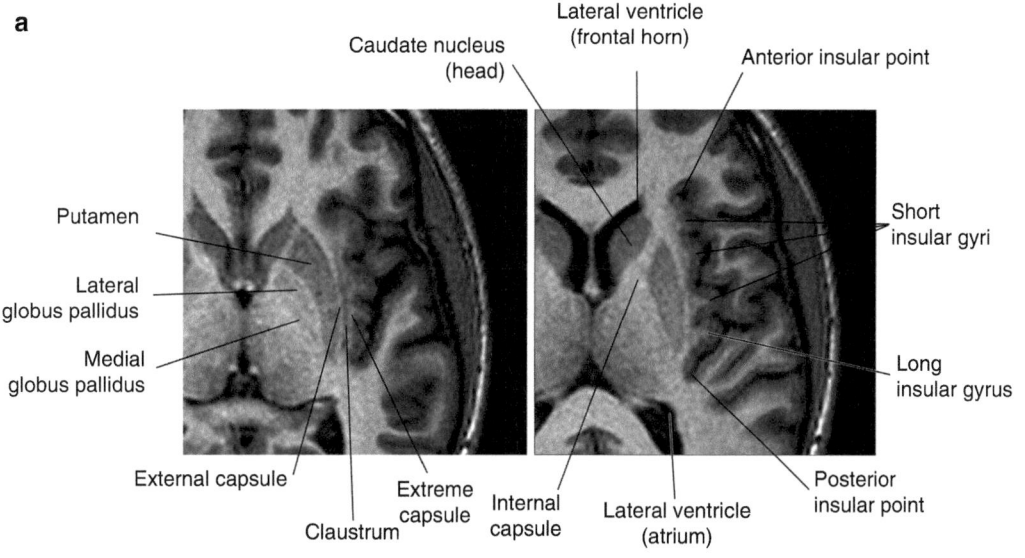

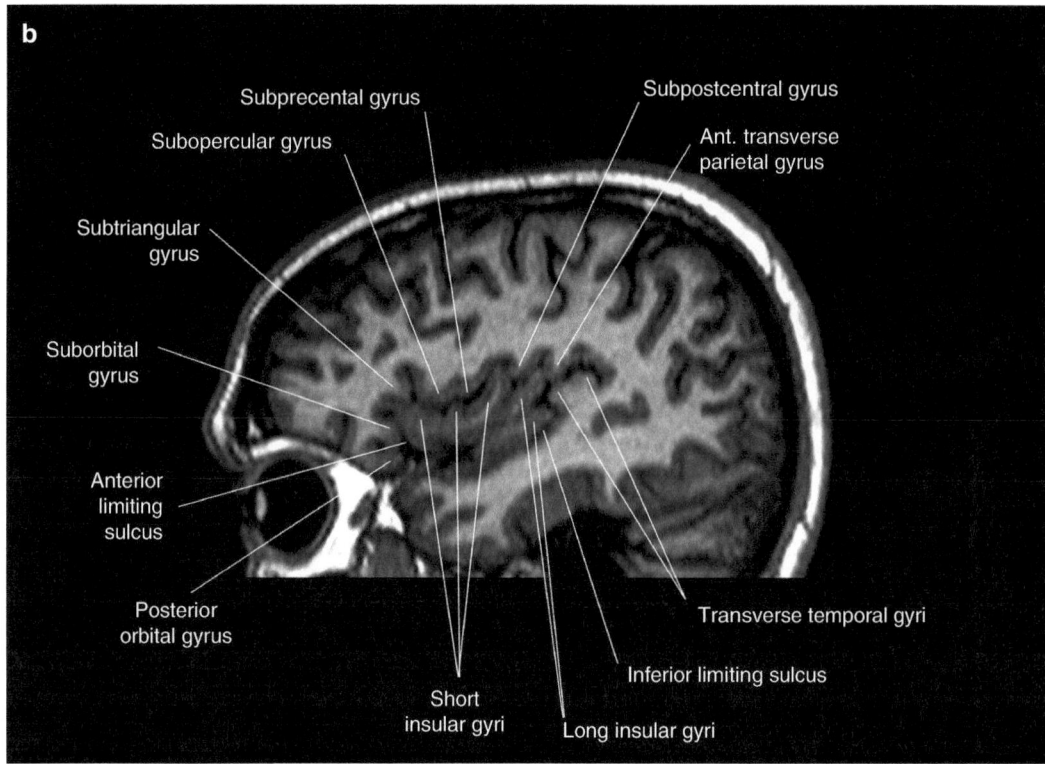

Fig. 2.3 Deep and superficial anatomical relationships of the insula in a T1-weighted magnetic resonance examination. (**a**) Axial view. The relationships with basal nuclei and the ventricular system are exposed. (**b**) Parasagittal view. The superficial relationships with the frontoparietal and temporal operculum are shown

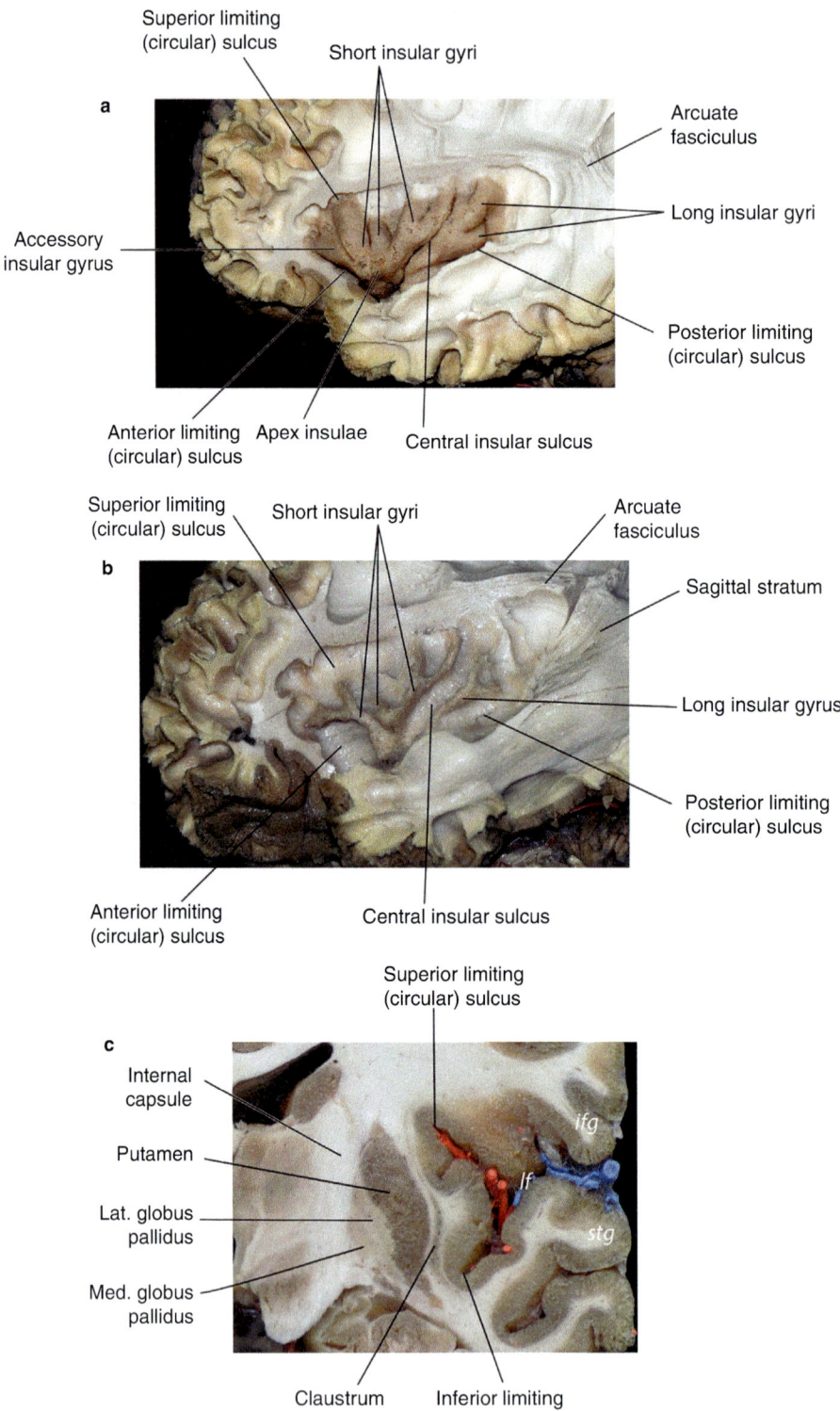

Fig. 2.2 Deep anatomical relationships of the insula. (**a**) The arcuate fasciculus surrounds a significant portion of the circular sulcus of the insula. (**b**) Extensions of the extreme capsule white matter reproduce the form of the insular sulci and gyri. (**c**) Deep to the extreme capsule, the insula maintains an indirect relationships with the claustrum, the external capsule, and elements of the lentiform nucleus

2.7 Superficial Relationships

As mentioned, the fronto-orbital portion of the superior operculum covers the anterior aspect of the insula. Through the lateral fissure, it is in relation with the posterior portion of the posterior orbital gyrus, a small portion of the lateral orbital gyrus, and the orbital portion of the inferior frontal gyrus. In the medial surface of the fronto-orbital operculum, two small gyri, the (superior and inferior) suborbital gyri are continuous with the cortex of the insula in the region of the accessory and short anterior gyri.

Similarly, the medial aspect of the triangular portion contains the subtriangular gyrus, which is continuous with the cortex of the anterior short insular gyrus and covers it laterally. The suborbital gyrus covers the anterior short gyrus anteriorly and the subopercular gyrus covers it posteriorly [4].

The subopercular gyrus also covers the short insular sulcus and part of the middle short insular gyrus. The subprecentral gyrus covers the rest of the middle short insular gyrus and the precentral sulcus of the insula. Since in most cerebral hemispheres the central sulcus of the cerebral hemisphere does not reach the lateral fissure, the subcentral gyrus recovers part of the central sulcus of the insula. As a consequence, the posterior short insular gyrus is covered anteriorly by the subprecentral gyrus and posteriorly by the subcentral and subpostcentral gyri.

Also, in the frontoparietal operculum, three small transverse parietal gyri (anterior, middle, and posterior) are located on the medial (internal) opercular surface. The anterior one covers part of the postcentral insular sulcus and the superior portion of the two long insular gyri. The anterior transverse parietal gyrus converges with the anterior transverse temporal gyrus of Heschl close to the posterior insular point.

In the temporal operculum, the superior temporal gyrus and the lower portion of the supramarginal gyrus cover the insula. The polar planum covers the limen of the insula and its inferior surface. About two-thirds of the inferior limiting sulcus is adjacent to the polar planum [4]. The remaining third closely follows the anterior transverse temporal gyrus of Heschl.

2.8 Preinsular Area

In the inferior aspect of the brain, when examining the anterior portion of the lateral fissure, a pronounced curvature is observed. It is present at the level of the limen of the insula, where the fissure passes from the inferior aspect of the cerebral hemisphere to its lateral face. This curved portion of the fissure (viz., the falciform sulcus) anteriorly crosses the limen of the insula (the falciform fold). Anterior to this, we see the diagonal band of Broca and the lateral olfactory stria, landmarks that limit the anterior perforated substance.

The trunk of the lateral fissure thus presents as a preinsular segment (Sylvian vallecula), and its roof contains the anterior perforated substance. The trunk of the lateral fissure follows the contours of the posterior border of the lesser wing of the sphenoid bone.

2.9 Retroinsular Area

We consider this as part of the lateral fissure that is located posterior to the insula and specifically, behind the inferior limiting sulcus. It is a deep depression in which are situated the posteriormost branches of the middle cerebral artery.

Immediately behind the insula, a *pli de passage* is observed, often well developed and of varying depth, running obliquely from inferior to superior and from anterior to posterior. It corresponds to an extension of the transverse temporal gyrus of Heschl behind the insula until the anterior transverse parietal gyrus as a temporoparietal connection. This cortical fold behaves as a significant anastomosis between the superior temporal gyrus and the supramarginal gyrus. It is usually unique in its origin, but it may divide in its trajectory into two (or more) secondary gyri that connect themselves with similar parenchymal extensions from the parietal lobe—the anterior, middle, and posterior transverse parietal gyri.

Such cortical folds do not belong to the insula because they are clearly separated from it by the inferior limiting sulcus. It is interesting to note that the entire lenticular nucleus is also located anterior to that sulcus, which makes it an important landmark. As a consequence, this nucleus maintains an anatomical relationship with the insula, but not with the retroinsular area.

Conclusion

The study of the descriptive and topographic anatomy of the insula, its internal organization, and its anatomical relationships allows an overview of concepts that are important in a number of fields, such as neuroimaging and neurosurgery. An adequate nomenclature is important for better communication, while a good understanding of its morphology is one of the bases for development of precise radiological interpretations and operative techniques. These concepts must be accompanied by the study of the functional aspects of the insular lobe.

References

1. Reil J. Die Sylvische Grube. Arch Phys Ther. 1809;9:195–208.
2. Monro A (1783) Observations on the structure and functions of the nervous system. Edinburgh
3. Vicq d'Azyr F. Traité d'Anatomie et de physiologie. Paris; 1786.
4. Ture U, Yasargil DC, Al-Mefty O, Yasargil MG. Topographic anatomy of the insular region. J Neurosurg. 1999;90(4):720–33.
5. Ribas GC. The cerebral sulci and gyri. Neurosurg Focus. 2010;28(2):E2.
6. Excerpta Medica Foundation. Nomina anatomica. 4th ed. Amsterdam: Excerpta Medica; 1975.
7. Guenot M, Isnard J, Sindou M (2004) Surgical anatomy of the insula. Adv Tech Stand Neurosurg 29:265–288
8. Rhoton AL Jr. The cerebrum. Neurosurgery. 2002;51(4 Suppl):S1–51.
9. Ribas GC, Oliveira E. The insula and the central core concept. Arq Neuropsiquiatr. 2007;65(1):92–100. doi:S0004-282X2007000100020 [pii]
10. Testut L (1899) Traité d'anatomie humaine. Tome deuxième—angéiologie, système nerveux central. Octave Doin, Paris,
11. Tanriover N, Rhoton AL Jr, Kawashima M, Ulm AJ, Yasuda A. Microsurgical anatomy of the insula and the Sylvian fissure. J Neurosurg. 2004;100(5):891–922. https://doi.org/10.3171/jns.2004.100.5.0891.
12. Ludwig E, Klingler J. Atlas cerebri humani. Basel: S. Karger A.G; 1956.
13. Fernandez-Miranda JC, Rhoton AL Jr, Alvarez-Linera J, Kakizawa Y, Choi C, de Oliveira EP. Three-dimensional microsurgical and tractographic anatomy of the white matter of the human brain. Neurosurgery. 2008;62(6 Suppl 3):SHC989–1026. discussion SHC1026-1028
14. Fernandez-Miranda JC, Rhoton AL Jr, Kakizawa Y, Choi C, Alvarez-Linera J. The claustrum and its projection system in the human brain: a microsurgical and tractographic anatomical study. J Neurosurg. 2008;108(4):764–74.

Surgical Anatomy of the Insula

3

Christophe Destrieux, Igor Lima Maldonado, Louis-Marie Terrier, and Ilyess Zemmoura

3.1 Introduction

Deeply located at the bottom of the lateral fossa, the insula has strong relationships with [1] the opercula forming the banks of the lateral fissure, [2] white matter tracts, and [3] the middle cerebral artery and related veins.

3.2 Sulco-Gyral Relationships [1–3]

The *opercula* (Latin, meaning "little lid") are made of the inferior border of the frontal and parietal lobes and the superior border of the temporal lobe. They cover and hide the insula, deeply located in the lateral fissure.

The *lateral fissure* is subdivided into three segments (Fig. 3.1): the middle one parallels the main axis of the temporal lobe and separates the opercular part of the inferior frontal and subcentral gyri (frontoparietal operculum) from the superior temporal gyrus (temporal operculum); the anterior segment of the lateral fissure anteriorly splits into a vertical and a horizontal ramus; finally the posterior segment of the lateral fissure ascends posterodorsally within the anterior part of the inferior parietal lobule (supramarginal gyrus).

The *frontoparietal operculum* is bordered anteriorly by the horizontal ramus of the lateral fissure and posteriorly by its posterior ascending segment. It includes (Fig. 3.1), from anterior to posterior [1, 4–6], the triangular and opercular parts of the inferior frontal gyrus and the subcentral gyrus.

– The *inferior frontal gyrus* (or F3) is limited ventrally by the lateral fissure, posteriorly by the inferior precentral and anterior subcentral sulci, and dorsally by the inferior frontal sulcus. The inferior precentral sulcus—sometimes fused with its superior homologue—runs parallel and anterior to the central sulcus. It more or less reaches the lateral fissure and is perpendicularly connected to the inferior frontal sulcus. This one runs anteriorly, parallel to the middle segment of the lateral fissure, and finally curves anteroventrally to become the lateral

---

C. Destrieux (✉) · L.-M. Terrier · I. Zemmoura
UMR 1253, iBrain, Université de Tours, Inserm, Tours, France

CHRU de Tours, Service de Neurochirurgie, Tours, France
e-mail: christophe.destrieux@univ-tours.fr

I. Lima Maldonado
UMR 1253, iBrain, Université de Tours, Inserm, Tours, France

Departamento de Biomorfologia, Universidade Federal da Bahia, Laboratório de Anatomia e Dissecção, Instituto de Ciências da Saúde, Salvador, Bahia, Brazil

Serviço de Neurocirurgia, Unidade Neuro-Músculo-Esquelética, C.H.U. Prof. Edgard Santos, Salvador, Brazil

LE STUDIUM Loire Valley Institute for Advanced Studies, Orléans, France

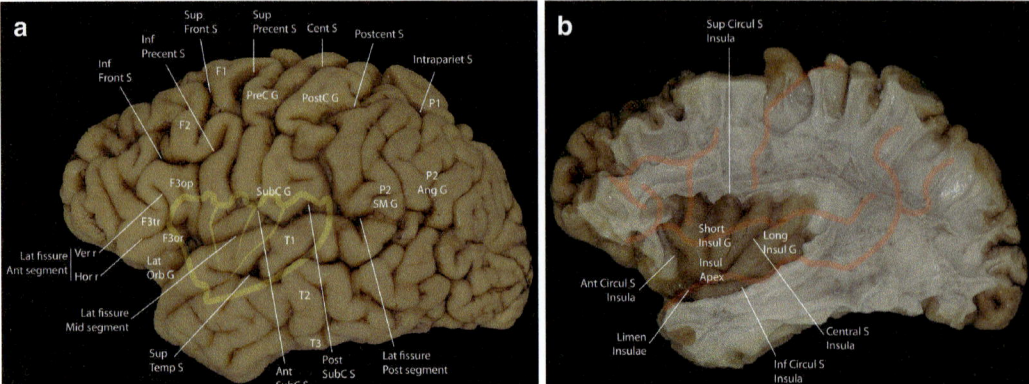

Fig. 3.1 Two steps of dissection of a left human hemisphere, lateral views. (**a**) Cerebral cortex after removal of the arachnoid, superficial vessels, and pia mater. The insula is hidden by the opercula. The lateral compartment of the lateral fissure is subdivided into three segments: middle (*Lat Fissure Mid seg*), slightly ascending posteriorly, between the temporal and frontal opercula; posterior (*Lat Fissure Post seg*), ascending and surrounded by the supramarginal gyrus (*P2SM G*); and anterior (*Lat Fissure Ant seg*), which divides into a horizontal and a vertical ramus limiting the pars triangularis (*F3tr*) from the pars opercularis (*F3op*) and orbitalis (*F3or*) of the inferior frontal gyrus. The frontoparietal operculum is made of *F3tr*, *F3op*, subcentral gyrus (*SubC G*), and part of *P2SM G* located dorsal and anterior to the *Lat Fissure Post seg*. The temporal operculum includes the remaining part of *P2SM G* ventral and posterior to the *Lat Fissure Post Seg*, continued by the superior temporal gyrus (*T1*). The fronto-orbital operculum is smaller and reunites *F3or* and the posterior part of the lateral orbital gyrus (*Lat Orb G*). The circular and central sulci of the insula (see (**b**) for legends) are projected in yellow onto the cortical surface, showing that F3tr points toward the anterior insula. The junction point of *Lat Fissure Mid seg* and *Lat Fissure Post seg* projects at the level of the posterosuperior corner of the insula. Finally, the central sulcus of the insula has about the same orientation as the central sulcus (*Cent S*) but may begin a few millimeters anterior or posterior to its ventral tip. (**b**) Dissection showing the insula after the opercula were removed. The insula has a triangular limit, the circular sulcus of the insula, made of anterior (*Ant Circul S insula*), superior (*Sup Circul S insula*), and inferior (*Inf Circul S insula*) segments. The *limen insulae* is located at the junction of *Ant Circul S insula* and *Inf Circul S insula*. The central sulcus of the insula (*Central S Insula*) subdivides the insula in long insular gyri (*Long Insul G*) posteriorly, and short insular gyri (*Short Insul G*) anteriorly. The lateral fissure, inferior precentral-inferior frontal sulci, the superior temporal sulcus, and central sulcus are projected in orange onto the insula to show that the two ventral thirds of the insula are hidden by the temporal operculum. Ant SubC S anterior subcentral sulcus, F1 superior frontal gyrus, F2 middle frontal gyrus, Inf Front S inferior frontal sulcus, Inf Precent S inferior precentral sulcus, Insul apex insular apex, Intrapariet S intraparietal sulcus, P1 superior parietal lobule, P2 Ang G sngular gyrus (posterior part of the inferior parietal lobule), Post SubC S posterior subcentral sulcus, Postcent G postcentral gyrus, Postcent S postcentral sulcus, Precent G precentral gyrus, Sup Front S superior frontal sulcus, Sup Precent S superior precentral sulcus, Sup Temp S superior temporal sulcus, T2 middle temporal gyrus, and T3 inferior temporal gyrus

orbital sulcus. The inferior frontal gyrus is subdivided by both rami of the anterior segments of the lateral fissure: its triangular part is located between them, whereas the opercular part is posterior to the vertical ramus and the orbital part is anteroventral to the horizontal ramus.

- The subcentral gyrus is the inferior frontoparietal *pli de passage* connecting the pre- and postcentral gyri. It curves around the ventral tip of the central sulcus and usually separates its inferior tip from the lateral fissure. The subcentral gyrus is thus located just posterior to the inferior frontal gyrus, from which it is limited by the anterior subcentral sulcus, a small notch originating from the lateral fissure. Similarly the subcentral gyrus is posteriorly limited from the supramarginal gyrus by the posterior subcentral sulcus. The supramarginal gyrus is the anterior part of the inferior parietal lobule (P2), its posterior part being the angular gyrus; the supramarginal gyrus curves around the posterior segment of the lateral fissure and is thus continuous with the subcentral gyrus dorsally and the superior temporal gyrus ventrally.

The *fronto-orbital operculum* covers the anterior part of the insula and is sometimes included in the frontoparietal operculum. It is made of the orbital part of the inferior frontal gyrus (ventral to the horizontal ramus of the anterior segment of the lateral fissure) and the posterior part of the lateral orbital gyrus.

The superior temporal gyrus (T1) forms the *temporal operculum*. It is bordered by the lateral fissure dorsally and by the superior temporal sulcus ventrally. The latter is a deep, continuous sulcus, which slightly ascends posteriorly and often splits into two posterior rami, one continuing its initial course and the second having a more ascending trajectory. These posterior rami of the superior temporal sulcus are surrounded by the angular gyrus, which is the posterior part of the inferior parietal lobule (P2).

The superior aspect of the superior temporal gyrus (T1) is subdivided into three distinct areas by the transverse temporal sulcus. The latter is posteromedially oriented and joins the lateral fissure, at the superficial aspect of the brain, to the circular sulcus of the insula, close to the junction point of its superior and inferior segments (see below). The *planum temporale*, located posterior to the transverse temporal sulcus, is the most posterior part of the superior aspect of the superior temporal gyrus. It is a flat area, posteriorly continuous with the anterior part of the supramarginal gyrus. The *temporal transverse gyrus*, containing the primary auditory area, is a strip of cortex having sometimes a double pattern; it follows the same posteromedial course as the temporal transverse sulcus, which is its posterior limit. Finally, the superior aspect of the superior temporal gyrus ends anteriorly as a second flat area, the *planum polare*, reaching the temporal pole.

Each of the opercula has three aspects (Fig. 3.2): lateral, visible from a surface inspection of the brain, dorsal (for the temporal operculum) or ventral (for the frontoparietal operculum), and medial, facing the insula. The junction point between the insula and the medial aspect of the opercula is the circular sulcus of the insula (Fig. 3.1). It is a triangle made of three segments that border the insula: superior, limiting the insula from the frontoparietal operculum; inferior, between the insula and the temporal operculum; and anterior, between the insula and the fronto-orbital operculum. In other words, the lateral fissure, limiting the frontoparietal operculum from the temporal one, appears as a straight cleft on a coronal slice (Fig. 3.2), with a latero-medial direction (opercular cleft); it medially splits into two branches, the insular clefts [7], oriented dorsomedially and ventromedially, limiting the medial aspect of the opercula from the insula. Taken together, the opercular and insular clefts have the shape of a "Y" or a "T" lying on its side on a coronal section. They limit a space, the lateral fossa, located between the medial aspects of the opercula laterally and the insula medially. The lateral fossa opens laterally through the lateral fissure. In the dorso-ventral direction, the fossa is wider at its anterior part, since the height of the insula decreases posteriorly.

3.3 White Matter Relationships

Several white matter tracts surround the insula (Fig. 3.3): one is dorsolateral and two are ventromedial (inferior fronto-occipital (IFOF) and uncinate (UF) fasciculi.

The *SLF* turns around the posterior part of the insula to connect the frontoparietal and temporal opercula [8–10]. It includes two superficial components (anterior or horizontal and posterior or vertical) and a deep one (arcuate fasciculus). The *anterior horizontal segment* is dorsal and lateral to the insula, just medial to the short association U-fibers of the frontoparietal operculum. It connects the supramarginal gyrus and posterior part of the superior temporal gyrus to the subcentral gyrus. The *posterior vertical segment* has weaker relationships to the insula: it is located posterior to the insula and is also superficial, just medial to the short U-fibers. It connects the supramarginal gyrus and posterior part of the middle and superior temporal gyri. Finally, the *deep or arcuate segment* is medial to the two previous ones and connects the posterior part of the superior, but also middle and inferior temporal gyri to the ventral precentral area, to the pars opercularis of the inferior frontal gyrus, and to the posterior part of the middle frontal gyrus. It is thus C-shaped and curves around the posterior angle of the insula, limited by the intersection of the superior and inferior segments of the circular sulcus of the insula.

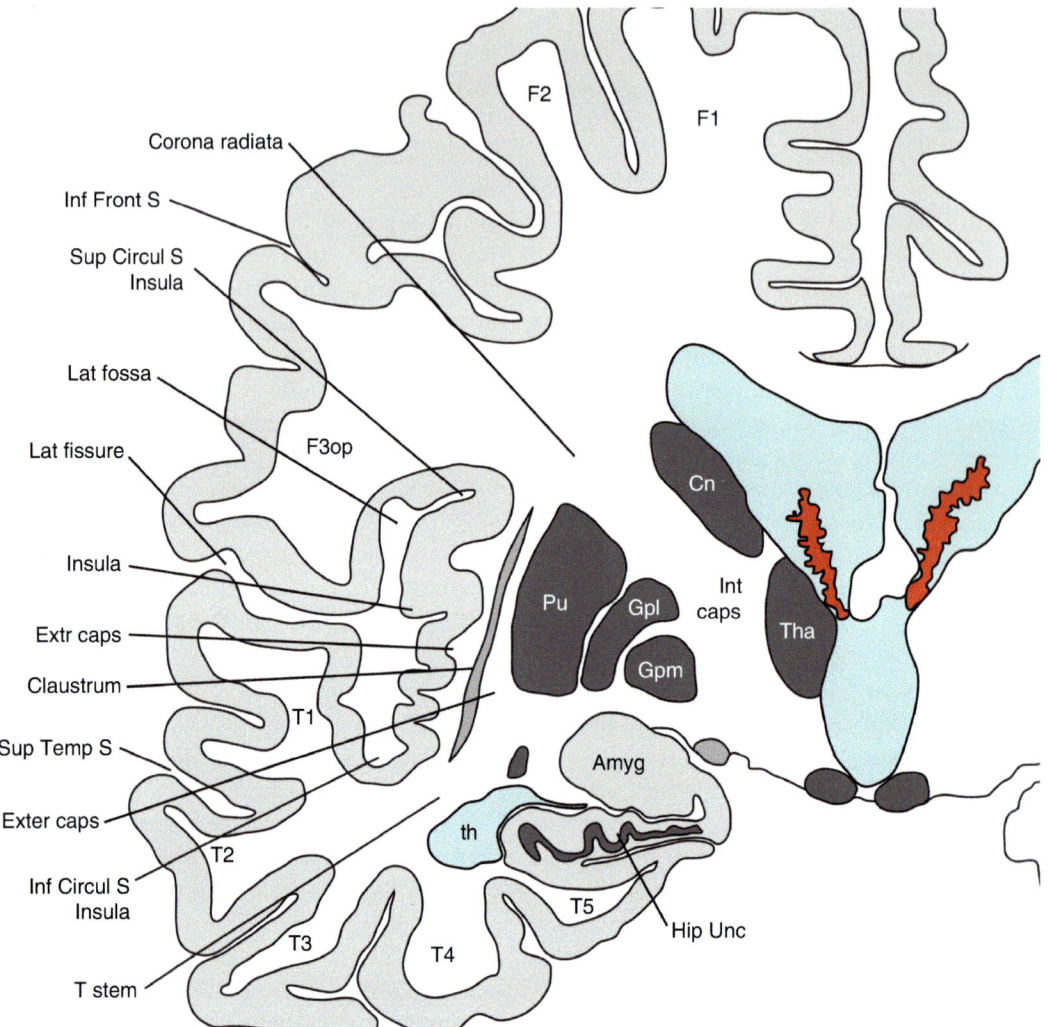

Fig. 3.2 Schematic coronal slice at the level of the hippocampal uncus. The lateral fossa (*Lat fossa*) is the space bordered by the *insula* medially and the opercula laterally, which contains branches of the middle cerebral artery. It is continued by the lateral segment of the lateral fissure (*Lat fissure*), which opens at the superficial aspect of the brain. The space located between the opercula is also known as the opercular cleft, whereas the ones located between the opercula and insula are known as the insular clefts. On this slice, the frontal and temporal opercula are made of the opercular part of the inferior frontal gyrus (*F3op*), and the superior temporal gyrus (*T1*), respectively. They are limited from the insula by the superior (*SupCircul S Insula*) and inferior circular sulci of the insula (*Inf Circul S Insula*). Several structures are found deep to the insula, from lateral to medial: the extreme capsule (*Extr caps*), the *claustrum*, the external capsule (*Exter caps*), the putamen (*Pu*), the lateral (*Gpl*) and medial (*Gpm*) parts of the globus pallidus and the internal capsule (*Int caps*). The *Extr caps* and *Exter caps* fuse at the inferior limit of the *claustrum* to become the temporal stem (*T stem*), located between the inferior circular sulcus of the insula (*Inf Circul S Insula*) and the temporal horn of the lateral ventricle (*th*). The three capsules and callosal fibers fuse at the superior aspect of the putamen and claustrum to form the *corona radiata*, making the *Int caps* more vulnerable at the superior aspect of the *insula*. *Amyg* Amygdala, *Cn* Caudate nucleus (body), *F1* Superior frontal gyrus, *F*: Middle frontal gyrus, *Hip unc* Hippocampal uncus, *Inf Front S* Inferior frontal sulcus, *Sup Temp S* Superior temporal sulcus, *T2* Middle temporal gyrus, *T3* Inferior temporal gyrus, *T4* Fusiform gyrus, *T5* Parahippocampal gyrus, *Thal* Thalamus

The *white matter located medial to the insula* is organized in three capsules (extreme, external, and internal) separated by gray matter (claustrum and lenticular nucleus).

- The *extreme capsule* is located just medial to the insular cortex. It may be subdivided into dorsal and ventral components. The *dorsal extreme capsule* contains short insulo-insular and long insulo-opercular association fibers that will be described in the chapter dedicated to insular projections. The *ventral extreme capsule* contains the trunk of the UF and IFOF [5, 11], a densely packed group of fibers having a global anteroposterior direction.
- The *claustrum* is a thin layer of gray matter lying between the extreme (laterally) and external (medially) capsules. Its dorsal part is compact, whereas its ventral part is made of small islands of gray matter intermingled with the fibers of the trunk of the IFOF and UF.
- The *external capsule*, which is medial to the claustrum, is also subdivided into dorsal and ventral parts: the *dorsal external capsule* contains radiate claustro-cortical fibers, whereas the trunk of the UF and IFOF forms the *ventral external capsule*.
- The *lenticular nucleus* is located medial to the external capsule and is triangular on a coronal slice: its lateral base is formed by the putamen, medially continued by the lateral and medial parts of the pallidum.
- Finally, the *internal capsule* lies medial to the lenticular nucleus and follows a ventromedial direction to reach the mesencephalon. It is subdivided into anterior and posterior limbs joined by a genu [12]. The *anterior limb*, located between the caudate and lenticular nuclei (lenticulo-caudate portion), is anterolaterally oriented on a horizontal slice. The fibers of the *posterior limb* of the internal capsule are fanning between the thalamus and caudate nucleus. Due to the rotation of the caudate nucleus around the thalamus, these fibers have a radiate orientation: some are mainly running dorsally and appear above the putamen (lenticulo-thalamic portion), some are posteriorly oriented (retrolentiform portion), and some are running ventrally, between the lenticular nucleus and the tail of the caudate nucleus (sublentiform portions). Fibers of the corticospinal tract run in the posterior limb of the internal capsule and project medial to the long insular gyri, whereas the corticonuclear tract, located in the *genu*, runs medial to the posterior short insular gyrus [13].

The internal, external, and extreme capsules and callosal fibers fuse dorsal to the superior limit of the claustrum and putamen to form the *corona radiata*. Similarly, the *stratum sagittale*, which may be regarded as the equivalent of the *corona radiata* for the temporal and occipital area, contains fibers of the sublentiform portion of the internal capsule (including optic and auditory radiations), fibers of the anterior commissure, and fibers of the IFOF. The distinction of these different contingents of fibers during dissection is a matter of debate in the anatomical literature; some authors are describing each of these fasciculi separately [9, 14, 15], while others are claiming that their fibers are so intermingled that any individual dissection is impossible, the stratum sagittale being only described as a whole [12].

The anteroventral part of the insula is in close relationships with the *trunks of the UF and IFOF* [5, 11], which occupy the ventral part of the extreme and external capsules. IFOF and UF are both made of two fans connected by a trunk.

- The *IFOF* joins the frontal to the parieto-occipital lobes and was described as made of two layers [15, 16]; its superficial dorsal one joins the inferior frontal gyrus (pars opercular, triangular, and orbicular) to the superior parietal lobule and superior and middle occipital gyri. Its deep ventral layer runs from the orbital gyri, the middle frontal gyrus, and the frontal pole to the inferior occipital gyrus and ventral aspect of the middle occipital gyrus and temporobasal cortex.
- The *UF* trunk is located ventrolateral to the IFOF trunk. The temporal portion of the UF fans in the anterior portion of the three temporal gyri, anterior to the amygdala; its frontal part reaches the gyrus rectus, orbital cortex, and subcallosal area [17].

Trunks of both the UF and IFOF extend from the external and extreme capsules to the *temporal stem* (Fig. 3.4). The latter is the white matter connecting the temporal lobe to the rest of the brain, between the inferior circular sulcus of the insula and the temporal horn of the lateral ventricle [17, 18]. It extends anteroposteriorly from the coronal plane of the amygdala to that one of the lateral geniculate body [18] and is continuous on a coronal section with the ventral part of the external and extreme capsule. It contains the UF and IFOF trunk anteriorly and the anterior commissure and optic radiations posteriorly [17–19]. The anterior commissure has a ventrolateral course from the midline and crosses the inferior limit of the insula to enter the temporal lobe via the temporal stem [19].

3.4 Vascular Relationships

The lateral fossa is the space located between the insula medially and the medial aspect of the opercula laterally. It is continued by the lateral part (or operculo-insular, [5]) and basal part (or stem [5]) of the lateral fissure: the lateral part of the lateral fissure was previously described and corresponds to the opening of the lateral fossa at the surface of the brain; the basal aspect of the lateral fissure is the cleft located between the ventral aspects of the frontal and temporal lobes; it runs from the limen insulae, where it laterally communicates with the lateral fossa, to the anterior perforated space, located medially.

Fig. 3.3 Step-by-step Klingler's dissection of the peri-insular white matter tracts. The red arrows point the approximate level of dissection on a coronal schematic slice. (**a**) Lateral segments of the arcuate fasciculus. After removing the cortex and short association "U" fibers, the two most superficial segments of the arcuate fasciculus appear: the vertical posterior segment (*SLF vert*) is posterior to the insula and joins the supramarginal gyrus to the posterior part of the superior and middle temporal gyri; the horizontal anterior segment (*SLF hor*), dorsal to the insula, joins the supramarginal gyrus and posterior part of the superior temporal gyrus, to the subcentral gyrus. Due to the partial removal of the frontoparietal (*Fr-Par operc*), temporal (*Temp operc*) and fronto-orbital (*Fr-Orb operc*) opercula, the *insula* appears. (**b**) Deep long segment of the SLF, or arcuate fasciculus (*SLF arc*). After the superficial segments of the SLF were removed, its deep or arcuate segment appears. It curves around the posterior tip of the insula and joins the superior, middle and inferior temporal gyri to the ventral precentral area, pars opercularis of the inferior frontal gyrus and posterior part of the middle frontal gyrus. The insula clearly appears, limited from the opercula by its circular sulcus (*Circul S insula*), and subdivided by the central sulcus of the insula (*Central S insula*) into long (*Long insul G*) and short insular (*Short insul G*) gyri. (**c**) Extreme capsule. The extreme capsule appears after the insular cortex is removed and contains the superficial insulo-insular "U" fibers. Its dorsal part (*Extr caps* dorsal) also includes insulo-opercular projection fibers, having a radiating pattern, whereas its ventral part (*Extr caps* ventral) is made of the trunks of the uncinate (*UF*) and Inferior Fronto-occipital fasciculi (*IFOF*). (**d**) Claustrum and external capsule. The *claustrum* is subdivided into a denser dorsal part (*Claustrum dorsal*), medial to the dorsal *Extr caps*, and a ventral part (*Claustrum ventral*) made of small islands of gray matter intermingled with the trunks of the *IFOF* and *UF*. The ventral part of the external capsule (*Exter caps ventral*) is continuous with *Extr caps ventral* and is made of the trunks of the *IFOF* and *UF*. Its dorsal part (*Exter caps dorsal*) is made of radiate claustro-cortical fibers. The putamen (*Pu*) appears just medial to *Exter caps dorsal* after a few fibers are removed. (**d**) Putamen (*Pu*). The *Pu* appears after removing the *Exter caps*. (**e**) The uncinate fasciculus (*UF*). The *UF* runs ventral within the extreme and external capsules and connects the anterior portion of the three temporal gyri, anterior to the amygdala, to the gyrus rectus, orbital cortex and subcallosal area. The trunk of the *IFOF* is dorsal to the *UF*; it connects the frontal to the parieto-occipital lobes. Dorsal to the putamen, the extreme, external and internal capsule fuse to form the *corona radiata*. (**f**) Internal capsule (*Int caps*). The *Pu* and globus pallidus are removed and the *Int caps* appears medial to them. It is continued dorsally by the *corona radiata* and posteriorly by the stratum sagittale (*SS*). The *SS* may be regarded as the equivalent of the corona radiata for the temporal and occipital area and contains fibers of the sublentiform portion of the internal capsule (including optic and auditory radiations), fibers of the anterior commissure, and fibers of the *IFOF*.

Ventral to the anterior perforated space and lateral to the optic chiasm (Fig. 3.5), the internal carotid artery bifurcates into the anterior (ACA) and middle cerebral arteries (MCA), the latter being about 4 mm in diameter, roughly twice that of the ACA [20].

- The M1 or sphenoidal segment of the MCA runs laterally, parallel, and posterior to the sphenoidal ridge. It is located in the basal part of the lateral fissure and reaches the limen insulae. It divides into two (78% of hemispheres), three (12%), or multiple (10%) trunks. It is therefore described as made of pre- and post-bifurcation segments [20], which continue their course parallel to each other. The respective size of these trunks is variable, and, in case of bifurcation, equivalent trunks are observed in about 2 out of 8 hemispheres, whereas superior dominant (3 out of 8) or inferior dominant (3 out of 8) patterns are more common [20]. The M1 segment perforators, the anterolateral central (or anterolateral lenticulostriate) arteries, mainly originate from the pre-bifurcation segment of M1 and ascend to reach the anterior perforated space. Depending on their origin, they are subdivided into lateral, medial, and sometimes intermediate groups. The lateral and intermediate groups run through the putamen to vascularize the head and body of the caudate and superior part of the internal capsule. The

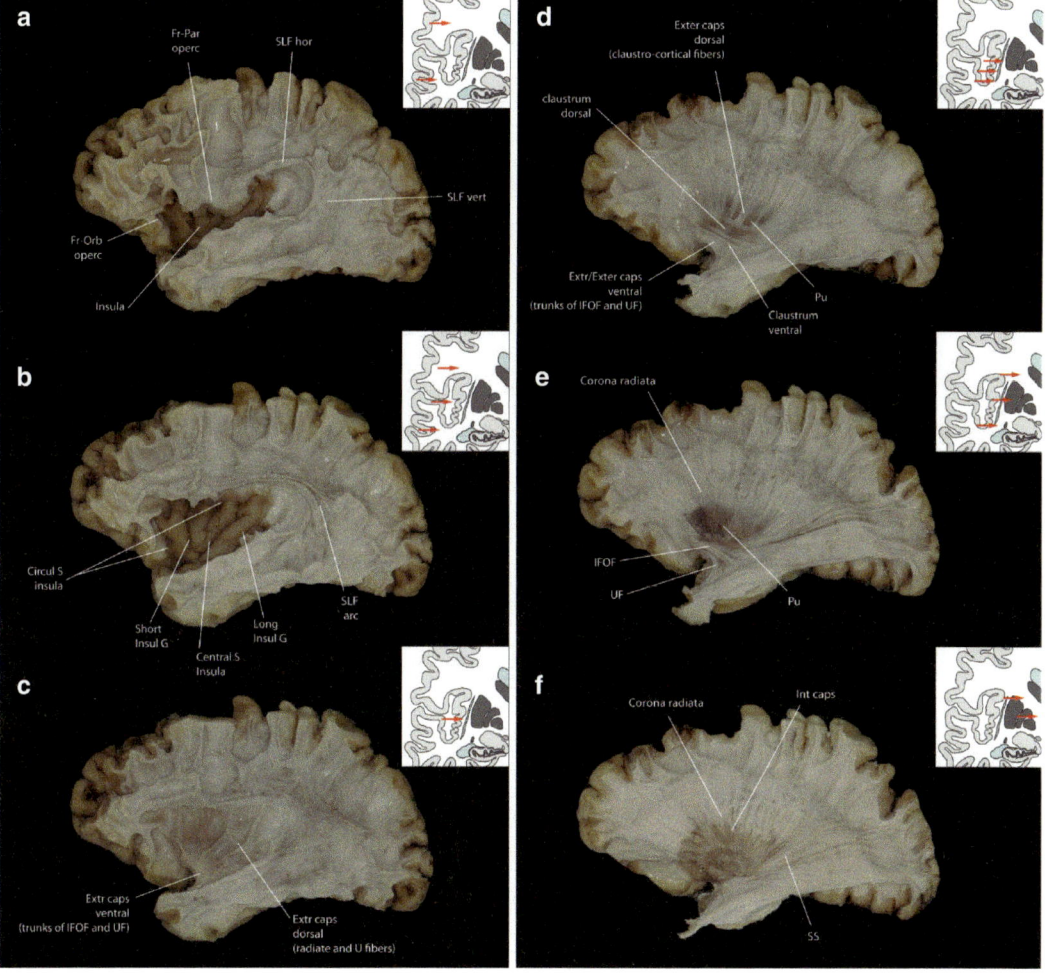

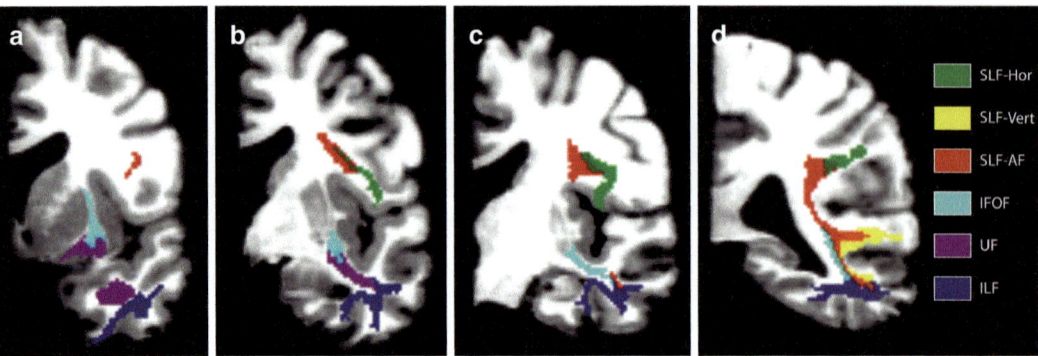

Fig. 3.4 Peri-insular white matter tracts. White matter tracts obtained from the dissection using Klingler's method were scanned at each step of dissection using a 3D scanner laser, 3D-reconstructed and ported into the reference space of the dissected specimen postmortem MRI (see [33] for details). The figure represents ex vivo MRI coronal slices of a left hemisphere, from the anterior insula (A) to a slice just retroinsular (D). Main peri-insular tracts are superimposed in color onto these MRI images: the horizontal segment of the superior longitudinal fasciculus (*SLF-Hor*, green) is located dorsal and lateral to the insula; the vertical segment of the same fasciculus (*SLF-Vert*, yellow) is posterior to the insula; the deep segment of the SFL, or arcuate fasciculus (*SLF-AF*, red) turns around the posterior insula to extend into the posterior temporal and ventral frontal lobes; the uncinate (*UF*, purple) and inferior fronto-occipital fasciculi (*IFOF*, light blue) trunks are located in the ventral external and extreme capsules. *ILF* Inferior longitudinal fasciculus (dark blue)

medial group vascularizes part of the globus pallidus, anterior limb of the internal capsule, and head of the caudate [20].

- At the level of the limen insulae, the MCA turns posterodorsally, with an angle of approximately 90°, to enter the lateral fossa (Fig. 3.6). This sharp angle, or genu, clearly visible on a frontal angiogram, is the limit between the M1-sphenoidal and M2-insular segments of the MCA. The trunks of the MCA divide here in a mean of eight stem arteries [5–11] fanning at the surface of the insula [7]. These stem arteries divide into two or more cortical arteries that reach the circular sulcus of the insula, considered as the distal limit of the M2 insular segment of the MCA. This division of the stem arteries in cortical arteries usually occurs before or at the level of the circular sulcus of the insula [7, 20]. Due to the shape of the insula, the M2 segment has a global triangular distribution, visible on lateral angiograms. Most of the arteries vascularizing the insula originate from the M2 segment [13]; the antero-dorsal part of the insula (short insular, accessory and transverse gyri, anterior circular sulcus, apex) is mainly vascularized by the superior trunk of the MCA and its branches; the posteroventral insula (posterior long insular gyrus, limen) is supplied by the inferior trunk and branches of the MCA; the central part of the insula (central sulcus of the insula and anterior long insular gyrus) gets a double vascularization from both trunks. Finally early branches, originating from M1 before its bifurcation, can supply every part of the insula except its central sulcus. A detailed description of the arterial supply of each insular gyrus is given in [7], other descriptions with a more extensive territory for the superior trunk being proposed by other authors [6, 13]. Most insular arteries are short [6, 21] and only supply the insular cortex and underlying extreme capsule; 10% are medium sized and vascularize also the claustrum and external capsule; finally 3–5% of insular arteries are long and reach deeper structures, as the corona radiata [13]. Three to eleven such long insular perforators are found per hemisphere; in about 80%, they originate from branches of the superior trunk at the level of M2 segment

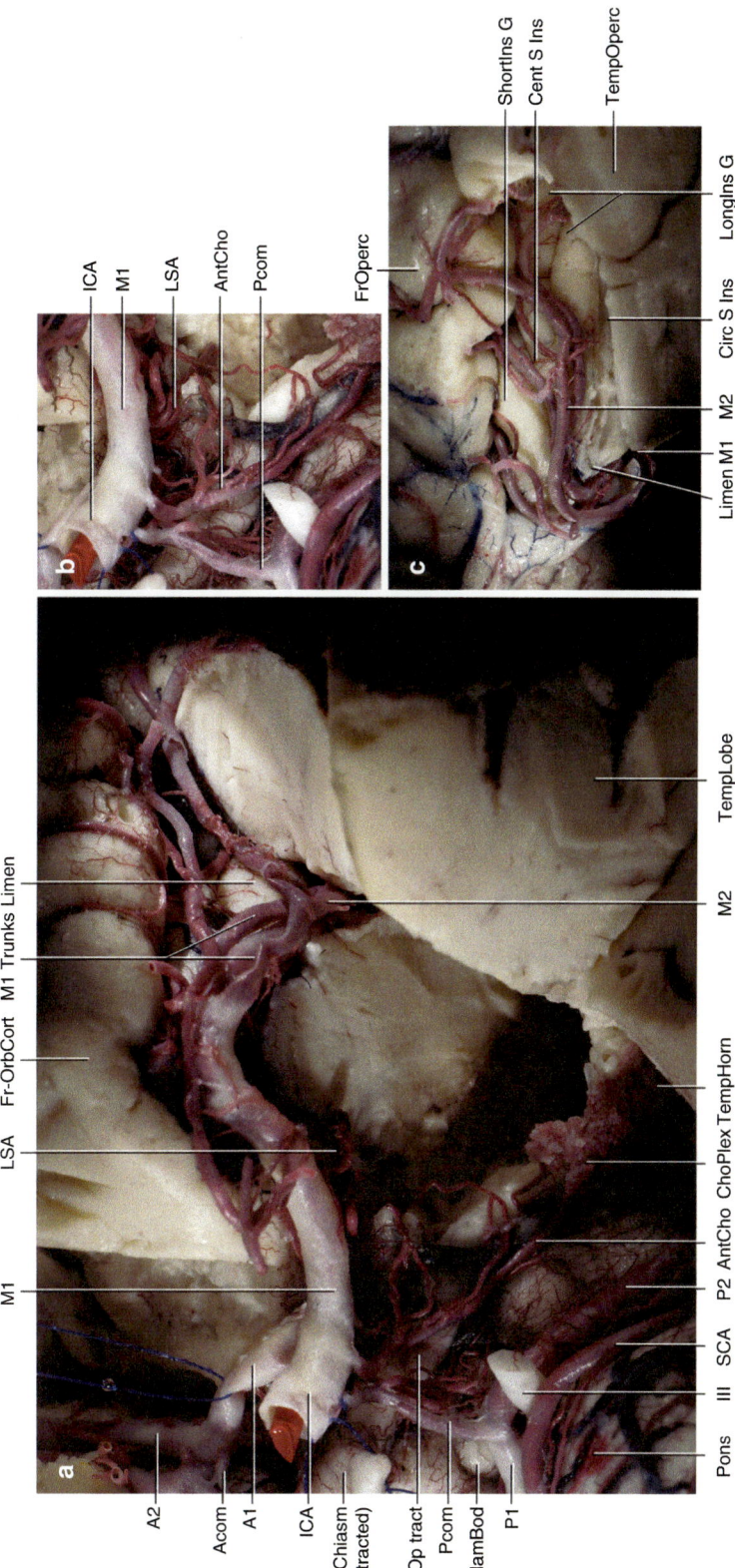

Fig. 3.5 M1 and M2 segments of the Middle cerebral artery. (**a**) Ventral view of a left hemisphere after part of the temporal lobe (*TempLobe*) was opened to show the temporal horn (*TempHorn*) of the lateral ventricle and choroid plexus (*ChoPlex*). From proximal to distal, the internal carotid artery (*ICA*) gives the posterior communicating artery (*Pcom*), the anterior choroidal artery (*AntCho*). It finally divides lateral to the optic chiasm (*Chiasm*) in precommunicating (*A1*) segment of the anterior cerebral artery and sphenoidal segment (*M1*) of the middle cerebral artery. The later runs in the basal compartment of the lateral fissure where it divides in trunks (*M1Trunks*). The prebifurcation part of M1 gives perforators, the anterolateral central (or lenticulate, *LSA*) arteries. After passing the limen insulae (*Limen*), M1 is continued by the insular segment (*M2*) of the middle cerebral artery. (**b**) Detailed view of the distal branches of the *ICA* and *LSA*. (**c**) Lateral view of the left hemisphere after the temporal operculum (*TempOperc*) was partially removed and the frontoparietal operculum (*FrOperc*) was retracted to show the *limen*, the long insular gyri (*LongIns G*), the central insular sulcus (*Cent S Ins*), and the most posterior short insular gyrus (*ShortIns G*). At the level of the *limen insulae*, the trunks of the middle cerebral artery follows a 90° angle that marks the *M1–M2* limit. Arteries belonging to the M2 segment fan and divide at the surface of the insula, up to the circular sulcus of the insula (*CircSIns*). *A2* Post communicating segment of the anterior cerebral artery, *Acom* Anterior communicating artery, *Fr-OrbCort* Fronto-orbital cortex, *III* Oculomotor nerve, *MamBod* Mammillary body, *Op tract* Optic tract, *P1* Precommunicating segment of the posterior cerebral artery, *P2* Postcommunicating segment of the posterior cerebral artery, *SCA* Superior cerebral artery

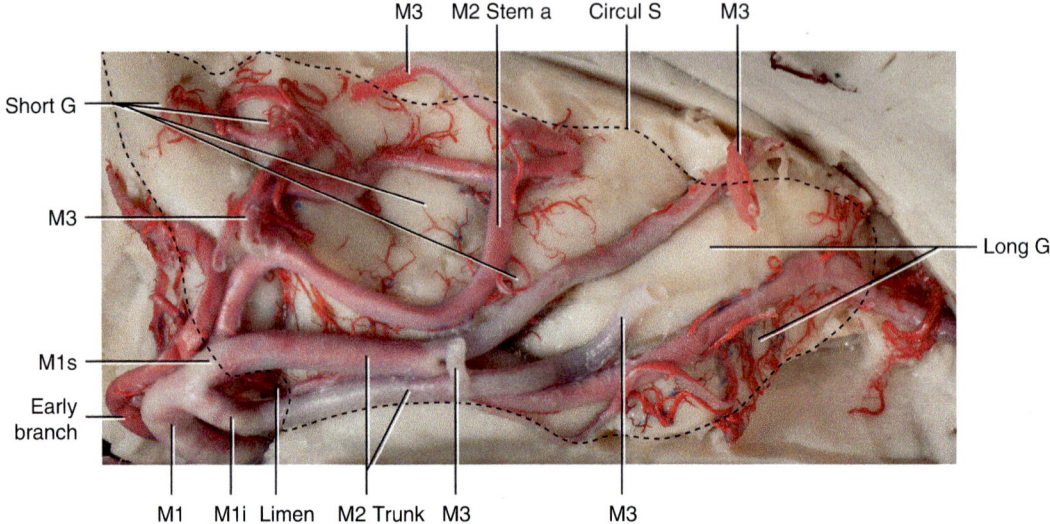

Fig. 3.6 Insular (M2) segment of the middle cerebral artery. Lateral view of a left hemisphere, after the opercula were removed to expose the short (*Short G*) and long (*Long G*) gyri of the insula. The sphenoidal (*M1*) segment of the MCA divides in a superior (*M1s*) and an inferior (*M1i*) trunks, which pass the limen insulae to enter the lateral segment of the lateral fossa. An early branch originates proximal to this bifurcation to reach the anterior part of the insula and the fronto-orbital operculum. The insular segment of the MCA (M2) is the one running at the surface of the insula, from the *limen* to the circular sulcus of the insula (*Circul S*, dash line). It is made of the main trunks (*M2 Trunk*), which divides in stem arteries (*Stem a*). The latter then divide again, close to the *Circul S* in cortical arteries. Distal to the *Circul S*, the MCA branches are known as the opercular segment (*M3*) which follow the medial and then horizontal aspects of the opercula to reach the lateral aspect of the brain

(51.6%), M2–M3 junction (37.4%), or proximal M3 (11%) [13]. These long perforators enter the dorsal part of the insula, close to the superior circular sulcus: at the level of the central insula in 36%, of the long insular gyri in 29%, and the anterior part of the insula in 35% [13, 21]. These perforators have an ascending direction, orthogonal to the insula, toward the body of the lateral ventricle. They then reach deep targets: arcuate fasciculus for all of them, corticonuclear tract for those penetrating the dorsal part of the short insular gyri, corticospinal tract for those penetrating the dorsal part of the long insular gyri [21].
- At the level of the circular sulcus of the insula, the MCA branches follow a hairpin curve to reach the medial aspect of the opercula. This 180° curve, which is more obvious for the M2 branches having an ascending course, is visible on angiograms and marks the limit between the M2-insular and M3-opercular segments. Once arrived at the junction of the medial and horizontal part of the opercula (superior for the temporal one, and inferior for the frontoparietal one), M3 changes again its direction to run laterally with a new turn of about 90° before reaching the lateral limit of the lateral fissure, which marks the limit between the M3-opercular and M4-cortical segments.
- From this point, the branches of the M4 segment exit the lateral fissure and spread over the surface of the cortex to reach their terminal destination. The territory of the MCA is subdivided into 12 different areas served by cortical arteries with a variable branching pattern from the stem arteries: orbitofrontal, prefrontal, precentral, central, anterior parietal, posterior parietal, angular, temporo-occipital, posterior temporal, middle temporal, anterior temporal, and temporopolar arteries [7, 20].

3.5 Venous Relationships

The venous pattern of the insula is variable, but four insular veins are commonly described:

- The *anterior insular vein* runs close to the anterior circular sulcus with a ventro-posterior direction, toward the limen insulae. It drains the anterior circular sulcus and the anterior short insular gyrus.
- The *precentral insular vein* drains the middle short insular gyrus and apex and follows the insular precentral sulcus with a ventro-posterior direction.
- The *central insular vein* courses along the central sulcus of the insula to drain the posterior short and anterior long gyri, central insular sulcus, and limen insulae.
- The *postcentral vein* follows the postcentral insular sulcus and gets blood from the posterior long gyrus, posterior circular sulcus, and limen insulae [7].

Classically, these insular veins join together near the limen insulae to form the deep middle cerebral veins (DMCV), which courses in the basal segment of the lateral fissure. At the level of the anterior perforated space, it anastomoses with the anterior cerebral vein to form the anterior part of the basal vein. The drainage of the insular veins was shown [7] to be more complex, some of them having a partial or total drainage toward the superficial middle cerebral vein (SMCV or superficial Sylvian vein). The SMCV parallels the lateral segment of the lateral fissure, usually coursing a few millimeter ventral to it. At the anterior tip of the lateral part of the lateral fissure, it turns medially to reach its basal part before ending in the sphenoparietal or more rarely in the cavernous or sphenopetrosal sinuses. On the way it drains superficial cortical veins, divided into frontosylvian, parietosylvian, and temporosylvian groups but also some of the insular veins, especially the anterior and precentral ones.

3.6 Surgical Consequences

Due to its deep location, surgical approaches of the insula have to deal with the opercula and the vessels contained in the lateral fossa. The transsylvian and subpial approaches will be detailed chapter XXX, but a few surgically relevant details are pointed here.

3.6.1 Transsylvian Approach

Identifying the lateral fissure is usually trivial, but in case of large expansive processes or malformation (for instance, megalencephaly), the usual anatomical landmarks can be left. The lateral fissure can then be identified as the sulcus where numerous arteries (M4 segment) exit with ascending and descending courses over its ventral and dorsal lips, to reach their terminal cortical territories.

The opening of the lateral fossa can be tailored depending of the area of the insula to be reached; the area of the limen insulae and basal compartment of the lateral fissure can be reached to access the ICA, MCA bifurcation, arterial circle, and sphenoidal ridge. A reliable landmark to open the lateral fissure in this area is the inferior tip of the triangular part of the inferior frontal gyrus, the approach being conducted ventrally from this point. More posterior approaches can benefit from the projection of the superficial sulci onto the insula (Fig. 3.1). Briefly:

- The dorsal end of the vertical ramus of the anterior segment of the lateral fissure projects onto the anterior insula, close to the superior circular sulcus [7].
- The central sulcus of the insula has about the same orientation as the central sulcus. Its dorsal tip can face the ventral tip of the central sulcus in 60% [13] or be located a few millimeters anterior or posterior to it.
- The middle horizontal segment of the lateral fissure corresponds to the juxtaposition of the

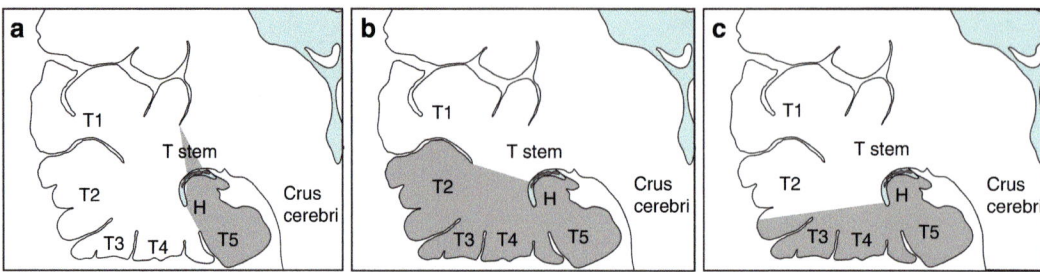

Fig. 3.7 Surgical consequences of amygdalohippocampectomy. The transsylvian approach (**a**) preserves most of the neocortex as compared to trans-superior temporal sulcus (**b**) or trans T2 (**c**) approaches. It nevertheless implies to cut the anterior part of the temporal stem (*T stem*). *H* Hippocampus, *T1* Superior temporal gyrus, *T2* Middle temporal gyrus, *T3* Inferior temporal gyrus, *T4* Fusiform gyrus, *T5* Parahippocampal gyrus

frontoparietal and temporal opercula, whereas its posterior segment is surrounded by the supramarginal gyrus. In other words, the middle segment is the only one facing the lateral fossa and insula, and the junction of the middle horizontal and posterior ascending segments of the lateral fissure is a good landmark for the posterodorsal limit of the insula (junction of the superior and inferior circular sulci.

Whatever the considered part of the lateral fissure, retraction of the opercula has to be as limited as possible, especially in the subcentral area to avoid postoperative contralateral facial palsy but also in the superior temporal and inferior frontal and inferior precentral gyri in the dominant hemisphere to preserve language areas.

Once the lateral fossa is opened, if the trunks, stem, and cortical arteries of the MCA need to be dissected, two main areas have to be especially considered because of perforators:

– The area located *medial to the limen insulae* (basal compartment of the lateral fissure) contains perforators (anterolateral central arteries) issued from the M1 segment of the MCA. The resection of a tumor in this area has to remain lateral to the most lateral of these arteries to avoid postoperative ischemia of the internal capsule and basal ganglia. The level of this most lateral M1 perforator can be considered as the lateral limit of the anterior perforated space. The area located between the anterior perforated space and the limen insulae, or limen recess, is free of perforators and is usually about 15 mm width [7].

– The second area with important relationships to perforators is the superior part of the insula, close to the superior circular sulcus. As shown by [13, 21] long perforators of the M2/M3 segments of the MCA penetrate the insula in this area and may reach deep structures: arcuate fasciculus for all of them, corticonuclear tract for those penetrating the dorsal part of the short insular gyri, and corticospinal tract for those penetrating the dorsal part of the long insular gyri. These perforators have of course to be respected during the dissection of the MCA branches in the lateral fossa.

The transsylvian approach was proposed to reach the temporal horn of the lateral ventricle (Fig. 3.7) to perform selective amygdalohippocampectomy [22]. In this approach, an incision of about 2 cm is performed along the anterior part of the inferior circular sulcus of the insula to reach the temporal horn of the ventricle at a depth of about 0.5 mm [7]. Such an approach has the advantage to avoid neocortical resection but implies the manipulation of MCA branches and, above all, the disconnection of the anterior part of the temporal stem, with possible lesions to the uncinate fasciculus, anterior commissure, and anterior part of the temporal loop of the optic radiations [19, 23].

3.6.2 Transopercular Approach

This approach begins by the partial resection of the temporal or/and frontoparietal opercula after the superficial pia mater was cut parallel to the inferior or superior lip of the lateral fissure, respectively. The extension of this resection depends on the part of the insula to be reached, a temporal approach being for instance privileged in the case of a temporo-insular glioma. The projection of the sulci of the lateral aspect of the hemisphere onto the insula shows that about 2/3 of the insula is located medial to the temporal operculum, the remaining dorsal third being hidden by the frontal operculum. This explains why a temporal resection is usually preferred when a subpial approach is proposed for a glioma limited to the insula [24].

The resection is performed subpially, the pia mater being followed from the lateral, to the inferior or superior, and then medial aspects of the operculum to reach the inferior or superior circular sulcus, giving access to the insula. In this approach the vascular dissection is minimal, branches of the MCA remaining embedded by the pia mater covering the walls of the lateral fossa. It nevertheless implies the resection of possibly highly functional areas involved in language (inferior frontal and inferior precentral gyrus, superior temporal gyrus in the dominant hemisphere) or motor function (subcentral area). For this reason, a cortical and subcortical mapping using direct electrical stimulation (DES) and awake surgery has to be performed prior and during the opercular resection [24]. For instance [25], *dysarthria* is induced by stimulation of the ventral sensorimotor cortex (vSMC) on both hemispheres (ventral part of the pre- and postcentral gyri and subcentral gyrus [26]) or by stimulation of the anterior horizontal segment of the SLF; it connects to the supramarginal gyrus (articulatory loop). A *total motor arrest* can be induced by DES of the vSMC or pars opercularis of the inferior frontal gyrus. *Phonological disorders* can be elicited by DES of the dorsal phonological stream: arcuate fasciculus around the posterodorsal part of the insula and areas it connects in the frontal (opercular part of the inferior frontal gyrus, posterior third of the middle frontal gyrus) and temporal lobes (posterior third of the middle and inferior temporal gyri). *Semantic disturbances* (or at maximum, anomia) are induced by stimulation of cortical components of the ventral semantic stream in the dominant hemisphere:

pars opercularis of the inferior frontal gyrus, posterior part of the superior and middle frontal gyri, and superior temporal sulcus [27].

3.6.3 Resection of the Insula

During tumor resection, several areas of the insula need to be approached with caution.

First, as previously mentioned, the entry area of the long perforators is located at the *superior aspect of the insula*. These vessels have to be respected in that they vascularize deep white matter.

The *ventral aspect of the extreme and external capsules*, located just medial to the ventral insula, also has to be respected because it contains the trunks of the IFOF and UF. The role of the UF remains unclear in the literature, but its preservation seems not to be crucial for language [27, 28]. In the dominant hemisphere, the situation is completely different for the IFOF, which completes the semantic network mentioned above, at the cortical level. DES of the IFOF consistently induces transient verbal (semantic paraphasia, anomia) [29, 30] and nonverbal semantic disturbances (impossibility to associate images belonging to the same semantic field during for instance the pyramid and palm tree test) [31]. It also produces a certain degree of unawareness of the induced deficits [31] and perseveration [32]. Its lesion induces non-compensable permanent deficit, pleading for a systematic use of DES during resection of insular lesions.

The *internal capsule*, including fibers of the corticonuclear and corticospinal tracts, is separated from the insula by the extreme capsule, claustrum, external capsule, and lenticular nucleus. The internal capsule is thus relatively preserved from a direct injury during insular resection, except at the superior border of the putamen, where the extreme, external, and internal capsules fuse to become the corona radiata. As a consequence the white matter located medial to the superior part of the insula needs to be frequently tested using DES to control extension of the resection in this direction.

Conclusion
Surgical approach of the insula remains quite challenging due to its deep location, medial to

the opercula, and also a rich environment made of vessels and white matter tracts. It is nevertheless possible, bearing in mind a few anatomical information, important to avoid postoperative deficits.

Acknowledgments We thank Daniel Bourry, photographer at the University of Tours, for his help in picture acquisition and post processing.

References

1. Duvernoy HM. The human Brain. Surface, blood supply, and three-dimensional sectional anatomy. 2nd ed. Wien, New York: Springer; 1999. 491 p.
2. Destrieux C, Terrier LM, Andersson F, Love SA, Cottier J-P, Duvernoy H, et al. A practical guide for the identification of major sulcogyral structures of the human cortex. Brain Struct Funct. 2017;222(4):2001–15.
3. Destrieux C, Fischl B, Dale A, Halgren E. Automatic parcellation of human cortical gyri and sulci using standard anatomical nomenclature. Neuroimage. 2010;53:1–15.
4. Guenot M, Isnard J, Sindou M, et al. Surgical anatomy of the insula. Adv Tech Stand Neurosurg. 2004;29:265–88.
5. Türe U, Yaşargil DC, Al-Mefty O, Yaşargil MG. Topographic anatomy of the insular region. J Neurosurg. 1999;90(4):720–33.
6. Varnavas GG, Grand W. The insular cortex: morphological and vascular anatomic characteristics. Neurosurgery. 1999;44(1):127–36; discussion 136–8.
7. Tanriover N, Rhoton AL Jr, Kawashima M, Ulm AJ, Yasuda A. Microsurgical anatomy of the insula and the sylvian fissure. J Neurosurg. 2004;100(5):891–922.
8. Martino J, De Witt Hamer PC, Berger MS, Lawton MT, Arnold CM, de Lucas EM, et al. Analysis of the subcomponents and cortical terminations of the perisylvian superior longitudinal fasciculus: a fiber dissection and DTI tractography study. Brain Struct Funct. 2013;218(1):105–21.
9. Fernandez-Miranda JC, Rhoton AL, Alvarez-Linera J, Kakizawa Y, Choi C, de Oliveira EP. Three-dimensional microsurgical and tractographic anatomy of the white matter of the human brain. Neurosurgery. 2008;62(6 Suppl 3):989–1026. discussion 1026-8.
10. Catani M, Jones DK, ffytche DH. Perisylvian language networks of the human brain. Ann Neurol. 2005;57(1):8–16.
11. Fernandez-Miranda JC, Rhoton AL, Kakizawa Y, Choi C, Alvarez-Linera J. The claustrum and its projection system in the human brain: a microsurgical and tractographic anatomical study. J Neurosurg. 2008;108(4):764–74.
12. Goga C, Türe U. The anatomy of Meyer's loop revisited: changing the anatomical paradigm of the temporal loop based on evidence from fiber microdissection. J Neurosurg. 2015;122(6):1253–62.
13. Delion M, Mercier P, Brassier G. Arteries and veins of the Sylvian fissure and insula: microsurgical anatomy. In: Schramm J, editor. Advances and technical standards in neurosurgery [Internet]. Cham: Springer International Publishing; 2016. p. 185–216. [cited 2017 Aug 13]. Available from: http://link.springer.com/10.1007/978-3-319-21359-0_7.
14. Sarubbo S, De Benedictis A, Milani P, Paradiso B, Barbareschi M, Rozzanigo U, et al. The course and the anatomo-functional relationships of the optic radiation: a combined study with 'post mortem' dissections and 'in vivo' direct electrical mapping. J Anat. 2015;226(1):47–59.
15. Martino J, Brogna C, Robles SG, Vergani F, Duffau H. Anatomic dissection of the inferior fronto-occipital fasciculus revisited in the lights of brain stimulation data. Cortex. 2009;46:691–9.
16. Sarubbo S, De Benedictis A, Maldonado IL, Basso G, Duffau H. Frontal terminations for the inferior fronto-occipital fascicle: anatomical dissection, DTI study and functional considerations on a multi-component bundle. Brain Struct Funct. 2013;218(1):21–37.
17. Ebeling U, von Cramon D. Topography of the uncinate fascicle and adjacent temporal fiber tracts. Acta Neurochir. 1992;115(3–4):143–8.
18. Kier EL, Staib LH, Davis LM, Bronen RA. MR imaging of the temporal stem: anatomic dissection tractography of the uncinate fasciculus, inferior occipitofrontal fasciculus, and Meyer's loop of the optic radiation. AJNR Am J Neuroradiol. 2004;25(5):677–91.
19. Peuskens D, van Loon J, Van Calenbergh F, van den Bergh R, Goffin J, Plets C. Anatomy of the anterior temporal lobe and the frontotemporal region demonstrated by fiber dissection. Neurosurgery. 2004;55(5):1174–84.
20. Rhoton AL. The supratentorial arteries. Neurosurgery. 2002;51(4 Suppl):S53–120.
21. Delion M, Mercier P. Microanatomical study of the insular perforating arteries. Acta Neurochir. 2014;156(10):1991–8.
22. Yasargil MG. Microneurosurgery: in 4 volumes. Vol. 1. New York Stuttgart: Thieme Stratton ; Georg Thieme Verlag; 1984. 4 v. in 6.
23. Destrieux C, Bourry D, Velut S. Surgical anatomy of the hippocampus. Neurochirurgie. 2013;59(4–5):149–58.
24. Duffau H, Capelle L, Lopes M, Faillot T, Sichez JP, Fohanno D. The insular lobe: physiopathological and surgical considerations. Neurosurgery. 2000;47(4):801–10. discussion 810–1.
25. Mandonnet E, Sarubbo S, Duffau H. Proposal of an optimized strategy for intraoperative testing of speech and language during awake mapping. Neurosurg Rev

[Internet]. 2016 19 [cited 2016 Jul 3]; Available from: http://link.springer.com/10.1007/s10143-016-0723-x
26. Bouchard KE, Mesgarani N, Johnson K, Chang EF. Functional organization of human sensorimotor cortex for speech articulation. Nature. 2013;495(7441):327–32.
27. Duffau H. The anatomo-functional connectivity of language revisited New insights provided by electrostimulation and tractography. Neuropsychologia. 2008;46(4):927–34.
28. Duffau H, Gatignol P, Moritz-Gasser S, Mandonnet E. Is the left uncinate fasciculus essential for language? A cerebral stimulation study. J Neurol. 2009;256(3):382–9.
29. Almairac F, Herbet G, Moritz-Gasser S, de Champfleur NM, Duffau H. The left inferior fronto-occipital fasciculus subserves language semantics: a multilevel lesion study. Brain Struct Funct. 2015;220(4):1983–95.
30. Duffau H, Herbet G, Moritz-Gasser S. Toward a pluricomponent, multimodal, and dynamic organization of the ventral semantic stream in humans: lessons from stimulation mapping in awake patients. Front Syst Neurosci [Internet]. 2013 Aug 26 [cited 2015 Aug 28];7. Available from: http://www.ncbi.nlm.nih.gov/pmc/articles/PMC3752437/
31. Moritz-Gasser S, Herbet G, Duffau H. Mapping the connectivity underlying multimodal (verbal and non-verbal) semantic processing: a brain electrostimulation study. Neuropsychologia. 2013;51(10):1814–22.
32. Khan OH, Herbet G, Moritz-Gasser S, Duffau H. The role of left inferior fronto-occipital fascicle in verbal perseveration: a brain electrostimulation mapping study. Brain Topogr. 2014;27(3):403–11.
33. Zemmoura I, Serres B, Andersson F, Barantin L, Tauber C, Filipiak I, et al. FIBRASCAN: a novel method for 3D white matter tract reconstruction in MR space from cadaveric dissection. Neuroimage. 2014;103:106–18.

Anatomy of the Insular Arteries

Servet Celik, Okan Bilge, Canan Yurttaş, and Mehmet Turgut

Abbreviations

A	Insular apex
ACA	Anterior cerebral artery
Ag	Accessory gyrus
AIP	Anterior insular point
ALg	Anterior long gyrus
ALs	Anterior limiting sulcus
Ang	Branch to angular gyrus
APr	Anterior parietal artery
aPx	Pial arterial plexus
ASg	Anterior short gyrus
ATm	Anterior temporal branch
Both	Transitional or mixed vascular area
CN	Caudate nucleus
Cn	Central sulcal artery
Cn1	Central sulcal artery 1
Cn2	Central sulcal artery 2
Cns	Central insular sulcus
Early IT	Early branch of middle cerebral artery suppling inferior trunk area
Early ST	Early branch of middle cerebral artery suppling superior trunk area
IC	Insular cortex
ICA	Internal cerebral artery
ILs	Inferior limiting sulcus
IO	Insuloopercular artery
IT	Inferior trunk of middle cerebral artery
L	Limen of the insula
LN	Lentiform nucleus
LOF	Lateral orbitofrontal artery
LPr	Long perforating artery
LV	Lateral ventricle
M1	First (sphenoidal) segment of middle cerebral artery
M2	Insular segment of middle cerebral artery
M3	Opercular segment of middle cerebral artery
M4	Cortical segment of middle cerebral artery
MCA	Middle cerebral artery
MPr	Middle perforating artery
MSg	Middle short gyrus
MTm	Middle temporal branch
Pal	Pallidum
PCA	Posterior cerebral artery
PIP	Posterior insular point
PLg	Posterior long gyrus
PLIC	Posterior limb of the internal capsule
PlTm	Polar temporal artery
PPr	Posterior parietal artery
Pr	Perforating insular artery
PreCn	Precentral sulcal artery
PreCnI	Precentral insular vein

S. Celik, M.D. (✉) · O. Bilge, M.D.
C. Yurttaş, M.D.
Faculty of Medicine, Department of Anatomy,
Ege University, Izmir, Turkey
e-mail: canan.yurttas@ege.edu.tr;
okan.bilge@ege.edu.tr; servet.celik@ege.edu.tr

M. Turgut, M.D.
Faculty of Medicine, Department of Neurosurgery,
Adana Menderes University, Aydın, Turkey

PreCns	Precentral insular sulcus
PreF	Prefrontal artery
PSg	Posterior short gyrus
PstCns	Postcentral insular sulcus
PTm	Posterior temporal branch
Put	Putamen
SLs	Superior limiting sulcus
SPr	Short perforating artery
Ss	Short insular sulcus
ST	Superior trunk of middle cerebral artery
Stm	Stem artery
Tg	Transverse gyrus
TmO	Temporooccipital branch
TrI	Terminal insular artery

4.1 Introduction

The human insula is regarded as the hidden fifth lobe, and as a result, macroscopic features also vascularity of insula are neglected in undergraduate and in even postgraduate medical education. The vasculature of the insula is of interest to many medical professionals and search areas such as; neuroanatomy, neuroimaging, neurosurgery, etc.

Arterial supply of insula affects its functionality, and it is associated with many diseases, such as vascular strokes or infarcts, arteriovenous malformations, varices and tumours of insula. Middle cerebral artery (MCA) is the main blood supplier of insula. Cortical branches of MCA lie over insula and give short, middle, and long perforating arteries to it. The insuloopercular artery (IO) and the terminal insular artery (TrI) have been discussed and depicted. Arterial supply regions of insular gyri and sulci are presented with illustrations. Arteries of the insula are systematized and refined based on the recent literature, in order for readers to better comprehend the vasculature. Knowledge of the vascular anatomy of the insula is necessary for medical trainees and professionals to understand functionality and pathogenesis of disorders and apply noninvasive and invasive medical diagnosis and treatments of the insula, especially neurosurgical procedures that necessitate exact and detailed anatomic knowledge of the insular vessels.

The lateral cerebral cistern (or the Sylvian cistern) is comprised of three parts, respectively, the fissure, opercular sulci, and the Sylvian fossa. The Sylvian fossa is the deepest part of cistern, hidden beneath the opercula. The insular vessels located in the deep part of the Sylvian cistern, namely, in the Sylvian fossa, over the surface of the insula [1]. We will examine insular vascularity the general organization of cerebral vasculature.

4.2 General Organization of the Cerebral Arteries

The brain is supplied by a complex anastomosis which it is called "circulus arteriosus" or "circle of Willis", located under the brain, composed of two internal carotid arteries (ICA) and two vertebral arteries. Vessels arising from the circle of Willis supply the brain. In general, the branches of ICA supply the forebrain, with the exception of the occipital lobe of the cerebral hemisphere. The vertebral arteries and their branches supply the brainstem, the occipital lobe and the cerebellum. Terminal branches of the ICA are the MCA and the anterior cerebral artery (ACA). The vertebral arteries join to form the basilar artery which divides into posterior cerebral arteries (PCA). The major arteries like the ACA, MCA, and PCA give off cortical branches. Major arteries and their cortical branches course within the subarachnoid space. The tentorium cerebelli divides the intracranial cavity into supratentorial and infratentorial compartment. Intracranial arteries are also classified in supratentorial and infratentorial groups according to their localization and supplying territory. The supratentorial group consists of the arteries that supply the hemispheres of the brain, and the infratentorial group is composed of arteries that supply the cerebellum and brain stem. The circle of Willis, major arteries and their cortical branches allow central arteries to supply deep parts of brain. These central arteries perforate the surfaces of the brain and are called perforating arteries [2, 3].

4.2.1 The Middle Cerebral Artery and Its Branches

There are many publications about termination patterns, cortical branches, and variations of the MCA or accessory MCA–these are not part of this chapter. The origin of the insular supply is mainly from the MCA and its cortical branches; therefore, we underline some features of the MCA especially related to insular supply.

The MCA originates from the ICA as a larger terminal branch. Its diameter at its origin is 3.9 mm. The location of its origin is at the end of the Sylvian fissure, below the anterior perforated substance, posterior to the tractusolfactorius and lateral to the optic chiasm. From its origin, the MCA routes laterally in the Sylvian stem, below the anterior perforated substance and about 1 cm posterior to the sphenoid ridge and parallel to it. When the MCA crosses below the anterior perforated substance, perforating branches, called lenticulostriate arteries, arise from it. At the level of, or immediately distal to the limen insula, the MCA or its main trunks turn about 90° and forms the genu (Fig. 4.1). The mean distance from the genu to the limen insula is 4.8 mm [3–6].

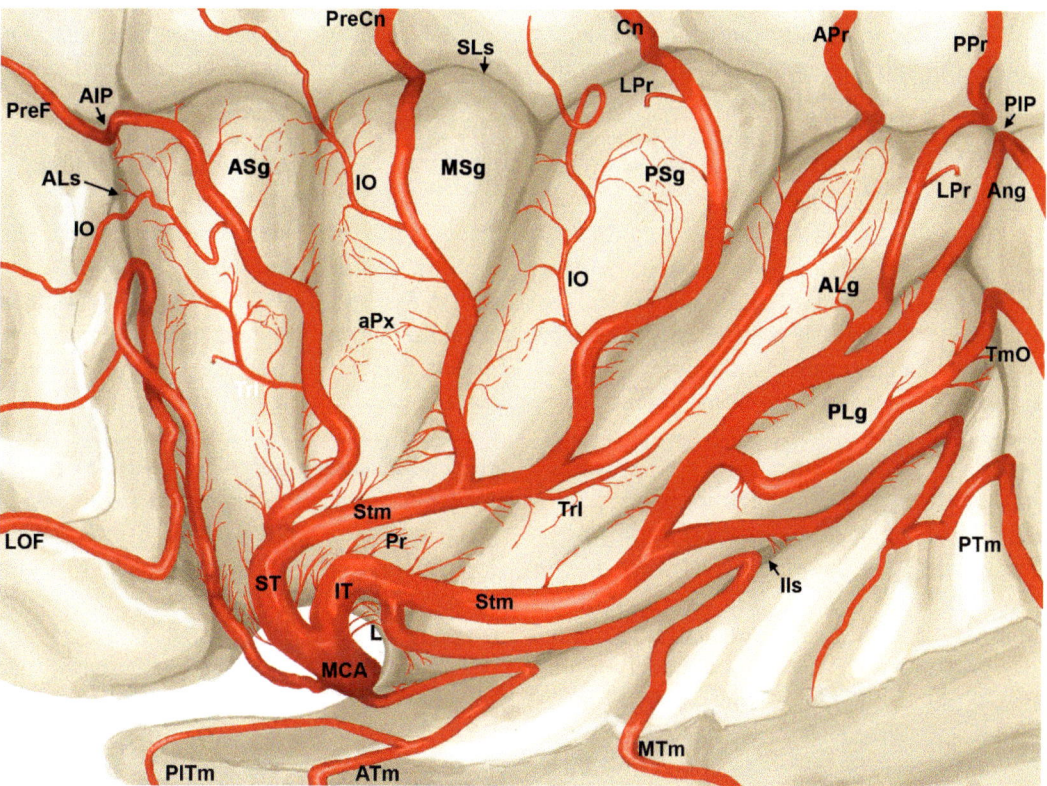

Fig. 4.1 An overview of the insular arteries following removal of overlying brain tissue. Stem arteries terminate before the superior limiting sulcus, lateral orbitofrontal artery, and the anterior temporal artery that gives rise to the polar temporal artery. *(Illustrated by Servet Celik, M.D.)*. *AIP* Anterior insular point, *ALg* Anterior long gyrus, *ALs* Anterior limiting sulcus, *Ang* Branch to angular gyrus, *APr* Anterior parietal artery, *ASg* Anterior short gyrus, *ATm* Anterior temporal branch, *aPx* Pial arterial pelxus, *ILs* Inferior limiting sulcus, *IO* Insuloopercular artery, *IT* Inferior trunk of middle cerebral artery, *L* Limen of the insula, *LPr* Long perforating artery, *MCA* Middle cerebral artery, *MSg* Middle short gyrus, *MTm* Middle temporal branch, *PIP* Posterior insular point, *PLg* Posterior long gyrus, *PlTm* Polar temporal artery, *PPr* Posterior parietal artery, *Pr* Perforating insular artery, *PSg* Posterior short gyrus, *PTm* Posterior temporal branch, *SLs* Superior limiting sulcus, *ST* Superior trunk of middle cerebral artery, *Stm* Stem artery, *TmO* Temporooccipital branch, *TrI* Terminal insular artery

The MCA and its cortical branches are considered clinically as a unit and divide into four segments. These are, respectively, sphenoidal (M1), insular (M2), opercular (M3) and cortical (M4) segments:

The sphenoidal (M1) segment: This segment begins from the ICA at the origin of the MCA and terminates at the main MCA branching point, which is located adjacent to the limen insula.

The insular (M2) segment: This segment extends from the limen insula to the superior limiting sulcus (SLs). At the level of the limen, the MCA or its main trunks turn about 90° more than those directly postero-superiorly above the insula. The MCA then divides into branches, over the insula and lateral cerebral surface.

The opercular (M3) segment: This part extends from the anterior limiting sulcus (ALs), SLs, and inferior limiting sulcus (ILs) to the lateral face of the frontoparietal opercula. The branches are on the medial surface of the opercula of the frontal, temporal and parietal lobes.

The cortical (M4) segment: This segment consists of the distributions of the branches over the supero-lateral surface of the hemisphere. The branches of this part begin from the outer lips of the operculum, extend to the cortical surface, and supply some of the lateral and basal surfaces of the cerebral hemisphere (Fig. 4.1) [3, 5, 7, 8].

4.2.1.1 Termination and Branches of the Middle Cerebral Artery

The branching of the MCA is classified as true and false branching. The true branching of the MCA is a pattern, which presents main trunks such as the superior trunk (ST) and inferior trunk (IT). When the MCA gives strong early branches before the true branching level, this pattern simulates main branching and is referred to as false branching. Common main branching patterns of the MCA are bifurcation, trifurcation, quadrification, and other patterns. In more than half of the hemispheres, the main bifurcation is located at the genu. The main bifurcation is distal to the genu in 27.5% of hemispheres, but it is proximal to the genu in the remaining hemispheres (15%) (Fig. 4.1). A true bifurcated MCA gives rise to ST and IT and a true trifurcated MCA gives rise to the middle trunk. Bifurcation and trifurcation of the MCA is seen in 66.6–92.7% and 7.0–28.6% of individuals respectively [5, 6, 9, 10].

4.2.1.2 Stem Artery of Middle Cerebral Artery

After branching, the ST, IT, or when present the middle trunks of the MCA give off secondary trunks that give rise to at least two cortical arteries termed "stem arteries" (Figs. 4.1, 4.2, and 4.3). Stem arteries vary in both the number and size of the area supplied. Most commonly, stem arteries have eight stem arteries per hemisphere. A stem artery generally gives rise to a range of one to five cortical arteries. The stem artery is located over the insula, and the cortical branches arise before the stem artery reaches the SLs. This means most of the stem arteries are from the M2 segment of the MCA [5, 6, 9, 10].

4.2.1.3 Cortical Branches of the Middle Cerebral Artery

A total of 12 cortical arteries arose from the trunks or stem artery of the MCA and leave the Sylvian fissure to reach the cerebral hemisphere and those supply the individual cortical areas. Cortical branches of the MCA are lateral orbitofrontal artery (LOF), prefrontal artery (PreF), precentral sulcal artery (PreCn), central sulcal artery (Cn), anterior parietal artery (APr), posterior parietal artery (PPr), branch to angular gyrus (Ang), temporo-occipital branch (TmO), posterior temporal branch (PTm), middle temporal branch (MTm), anterior temporal branch (ATm) and polar temporal artery (PlTm) (Fig. 4.1). The LOF, PreF, PreCn and Cn arteries usually arise from the ST, PlTm, ATm, MTm, PTm, TmO, and the Ang arteries generally arise from the IT. The APr and PPr arteries can arise from both the ST and IT, usually from the dominant trunk. The smallest cortical arteries stem from the anterior limit of the Sylvian fissure, while the largest cortical arteries stem from the posterior end of the Sylvian fissure. The insular arteries cross perpendicular to the SLs toward the medial surface of the operculum. There is no cortical artery lying parallel along the SLs (Figs. 4.2 and 4.3) [5, 6, 9–11].

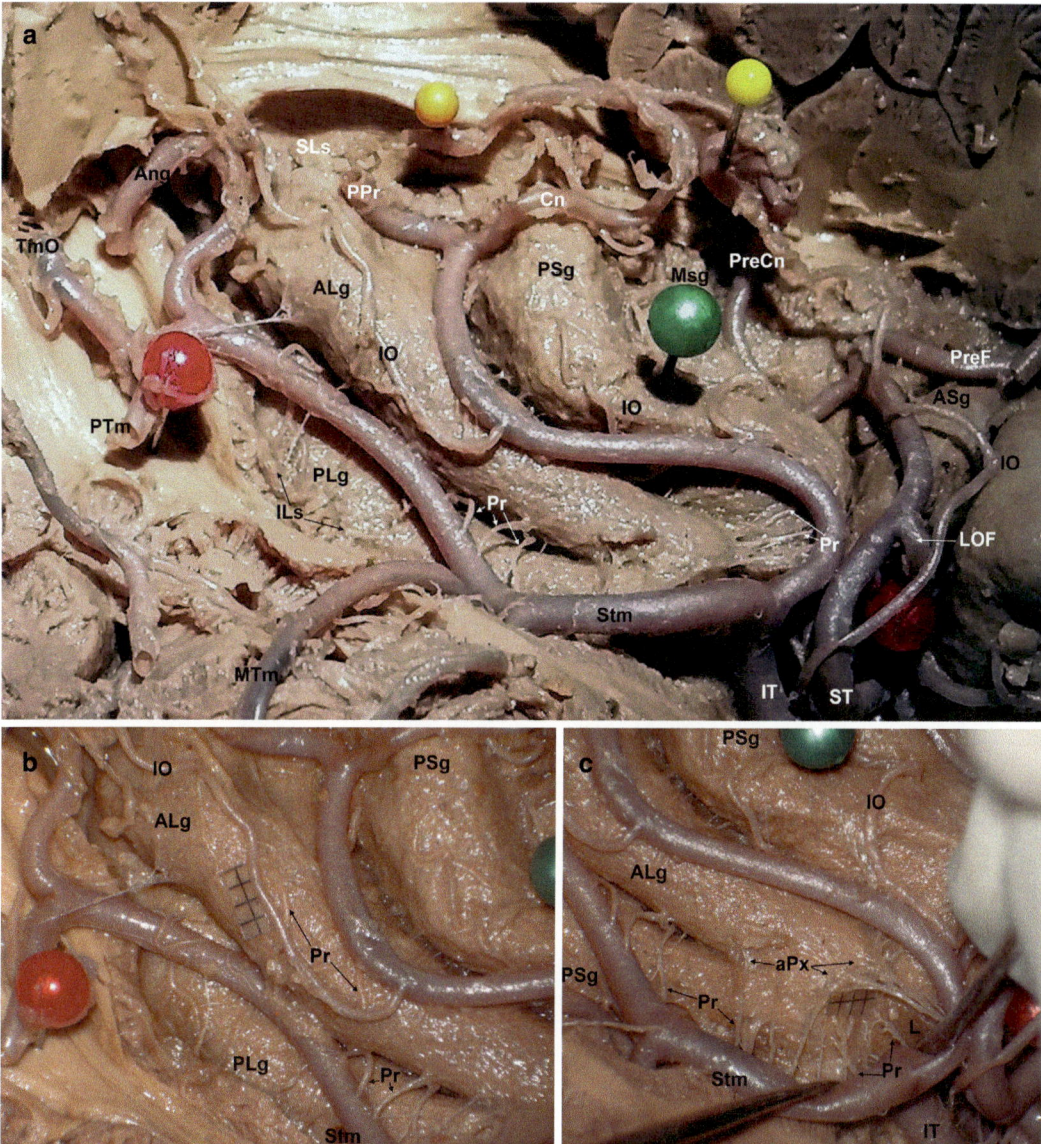

Fig. 4.2 (a) General overview of the insular arteries of the right hemisphere operculum removed. (b) Stem artery coursing near or in the insular sulcus and giving rise to insuloopercular artery. The insuloopercular artery lies over the anterior long gyrus and gives rise to long perforating arteries penetrating the anterior long gyrus. (c) The subapical and limen areas take perforating arteries from the IT and stem arteries. A square on a scale bar is 1 × 1 mm. *(With permission of Ege University Faculty of Medicine Department of Anatomy).* *ALg* Anterior long gyrus, *Ang* Branch to angular gyrus, *aPx* Pial arterial pelxus, *ASg* Anterior short gyrus, *Cn* Central sulcal artery, *ILs* Inferior limiting sulcus, *IO* Insuloopercular artery, *IT* Inferior trunk of middle cerebral artery, *LOF* Lateral orbitofrontal artery, *MSg* Middle short gyrus, *MTm* Middle temporal branch, *PLg* Posterior long gyrus, *PPr* Posterior parietal artery, *Pr* Perforating insular artery, *PreCn* Precentral sulcal artery, *PreF* Prefrontal artery, *PSg* Posterior short gyrus, *PTm* Posterior temporal branch, *SLs* Superior limiting sulcus, *ST* Superior trunk of middle cerebral artery, *Stm* Stem artery, *TmO* Temporooccipital branch, *Red head of pins*: Anterior and posterior Sylvian points of the Sylvian fissure, *Green head of pin*: Pin thorough interventricular foramen (Monro) pricked before dissection

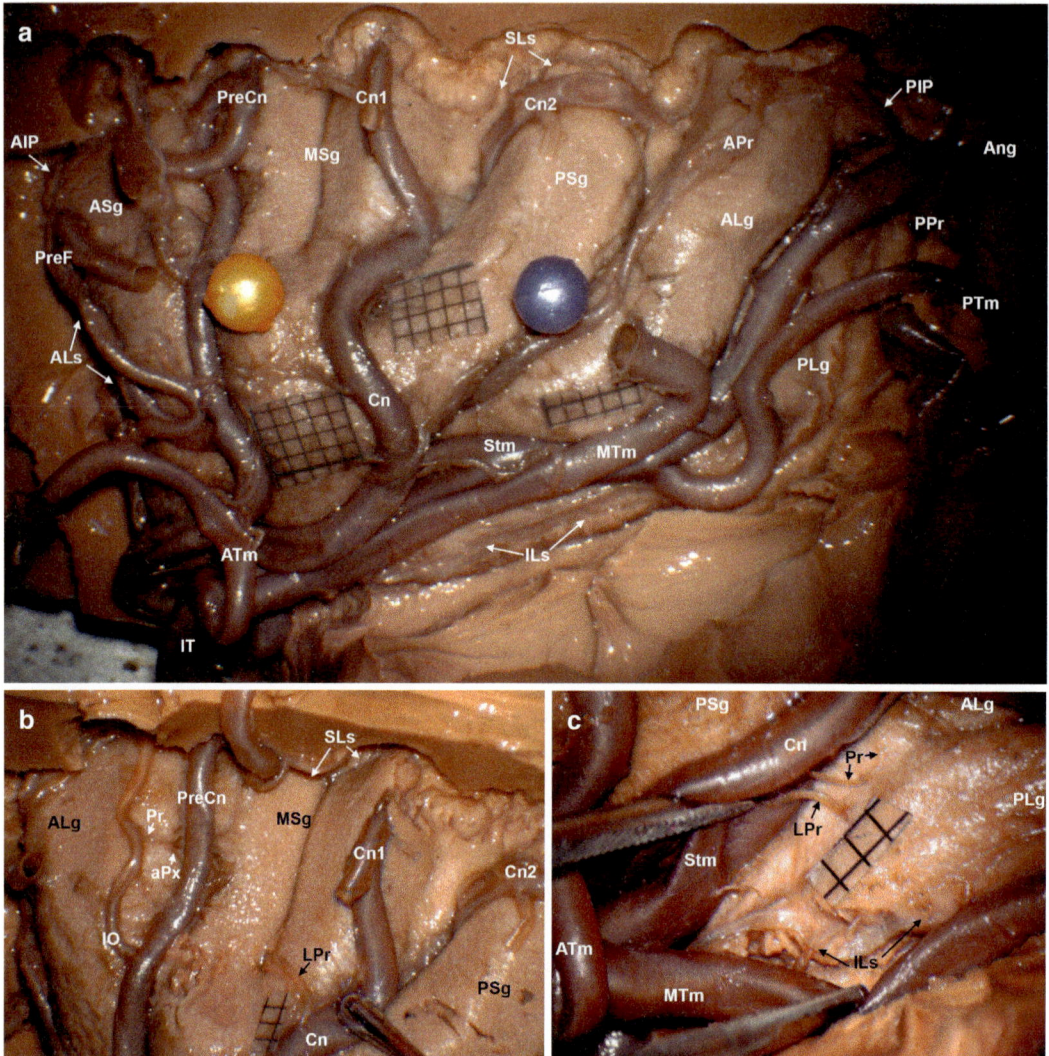

Fig. 4.3 (a) General view of the insular arteries on the left hemisphere operculum removed. (b) The insuloopercular artery arises from the precentral artery, which gives off perforating arteries and long perforating arteries from the central artery perforated perpendicularly from the middle short gyrus. (c) The long perforating artery aroused from stem artery perforates the junction of the long gyri. (d) Short and middle perforating arteries from the precentral artery supply limen, precentral sulcus, and anterior short gyrus. Pial arterial plexus of the limen level is presented. (e) Short and middle perforating arteries from the middle temporal artery supply inferior limiting sulcus and posterior long gyri. A scale bar is 1 × 1 mm. *(With permission of Ege University Faculty of Medicine Department of Anatomy)*. *AIP* Anterior insular point, *ALg* Anterior long gyrus, *Ang* Branch to angular gyrus, *aPx* Pial arterial pelxus, *ASg* Anterior short gyrus, *ATm* Anterior temporal branch, *Cn* Central sulcal artery, *Cn1* Central sulcal artery 1, *Cn2* Central sulcal artery 2, *ILs* Inferior limiting sulcus, *IO* Insuloopercular artery, *IT* Inferior trunk of middle cerebral artery, *LOF* Lateral orbitofrontal (frontobasal) artery, *LPr* Long perforating artery, *MSg* Middle short gyrus, *MTm* Middle temporal branch, *PIP* Posterior insular point, *PLg* Posterior long gyrus, *PPr* Posterior parietal artery, *Pr* Perforating insular artery, *PreCn* Precentral sulcal artery, *PreF* Prefrontal artery, *PSg* Posterior short gyrus, *PTm* Posterior temporal branch, *SLs* Superior limiting sulcus, *ST* Superior trunk of the middle cerebral artery, *Stm* Stem artery

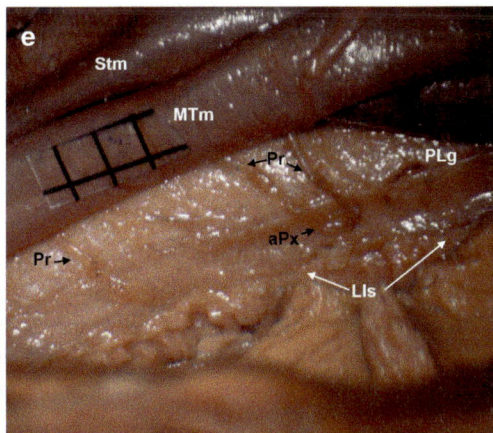

Fig. 4.3 (continued)

4.2.1.4 Early Branches of the Middle Cerebral Artery

The cortical arteries which stem from the main trunk proximal to the bifurcation or trifurcation are termed as "early branches" (Fig. 4.1). The early branches can be early frontal, early temporal, or both frontal and temporal. The number of early branches per hemisphere typically reported is one, but it can often be more than one [1, 5]. Early frontal branches have been reported 10–36.1%. The prevalence of the early temporal branch is 34.0–63.9% of hemispheres. Cortical branches of the MCA that appear as early branches are LOF, PlTm artery, and the ATm arteries [6, 8, 9].

4.2.1.5 Other Patterns of the Middle Cerebral Artery

The MCA can be accessory, duplicated, or fenestrated. The accessory MCA usually arises from the anterior cerebral or anterior communicating artery. The accessory MCA is observed in only 0.1–9.1%. The duplicated MCA presents in 0.3–7.1% of the cases and with origins from the ICA [6, 8, 9].

4.2.2 Parenchymal and Perforating Arteries of Brain

The main cerebral arteries and their cortical branches course along the surface of the sulci and gyri or deep within the sulci. Through its course, these arteries branche several times, and finally the pial arterial plexuses over the gyri and the sulci (Figs. 4.1 and 4.3b). The pial arterial plexuses give off small central branches, which finally perforate the brain from all surfaces to enter the parenchyma. Some branches of cortical arteries perforate brain surfaces without joining the pial arterial plexuses. The central arteries called "perforating arteries" or "perforans artery", because these perforate the surface of the brain [12].

Perforating arteries can be short, middle or long, and they have intraparenchymal and extraparenchymal segments. Before penetrating brain surface in a perpendicular direction, perforating arteries have extraparenchymal segments. The intraparenchymal segment of perforating arteries constitutes intracortical, subcortical and medullary arteries. Short perforators are intracortical arteries, which can give superficial, intermediate, or deep horizontal branches to the various cellular layers of grey matter. Some large arteries pass through the cortex, with or without branching, and vascularize the white matter as a medullary artery, which is also considered a long perforating artery. The medullary or long perforating arteries supply the superficial and deep white matter via short and long medullary branches, respectively. As medullary arteries course towards the body, the lateral ventricle presents two branching zones [13–15].

On the gyral surfaces the arterial network covers the venous network. The superficially located cortical arteries and arterioles are linked to the arachnoid by numerous bridges. Attachments of trabeculae are firm and these must be cut to free the vessels. The venous network underlying the arteries is generally free of arachnoid trabeculae. Inversely, veins draining the venous network separate from the pia mater and markedly adhere to the arachnoid, which makes the dissection of these veins difficult [12].

The vessels of the brain are located in subarachnoid space within the pia-arachnoid or the outmost layer of the cortex and are comprised of astrocytic end-feet. These vessels are surrounded by cerebrospinal fluid. Perforating arterioles lie within the Virchow-Robin spaces as they penetrate gyral or sulcal surfaces. The Virchow-Robin space is a continuation of the subarachnoid space, and the depth of space varies. The perforating arteries convert into intraparenchymal arterioles when they become almost completely surrounded by astrocytic end-feet in the brain parenchyma. Compared to the insular perforating arteries, the Virchow-Robin spaces were not observed around the insular veins [16–19].

Unlike the pial arterial vessels that have anastomoses among themselves, the intraparenchymal vessels, such as long perforating arteries, are essentially end arteries. Two neighboring vascular areas of end arteries of the brain form the boundary zone or watershed zone. The watershed zone between deep and superficial parenchymal perforating arteries is considered a deep boundary zone because the boundary zone is more superficial in the brain between the cortical branches of the ACA, MCA, and PCA. However, the larger perforating arteries may have web-like anastomoses at the precapillary level in the parenchyma, which means the nature of these vessels are as end arteries [17, 18, 20].

The extraparenchymal segment arteries receive perivascular innervation from the peripheral nervous system also known as "extrinsic" innervation, whereas intraparenchymal arterioles are "intrinsically" innervated from within the brain neuropil. Intraparenchymal arterioles supply the cerebral microcirculation, known as the neurovascular unit [21].

4.3 Insular Arteries

In a similar manner as the general arterial organization of the brain, insula has cortical and central branches. The MCA and its branches constitute cortical branches and these give off central branches, which are called "perforating arteries." "Perforans artery" in each hemisphere, an average of 96 insular perforating arteries supply the insula. The average diameter of insular arteries is 0.23 mm. The majority number of insular arteries (75–104) arose from the M2 segment. In 55% of hemispheres, the M1 segment gives off 1–6 insular arteries. In 25% of hemispheres, the M3 segment gives off 1–2 insular perforating arteries and supplies the region of either the SLs or ILs sulci. Central or perforating branches from other major arteries (ICA, ACA, PCA, basal artery) and segment M4 of MCA do not supply the insula [6–8, 22].

4.3.1 Classification of the Insular Perforating Arteries

The length of a perforating artery is described according to the deep structure accessed from the surface of the insula. There are three types of insular perforators (short, middle and long perforating arteries) (Figs. 4.1, 4.2, and 4.3):

Short perforating arteries (SPr): Small perforating arteries arise from all of the MCA trunks and cortical branches as they cross the insula. Approximately 85–90% of insular arteries are SPr, and they supply the insular cortex and extreme capsule.

Middle perforating arteries (MPr): MPr constitute approximately 10% of insular perforating arteries. In addition the same area of the SPr also supplies the claustrum and the external capsule. The number of SPr or MPr arteries, in 0.1–0.2 mm size, lying on the surface of the insula is 114.

Long perforating arteries (LPr): The LPr arteries per hemisphere are in the range of 4.55–9.9 mm. The LPr arteries make up less than 5% (3–5%) of all the insular perforators. The LPr arteries are bigger in diameter and length than the SPr and MPr arteries. The diameter of the LPr arteries are equal or greater than 0.3 mm and equal to or greater than 0.5 mm in 61.5% of cases. The LPr arteries in 0.3–0.5 mm diameter are in 79%, and the rest of the LPr arteries are longer than 0.5 mm [6–8, 22, 23].

The LPr arteries arose from the M2 segment in 51.6 %, from the junction M2-M3 segments in 37.4 %, and from the M3 segment of the MCA in 11 %. In most hemispheres (72.5%), LPr arteries arise from branches or directly from the ST of the MCA. The LPr arises from branches or directly from the IT of the MCA in 9.8% (Figs. 4.2 and 4.3). When the middle trunk is present, it gives off 7.7 % of the LPr arteries. The LPr arteries from the APr artery (26.4 %), from the Cn artery (17.6 %), from the PPr artery (17.6 %), or from the PreCn artery (13.2 %). The only cortical branches of the MCA that it does not give off the LPr and PlTm artery. Another study suggests that the Cn artery (20%), followed by the Ang, and the PPr arteries give off LPr arteries (Fig. 4.3c). The LPr arteries are bigger than 0.5 mm and originate mainly from the Ang and TmO arteries. In nearly all hemisphere 0.3–0.5 mm LPr arteries arise from Cn artery, following PPr and APr arteries [7, 8, 22].

4.3.2 Location and Penetration Site of the Long Perforating Arteries

The LPr arteries are most commonly located on the posterior half of the central insular sulcus and on the long gyri (Fig. 4.4). Delion et al. reported

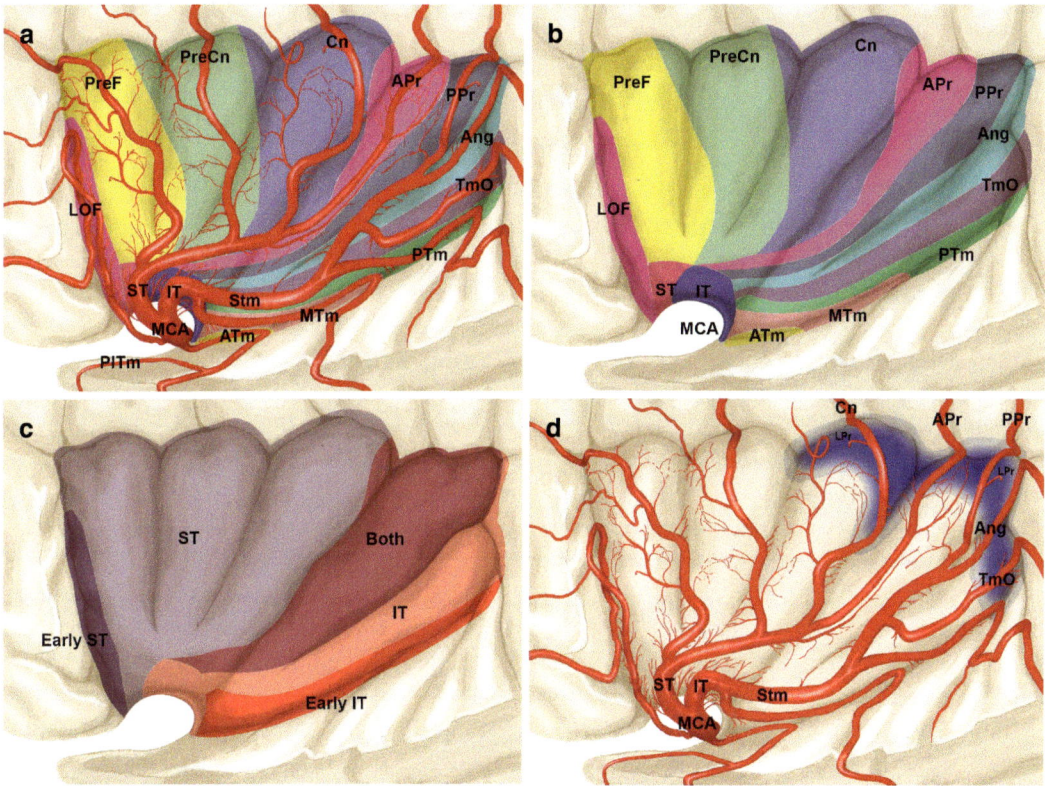

Fig. 4.4 Arterial supply of the insula. (**a**) Main trunks and cortical branches of the middle cerebral artery (MCA) projected over the areas supplied. (**b**) Territories supplied by the various arteries. (**c**) Common penetration sites of long perforating arteries. *(Illustrated by Servet Celik, M.D.). Ang* Branch to angular gyrus, *APr* Anterior parietal artery, *ATm* Anterior temporal branch, *Both* Transitional or mixed vascular area, *Cn* Central sulcal artery, *Early ST* Early branch of the middle cerebral artery supplying the superior trunk area, *Early IT* Early branch of middle cerebral artery suppling inferior trunk area, *IT* Inferior trunk of the middle cerebral artery, *LOF* Lateral orbitofrontal artery, *LPr* Long perforating artery, *MCA* Middle cerebral artery, *MTm* Middle temporal branch, *PlTm* Polar temporal artery, *PPr* Posterior parietal artery, *PreCn* Precentral sulcal artery, *PreF* Prefrontal artery, *PTm* Posterior temporal branch, *SLs* Superior limiting sulcus, *ST* Superior trunk of middle cerebral artery, *Stm* Stem artery, *TmO* Temporooccipital branch

that in one-third of the LPr arteries, penetration is in the central part (superior part of posterior short gyrus (PSg), SLs adjacent to the PSg, and SLs adjacent to the precentral sulcus); in another third of the cases, penetration is in the posterior part of the insula (in the insular long gyri and in the SLs adjacent to the insular long gyri); in the last third of the cases, penetration is in the anterior part of the insula. The branches larger than 0.3 mm in diameter most commonly arose from the Cn, Ang and PPr arteries and penetrated the posterior half of the Cns and ILs sulci and the long gyri. The LPr arteries located in ILs are especially in posterior half. For all of LPr arteries, 20% have a diameter greater than 0.5 mm and direct predominantly to the postero-superior part of the long gyri. Varnavas and Grand observed the LPr arteries at the junction of the superior and inferior limiting sulci in 25% of hemispheres. In contrast to previous reports, the LPr arteries were not only in the postero-superior portion of the insula but also postero-inferiorly [6–8, 22, 23].

4.3.3 Cortical and Subcortical Course of Long Perforating Arteries

After penetration, the LPr artery courses vertical to the insular surface plane and then slightly ascends toward the lateral ventricular body. Three types can be determine according to length of the LPr and the structures of which they supply (Fig. 4.5). In 36% of cases, the LPr artery reaches the corona radiata or deeper than the fibers of the corona radiata. Some of these groups crossed the fibers of the corona radiata and joined the ependym of the lateral ventricular body. In 27% of the hemisphere, LPr arteries end at the lateral aspect of the corona radiata. In 30% of hemisphere, LPr arteries terminate between the sagittal plane of the claustrum and the lateral fibers of the corona radiata, matched to the location of the arcuate fasciculus [7, 22].

The genu of the internal capsule projects over the PSg gyrus. LPr arteries penetrating into the superior part of the PSg supply the corticonuclear fibers of the corona radiata in 91% of cases. The posterior limb of the internal capsule projects over the long gyri of the insula. LPr arteries penetrating into the superior part of the long gyri supply the corticospinal fibres of the corona radiata in 45% of cases and just the external part of it in 45%. In total, 90% of this type of LPr arteries may supply the corticospinal tract. The putamen, globus pallidus, and the internal capsule were vascularized by the lenticulostriate arteries. The external capsule is the border of the territories supplied by the lenticulostriate arteries and the insular arteries. There is no anastomosis between the insular arteries and the lenticulostriate arteries [6, 7, 22, 24, 25].

4.3.4 Insuloopercular and Terminal Insular Branches

Insuloopercular artery (IO) is large caliber artery which it courses along the surface of the insula and then loops laterally, providing branches to both the insula and the medial surface of the operculum (Figs. 4.1, 4.2, and 4.3). The average number of IO artery is 3.5 per hemisphere (range 1–7) [6]. We determined in our dissection series insulopercular artery gives off long perforating branches (Fig. 4.2a).

Varnavas and Grand reported that terminal insular arteries (TrI), (diameter >0.7 mm), course over the insular gyri. Characteristically, the TrI is not found in the sulci (Fig. 4.1). In 67.9% of hemispheres, the terminal artery is related to the artery course in central insular sulcus. The occurrence of terminal arteries in right, left, and bilateral hemispheres are, respectively, 22.6%, 26.4% and 18.9%. TrI arteries are located in the same gyrus on both sides in 10% of cases [23].

Our observation from pervious dissections and literature is that insuloopercular and terminal arteries are similar. In the general description, IO arteries can be longer than TrI arteries and pass medial aspect of the operculum. The TrI arteries course only over the insular gyri. But IO artery may arise more distally than terminal branches and these pass along the medial aspect of operculum. The IO arteries can give rise to short, middle, and long perforating arteries to the insula (Fig. 4.2).

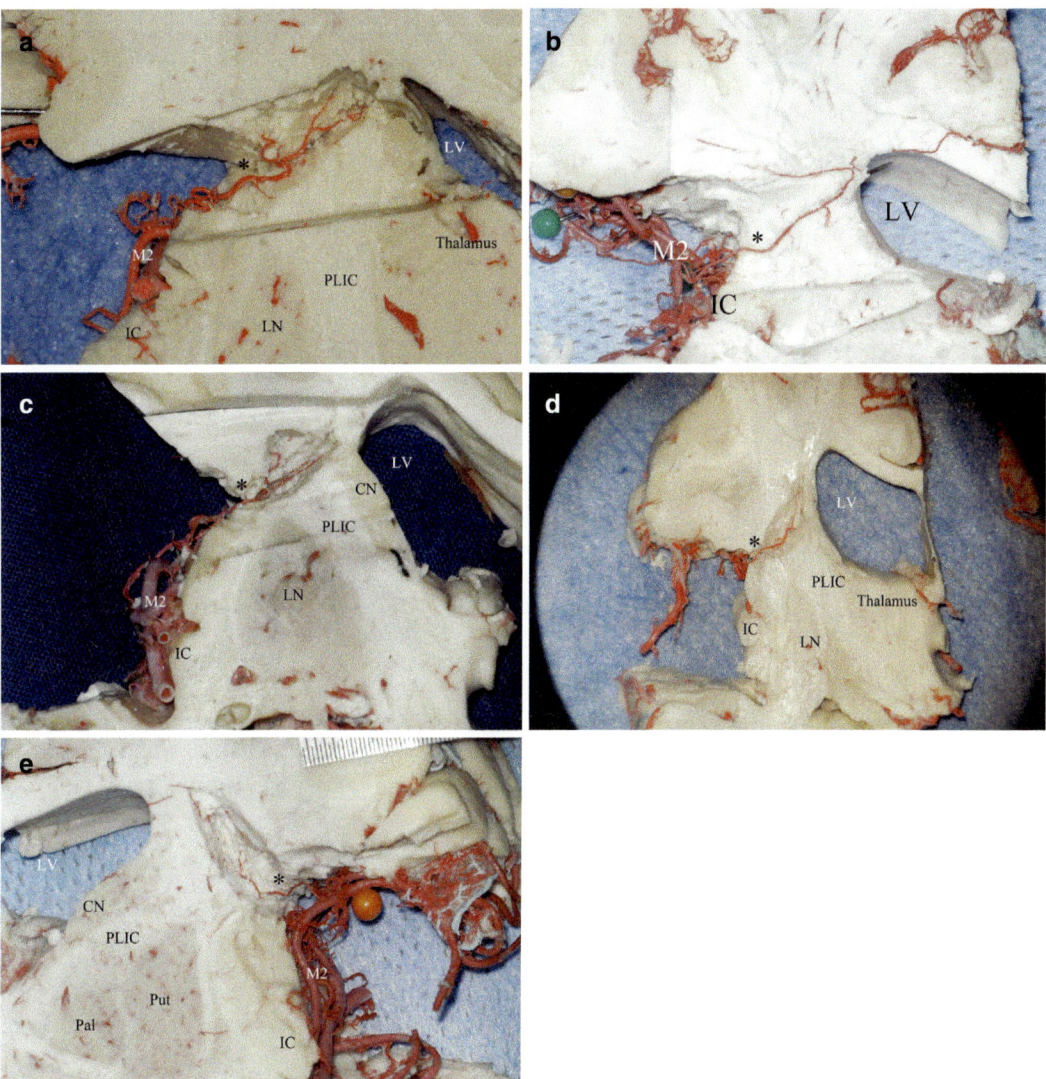

Fig. 4.5 Subcortical course of the long perforating arteries (LPr) at different depths are presented on coronal sections of the hemispheres. (**a**) The section passing through the precentral gyrus and the posterior limb of the internal capsule. The LPr penetrates into the superior limiting sulcus adjacent to the anterior long gyrus. And it supplies the motor fibers of the corona radiata. (**b**) The LPr enters through the posterior long gyrus. It exceeds the fibers of the corona radiata and reaches to ependyma of the body of the lateral ventricle. It may supply the corona radiata and the occipitofrontal tract. (**c**) The section passing through the posterior limb of the internal capsule. The LPr does not exceed the sagittal plane passing through the top of the lentiform nucleus. LPr supplies the external part of the corona radiata. (**d**) The LPr supplies the motor fibers of the corona radiata. (**e**) The LPr ends before the sagittal plane passing through the top of the lentiform nucleus. It may supply the corticonuclear tract and the arcuate fasciculus. (*With permission of Springer Publishing Company, Source: Delion M, Mercier P. Microanatomical study of the insular perforating arteries. ActaNeurochir. 2014;156(10):1991–7; discussion 7–8. doi:*https://doi.org/10.1007/s00701-014-2167-9.*, Figs. 5, 6, 7, 8 and 9 of article.*). *CN* Caudate nucleus, *IC* Insular cortex, *LN* Lentiform nucleus, *LPr* Long perforating artery, *LV* Lateral ventricle, *M2* Insular segment of middle cerebral artery, *Pal* Pallidum, *PLIC* Posterior limb of the internal capsule, *Put* Putamen

4.3.5 Supplier Arterial Pattern of Insula

The branches of the MCA give off different sized insular perforating arteries to supply the insula. Most of perforators are located in the M2 segment. The insular arteries originate from the MCA branches, and are as follows:

Superior trunk (ST): The ST originated from the stem arteries and cortical branches provide the sole supply to the accessory gyrus (Ag), transverse gyrus (Tg), and the three short gyri: the ALs, short insular sulcus (Ss), and the insular apex in almost all hemispheres. Namely, the anterior lobe is supplied by ST in nearly all hemispheres. Varnavas and Grand [23] determined the supply of the anterior lobe is by the ST in 50.9% of hemispheres, and the limen is supplied predominantly by ST. The ST and IT supplied the anterior long gyrus (ALg) and central insular sulcus (Cns) in similar percentages.

Inferior trunk (IT): In most cases, the IT and its branches supply the posterior long gyri (PLg) and the ILs sulci. The IT branches supply the ILs in 93% of the hemispheres, and the early branches contribute to this sulcus nearly half of the hemispheres. Tanriover et al. [8] found that the IT supplied the limen in almost all hemispheres, with minimal contributions from the ST and the early branches. The overlap area of arterial distributions of ST and IT does not present. But there is a "transition or mixed zone" of supply between the ST and IT. The Cns and ALg gyrus are considered as a "transition or mixed zone" by Tanriover et al. [8] because only this area receives branches from the ST or IT in nearly equal percentages (Fig. 4.4c) [6, 8, 23].

As the names imply, the cortical branches of the MCA are described by the cortical supply area. Therefore, describing them on the surface of insula without considering the cortical area is not easy. There are many studies about the MCA and the cortical branches, but little about the insular supply of them. The insular supply areas of each cortical branch are reviewed from the literature (Fig. 4.4):

The lateral orbitofrontal artery (LOF): It supplies the Ag, Tg, and ALs in the majority of the hemispheres.

The prefrontal artery (PreF): This artery supplies the anterior short gyrus (ASg) in most hemispheres, and in 77% of cases gives off branches at the apex of the insula. Finally it reaches the junction of the ALs and SLs, which it is the anterior insular point (AIP), and PreF is the only blood supply here.

The precentral artery (PreCn): It supplies the insular apex in 81% of the hemispheres, and it sends branches to the middle short gyrus (MSg) and the precentral insular sulcus (PreCns) in the majority of hemispheres.

The central sulcal artery (Cn): The Cns in 60% and the PSg in 78% of the hemispheres have branches from the Cn. It arises from the ST of the MCA in 86.8% of cases. The central insular sulcus, did not always travel within or along the entire length of the Cns.

The anterior parietal artery (APr): The Cns and posterior long gyrus (PLg) are supplied by the APr in more than half of the hemispheres. In almost half of the hemispheres, APr gives perforating arteries along the central insular sulcus to support the supply of the PSg and the ALg.

The posterior parietal artery (PPr): In 65% of cases, it contributes supply to the ALg and in 40% it supplies the Cns.

The angular artery (Ang): This artery arises from the IT of the MCA and most of the PLg are supplied by it. It courses postero-superiorly to reach the junction of the SLs and inferior ILs, of which it is the posterior insular point (PIP) and gives off perforating arteries here. In 49.1% of specimens, it does not give a branch to the insula.

The temporooccipital artery (TmO): It arises from a stem in common with the Ang artery in the majority of hemispheres and courses along the ILs in almost all hemispheres. This sulcus and PLg are supplied by TmO artery.

The posterior temporal artery (PTm): In almost all hemispheres, it supplies ILs. Nearly half of cases contribute to the vasculature of the PLg.

The middle temporal artery (MTm): In almost three quarters of cases, this artery sends branches to the ILs. It is the major artery that gives branches to the limen.

The anterior temporal artery (ATm): Similar to MTm, this artery supplies the ILs and limen. Varnavas and Grand [23] reported that it does not supply the insula.

Polar temporal artery: The arterial contributions to the insula from the PlTm artery have not been determined [6, 8, 23].

There are some verities of the insular supply:

The early branch: The early MCA branches have been described in almost in all hemispheres, and these can supply part of the insula, except the central insular sulcus. The sides commonly supplied by early branches are the ILs, ALs, and limen.

Accessory MCA: When present, it supplies the Ag in 8% and Tg in 4% of cases. Also it contributes little supply to ALs, prior to terminating in the orbitofrontal area.

M1 of MCA segment: This segment primarily gives rise to lenticulostriate arteries. The lenticulostriate arteries are classified as medial, middle, and lateral groups. Uncommon segments stem from lateral groups and extend more laterally and give insular branches to the limen and ASg [6, 8, 23, 26].

4.3.6 Arterial Supply of Insular Regions

Each insular sulcus and gyrus had a distinct pattern of supply from the branches of the MCA (Fig. 4.4):

Accessory and transverse gyri (Ag and Tg): In the majority of cases, the stem arteries and the cortical branches arise only from the ST by supplied by the Ag and Tg. In most hemispheres the LOF artery supplied both gyri with some contribution from the PreF.

Anterior limiting sulcus (ALs): The ST of the MCA gives branches to ALs in 77–100% of hemispheres. The study found that the vascularization of this sulcus by a single branch from the MCA was seen in 75.5% of hemispheres, with the rest of the hemispheres supplied by two branches from the MCA. In another study, the LOF artery supplies this sulcus in 77% of hemispheres, and the PreF artery contributed in 42% of hemispheres. In early a quarter of hemispheres, early branches provide blood to the sulcus. In more than half of the cases, at least half the length of the supplying artery courses within the sulcus.

Anterior short gyrus (ASg): This gyrus is always supplied by branches arising from the ST. The PreF artery most commonly (88%) supplied this gyrus. Supply of LOF and PreCn arteries are, respectively, 14% and 9% of hemispheres, and an early branch presents in 14%. In 56.6% of hemispheres, at least half the length of supplying arteries courses over the gyrus.

Short insular sulcus (Ss): It is always from the ST. In the majority of hemispheres, the sulcus contained a single supplying artery, and at least half of the length of it will course in the sulcus in 66% of cases.

Middle short gyrus (MSg): The branches arising from the ST supplied the MSg in 93–100% of the hemispheres. The PreCn artery most often supplies the MSg, followed by the PreF artery. Rarely, the middle trunk or an early branch contributes to its supply. Double supplying arteries were present in 60.8%, and these course over at least half of the length of the gyrus in 58.8% of cases.

Precentral insular sulcus (PreCns): In almost all hemispheres, it receives arteries from the ST of the MCA. Predominantly, the PreCn artery supplies the precentral sulcus, followed by the Cn artery. Three quarters of sulci have a single supplying artery, and at least half of its length course is within the sulcus in 57.7% of cases.

Posterior short gyrus (PSg): The branches of the ST supplied the posterior short gyrus in most hemispheres. Most commonly, the Cn artery supplies the gyrus, followed by the PreCn and APr arteries. In two-thirds of cases, double arteries are present on the surface of the gyrus, but at least half their length over the gyri [8, 23].

Alternation of supply between ST and IT is on the posterior lobe of insula (Fig. 4.4).

Central insular sulcus and anterior long gyrus (Cns and ALg): These structures are considered as a "transition or mixed zone" by Tanriover et al. [8] because only this area receives branches from the ST or IT in nearly equal percentages. The overlap supply area of ST and IT is not present. Similar alternation was reported by Varnavas and

Grand [23] with a lesser contribution ratio of IT. Almost all central sulci has a single cortical artery that at least half of its length was within the sulcus. The Cn and APr arteries supplied the central insular sulcus with nearly similar percentages. The APr and PPr arteries supplied the ALg in most of the cases.

Postcentral insular sulcus (PstCns): In almost two-thirds of hemispheres, branches from ST supply this sulcus and the rest supplied by branches of IT. Nearly all PstCns contain a single supplying artery, and in 68.6% it courses within the sulcus at least the half of its length. In the series of Tanriover et al., it is predominantly supplied by the arteries arising from the IT.

Posterior long gyrus (Plg): Branches arising from the IT supplied the PLg in 80% of the hemispheres. The Ang and TmO arteries exclusively supplied the PLg, in nearly half of cases, the PTm artery contributes to the supply. Inversely, in most of hemispheres the single contributor artery and dominance ST is reported.

Inferior limiting sulcus (ILs): The ILs received branches arising from the IT in 90% of the hemispheres and from the early branches in more than half of the hemispheres. Another series reported nearly equal supply of the ST and IT. The cortical arteries most often supplying the ILs are the TmO and PTm arteries, followed by the MTm artery. Almost all sulci have a single cortical artery that at least half of its length is within the sulcus [8, 23].

The source of suppling the apex and limen differs between the ST and IT (Fig. 4.4).

Insular apex (A): In the majority of hemispheres, cortical arteries arising from the ST supply the insular apex. The PreF and PreCn arteries supply the apex in nearly equal percentages. There is little contribution from early branches.

Subapical area, limen area (L): Varnavas and Grand reported that supply of area is by ST in 81.1% and both the ST and the IT in 18.9% of cases. Differently, Tanriover et al. determined predominantly the initial portion of the IT, proximal to the origin of the first cortical artery as supplying the limen insula in more than 80% of the hemispheres and received a contribution from the early branches in approximately one-third of the cases. The MTm supplies the limen area in 30% of the hemispheres, followed by ATm and Ang arteries in equal percentages [8, 23].

The junctions of the limiting sulci are important clinical determinants.

Anterior and posterior insular points (AIP and PIP): The junction of the ALs and SLs is AIP supplied by ST. The PreF artery is the only cortical branch that gives blood to the AIP. The junction of the inferior limiting sulcus (ILs) and SLs is PIP. In most of the hemispheres, the Ang artery supplies it, and it originates predominantly from the IT [8, 23].

Variations of the MCA and its branches can affect the origin and territory supplied to the insula.

References

1. Yasargil MG, Krisht AF, Ture U, Al-Mefty O, DCH Y. Microsurgery of insular gliomas part I: opening of the Sylvian fissure. Contemp Neurosurg. 2002;24(11):1–8.
2. Rhoton AL Jr. The cerebrum. Neurosurgery. 2002;51(suppl_1):S1–1-S1-52. https://doi.org/10.1097/00006123-200210001-00002.
3. Standring S. Chapter 19: Vascular supply and drainage of the brain. In: Standring S, editor. Gray's anatomy the anatomical basis of clinical practice, e-book, 2016, Elsevier Health Sciences; 2015. p. 280–290.
4. Marinkovic SV, Kovačević MS, Marinković JM. Perforating branches of the middle cerebral artery: microsurgical anatomy of their extra cerebral segments. J Neurosurg. 1985;63(2):266–71.
5. Rhoton AL Jr. The supratentorial arteries. Neurosurgery. 2002;51(suppl_1):S1-53–S1-120. https://doi.org/10.1097/00006123-200210001-00003.
6. Ture U, Yasargil MG, Al-Mefty O, Yasargil DC. Arteries of the insula. J Neurosurg. 2000;92(4):676–87. https://doi.org/10.3171/jns.2000.92.4.0676.
7. Delion M, Mercier P, Brassier G. Arteries and veins of the sylvian fissure and insula: microsurgical anatomy. Adv Tech Stand Neurosurg. 2016;43:185–216. https://doi.org/10.1007/978-3-319-21359-0_7.
8. Tanriover N, Rhoton AL Jr, Kawashima M, Ulm AJ, Yasuda A. Microsurgical anatomy of the insula and the Sylvian fissure. J Neurosurg. 2004;100(5):891–922. https://doi.org/10.3171/jns.2004.100.5.0891.
9. Cilliers K, Page BJ. Anatomy of the middle cerebral artery: cortical branches, branching pattern and anomalies. TurkNeurosurg. 2017;27(5):671–81. https://doi.org/10.5137/1019-5149.JTN.18127-16.1.
10. Yasargil MG. Operative anatomy. In: Yasargil MG, editor. Microneurosurgery in 4 volumes:

I-microsurgical anatomy of the basal cisterns and vessels of the brain, diagnostic studies, general operative techniques and pathological considerations of the intracranial aneurysms. Stuttgart: Georg Thieme Verlag; 1984. p. 5–168.
11. Afif A, Mertens P. Description of sulcal organization of the insular cortex. Surg Radiol Anat. 2010;32(5): 491–8. https://doi.org/10.1007/s00276-009-0598-4.
12. Yasargil MG. Organization of the cerebral microcirculation. In: Yasargil MG, editor. Microneurosurgery in 4 volumes: III A-AVM of the brain, history, embryology, pathological considerations, hemodynamics, diagnostic studies, microsurgical anatomy. Stuttgart: Georg Thieme Verlag; 1987. p. 284–337.
13. Duvernoy HM. Cortical blood vessels of the human brain. In: Yasargil MG, editor. Microneurosurgery in 4 Volumes: III A-AVM of the brain, history, embryology, pathological considerations, hemodynamics, diagnostic studies, microsurgical anatomy. Stuttgart: Georg Thieme Verlag; 1987. p. 338–49.
14. Marin-Padilla M. The human brain intracerebral microvascular system: development and structure. Front Neuroanat. 2012;6:1–14. https://doi.org/10.3389/fnana.2012.00038.
15. Okudera T, Huang YP, Fukusumi A, Nakamura Y, Hatazawa J, Uemura K. Micro-angiographical studies of the medullary venous system of the cerebral hemisphere. Neuropathology. 1999;19(1):93–111. https://doi.org/10.1046/j.1440-1789.1999.00215.x.
16. Bradac GB. Cerebral veins. In: Bradac GB, editor. Cerebral angiography normal anatomy and vascular pathology. 2nd ed. London: Springer; 2014. p. 109–34.
17. Cipolla MJ. Anatomy and ultrastructure. In: Cipolla MJ, editor. The cerebral circulation. Williston: Morgan & Claypool Life Sciences; 2010. p. 3–9. https://doi.org/10.4199/C00005ED1V01Y200912ISP002.
18. Kwee RM, Kwee TC. Virchow-Robin spaces at MR imaging. Radiographics. 2007;27(4):1071–86. https://doi.org/10.1148/rg.274065722.
19. Song CJ, Kim JH, Kier EL, Bronen RA. MR imaging and histologic features of subinsular bright spots on T2-weighted MR images: Virchow-Robin spaces of the extreme capsule and insular cortex. Radiology. 2000;214(3):671–7.
20. Takahashi S, Mugikura M. Intracranial arterial system: the main trunks and major arteries of the cerebrum. In: Takahashi S, editor. Neurovascular Imaging MRI & Microangiography. London: Springer-Verlag; 2010. p. 3–51.
21. Cipolla MJ. Perivascular innervation. In: Cipolla MJ, editor. The cerebral circulation. Williston: Morgan & Claypool Life Sciences; 2010. p. 13–6. https://doi.org/10.4199/C00005ED1V01Y200912ISP002.
22. Delion M, Mercier P. Microanatomical study of the insular perforating arteries. Acta Neurochir. 2014;156(10):1991–7.; discussion 7–8. https://doi.org/10.1007/s00701-014-2167-9.
23. Varnavas GG, Grand W. The insular cortex: morphological and vascular anatomic characteristics. Neurosurgery. 1999;44(1):127–36. discussion 36-8
24. Iwasaki M, Kumabe T, Saito R, Kanamori M, Yamashita Y, Sonoda Y, et al. Preservation of the long insular artery to prevent postoperative motor deficits after resection of insulo-opercular glioma: technical case reports. Neurol Med Chir (Tokyo). 2014;54(4):321–6. https://doi.org/10.2176/nmc.cr2012-0361.
25. Wen HT, Rhoton AL Jr, de Oliveira E, LHM C, Figueiredo EG, Teixeira MJ. Microsurgical anatomy of the temporal lobe: part 2-Sylvian fissure region and its clinical application. Neurosurgery. 2009;65(6):1–35. https://doi.org/10.1227/01.Neu.0000336314.20759.85.
26. Yasargil MG, Krisht AF, Ture U, Al-Mefty O, Yasargil DCH. Microsurgery of insular gliomas part II: opening of the Sylvian fissure. Contemp Neurosurg. 2002;24(12):1–5.

Anatomy of the Insular Veins

5

Servet Celik, Okan Bilge, Mehmet Turgut, and Canan Yurttaş

Abbreviations

AAr	Anterior arcuate branch of the central insular vein
AC	Anterior cerebral vein
Ag	Accessory gyrus
AI	Anterior insular vein
ALg	Anterior long gyrus
ALs	Anterior limiting sulcus
Am	Arachnoid membrane
An	Anastomosis of arcuate veins
Ang	Branch to angular gyrus
APo	Anterior pole of the insula
APr	Anterior parietal artery
aPx	Pial arterial plexus
ASg	Anterior short gyrus
ATm	Anterior temporal branch
B	Basal vein
Both	Transitional or mixed vascular area
cdH	Head of the caudate nucleus
CIV	Capsulo-insular vein
Cla	Claustrum
CN	Caudate nucleus
Cn	Central sulcal artery
CnI	Central insular vein
Cns	Central insular sulcus
Cs	Confluence of insular veins
CT	Common trunk of insular veins
D	Dura mater
DMC	Deep middle cerebral vein
EC	External capsule
FSy	Frontosylvian vein
GP	Globus pallidus
IA	Inferior anastomotic vein (Labbe)
IC	Internal capsule
ILs	Inferior limiting sulcus
IT	Inferior trunk of middle cerebral artery
L	Limen of the insula
LOF	Lateral orbitofrontal artery
LSSV	Lateral superior striate vein
LV	Lateral ventricle
MCA	Middle cerebral artery
MSg	Middle short gyrus
MT	Middle temporal vein
MTm	Middle temporal branch
PAr	Posterior arcuate branch of central insular vein
PI	Posterior insular vein
PLg	Posterior long gyrus
PlTm	Polar temporal artery
PPo	Posterior pole of the insula
PreCn	Precentral sulcal artery
PreCnI	Precentral insular vein
PreCns	Precentral insular sulcus
PreF	Prefrontal artery
PSg	Posterior short gyrus

S. Celik, M.D. (✉) · O. Bilge, M.D. · C. Yurttaş, M.D.
Department of Anatomy, Ege University School of Medicine, Izmir, Turkey
e-mail: servet.celik@ege.edu.tr; okan.bilge@ege.edu.tr; canan.yurttas@ege.edu.tr

M. Turgut, M.D.
Department of Neurosurgery, Adnan Menderes University School of Medicine, Aydın, Turkey

© Springer International Publishing AG, part of Springer Nature 2018
M. Turgut et al. (eds.), *Island of Reil (Insula) in the Human Brain*,
https://doi.org/10.1007/978-3-319-75468-0_5

PstCns	Postcentral insular sulcus
PTm	Posterior temporal branch
Pt	Putamen
SA	Superior anastomotic vein (Trolard)
SLs	Superior limiting sulcus
SMC	Superficial middle cerebral vein
Spf	Superficial rami of insular vein
Spf*	Superficial rami of insular vein; e.g., the postcentral insular vein
ST	Superior trunk of middle cerebral artery
SW-MRI	Susceptibility-weighted phase-sensitive magnetic resonance imaging
T	Study of Tanriover et al. 2004 [1]
TCV	Transverse caudate tributaries
Th	Thalamus
Th1	Group (1) thalamic veins that flow superomedially
Th2	Group (2) thalamic veins that flow superiorly
Th3	Group (3) thalamic veins that flow inferiorly
TSy	Temporosylvian vein
V	Study of Varnavas and Grand 1999 [2]
VIC	Vein of the internal capsule
VEC	Vein of the external capsule
vPx	Pial venous plexus
*	Tip of the gyrus
1v	One vein
2v	Two veins
≥3v	Three or more veins

5.1 Introduction

Information on the insular veins is less extensive than information on the insular arteries in the recent literature. Here we present our findings of cadaver dissections and observations of the insular veins. Venous drainage of the insula affects its functionality, and this drainage is associated with many diseases, such as arteriovenous malformations, varices and tumours of the insula. We note that the deep and superficial insular veins drain into the deep middle cerebral vein (DMC) and superficial middle cerebral vein (SMC). The capsulo-insular vein (CIV) and deep veins of the insula are discussed and depicted. Venous drainage regions of the insular gyri and sulci are presented with clear review illustrations. Also, the veins of the insula are systematized in the light of recent literature to provide information that will be clearly comprehended by readers.

5.2 General Organisation of the Cerebral Veins

The veins of the brain are classified as superficial and deep venous systems. The superficial venous system drains the cortical surfaces. The deep venous system drains the deep white and grey matter and collects into small veins that course through the walls of the ventricles and basal cisterns to drain into the internal cerebral, basal and great veins. The superficial and deep venous systems have parenchymal and cortical venous tributaries [3–5].

5.2.1 Deep and Superficial Parenchymal Veins

The parenchymal veins of the cerebral hemispheres can be divided into those draining superficially (superficial draining veins), coursing in a centrifugal direction to the superficial venous system, and those draining deeply (deep draining veins), coursing in a centripetal direction to the deep venous system. The superficial parenchymal veins consist of the superficial medullary veins, subcortical veins and intracortical veins. These veins drain 1–2 cm of the outer portion of the white matter and the cortex in a centrifugal direction to the superficial pial venous plexus, which is situated over the gyri. The small-sized pial veins join the superficial cortical veins that finally end in the dural sinuses. The deep parenchymal veins may be short or long. These veins consist of the deep medullary veins and they join the subependymal venous system. Before draining into the subependymal venous system, these veins have four zones of convergence (from outside to inside, respectively: superficial, candelabra, palmate and subependymal zones). Finally, the deep parenchymal veins collect and join the deep veins of the brain: the internal cerebral, the basal and the great cerebral vein of Galen. The deep system drains the remainder of the white and grey matter that is not drained by the superficial cortical veins. An anastomotic medullary vein is a bridging vein between a superficial medullary vein and a deep medullary vein. The parenchymal vein extending from the pia down to the subependymal region is called the transcerebral vein. Cerebral veins differ from other veins of the body in that they have no valves. The presence of anastomotic veins and the absence of valves make bidirectional blood flow in the brain possible [6–8].

5.2.2 The Superficial Middle Cerebral Vein (Superficial Sylvian Vein)

The group of veins in the superficial venous system located superficial to the insula is called the sphenoidal group. This group consists of the bridging veins that send their blood into the sinuses on the inner surface of the sphenoid bone. The major vein of the sphenoidal group is the SMC (Fig. 5.1). In

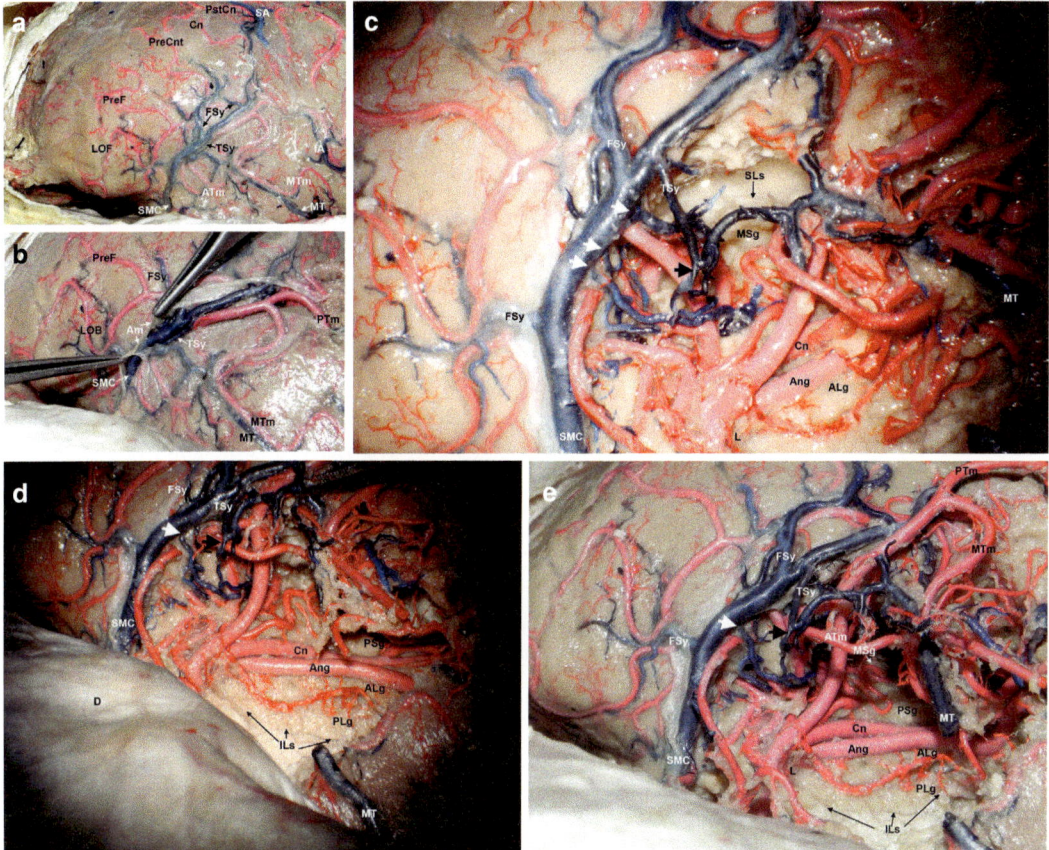

Fig. 5.1 (a–e) Three-dimensional configuration of insular veins and Sylvian vessels filled with coloured epoxy on modified Larssen solution-fixed cadaver brain. (**a**) General orientation view of vessel of Sylvian fissure. (**b**) Dissection of arachnoid membrane along the inferior border of the superficial middle cerebral vein (SMC) without incision of vessel. (**c**) Grey and white matter of operculum was suctioned to show the vascular tree three-dimensionally without disturbing the natural position of the vessel. (**d**) Inferior continuation of vessel that was incised and vessel pulled slightly upwards to show insular veins. (**e**) Final three-dimensional view of epoxy-filled insular vein without traction. *ALg* Anterior long gyrus, *Am* arachnoid membrane, *Ang* branch to angular gyrus, *ASg* anterior short gyrus, *ATm* anterior temporal branch, *Cn* central sulcal artery, *D* dura mater, *FSy* frontosylvian vein, *IA* inferior anastomotic vein (vein of Labbe), *ILs* inferior limiting sulcus, *L* limen of the insula, *LOF* lateral orbitofrontal artery, *MSg* middle short gyrus, *MT* middle temporal vein, *MTm* middle temporal branch, *PLg* posterior long gyrus, *PreCn* precentral sulcal artery, *PreF* prefrontal artery, *PSg* posterior short gyrus, *PTm* posterior temporal branch, *SA* superior anastomotic vein (vein of Trolard), *SLs* superior limiting sulcus, *SMC* superficial middle cerebral vein, *TSy* temporosylvian vein. *White arrow* Entry point of insular vein to the SMC. *Black arrow* Entry point of insular vein to the temporosylvian vein

our cadaver study, the SMC began at the posterior end of the lateral cerebral or the Sylvian fissure in most of the hemispheres examined (88.3%). Rarely, it was absent (4.7%) and it could be hypoplastic posterior to the central sulcus (7%). The SMC courses anteriorly and inferiorly along the lips of the Sylvian fissure. At the end of the Sylvian fissure, it penetrates the arachnoid membrane, and just below the medial part of the sphenoid ridge, it joins the sphenoparietal sinus or may join the cavernous sinus. The SMC receives the frontosylvian, parietosylvian and temporosylvian veins as it courses at the posterior ramus of the Sylvian fissure. It connects to the superior sagittal sinus via the superior anastomotic vein (vein of Trolard) and connects to the transverse or sigmoid sinus via the inferior anastomotic vein (vein of Labbe) (Fig. 5.1) [1, 3, 4].

5.2.3 The Deep Middle Cerebral Vein (Deep Sylvian Vein)

The deep venous system of the brain is classified into ventricular and cisternal groups. The ventricular group is composed of veins draining the walls of the lateral ventricles, and the cisternal group includes veins draining the walls of the basal cisterns. The cisternal venous group is divided into anterior, middle and posterior incisural groups according to relationships with the brainstem and tentorial incisura. The anterior incisural group is located in front of the brainstem, the middle incisural group is situated lateral to the brainstem and the posterior incisural group is located behind the brainstem. The anterior incisural region reaches upwards around the optic chiasm to the subcallosal area and laterally below the anterior perforated substance into the Sylvian fissure and over the surface of the insula [3, 4].

Generally, the union of insular veins forms the DMC near the limen insula (Figs. 5.2 and 5.3a). The DMC was single in most of the hemispheres examined (86%). Sometimes the DMC can be double and starts from the collection of the venous tributaries of the lateral perforator zone, without receiving any tributaries from the insular cortex. The DMC travels medially through the anterior perforated substance, where it unites with the anterior cerebral vein to form the anterior segment of the basal vein. In 53.5% of examined hemispheres, the DMC crossed the limen insula and passed between the proximal M1 segment of the middle cerebral artery (MCA) and the anterior perforated substance on the way to the basal vein. Prior to reaching the anterior perforated substance, the DMC usually receives the frontoorbital vein. As the anterior temporal vein passes the anterior perforated substance, the olfactory vein and the inferior striate vein, it may join the DMC. The uncal vein from the medial surface of the uncus crosses the anterior incisural region, and it can join the DMC. Historically, the DMC and anterior cerebral vein united to form the basal vein (Fig. 5.2). In our cadaver study, the DMC joined with the anterior cerebral vein to form the anterior segment (first or beginning) of the basal vein in 53.5% of hemispheres, and in the remainder of the hemispheres, the DMC and some of the tributaries of the initial segment of the basal vein formed a common stem, which coursed forwards to empty into the sphenoparietal sinus. The DMC and the anterior segment of the basal vein, or their tributaries, may be connected by a bridging vein to the sphenoparietal or cavernous sinus [1, 2, 4, 9].

5.3 Insular Veins

The insular veins draining the insula are included with the deep venous system because they drain predominantly through the DMC into the basal vein. The insular veins are part of the cisternal group of deep veins located in the anterior incisural region. The cerebral veins are classified into supratentorial and infratentorial groups and dural sinuses. The insular veins, one of the major groups contributing to the first part of the basal vein, are part of the supratentorial group [3, 4, 10].

The insular veins have many anatomical variations and asymmetries, but some common properties can be shown. As a general classification of the brain veins, the insular veins can collect the superficial cortical and deep veins.. The superficial insular veins course over the gyri and sulci of

5 Anatomy of the Insular Veins

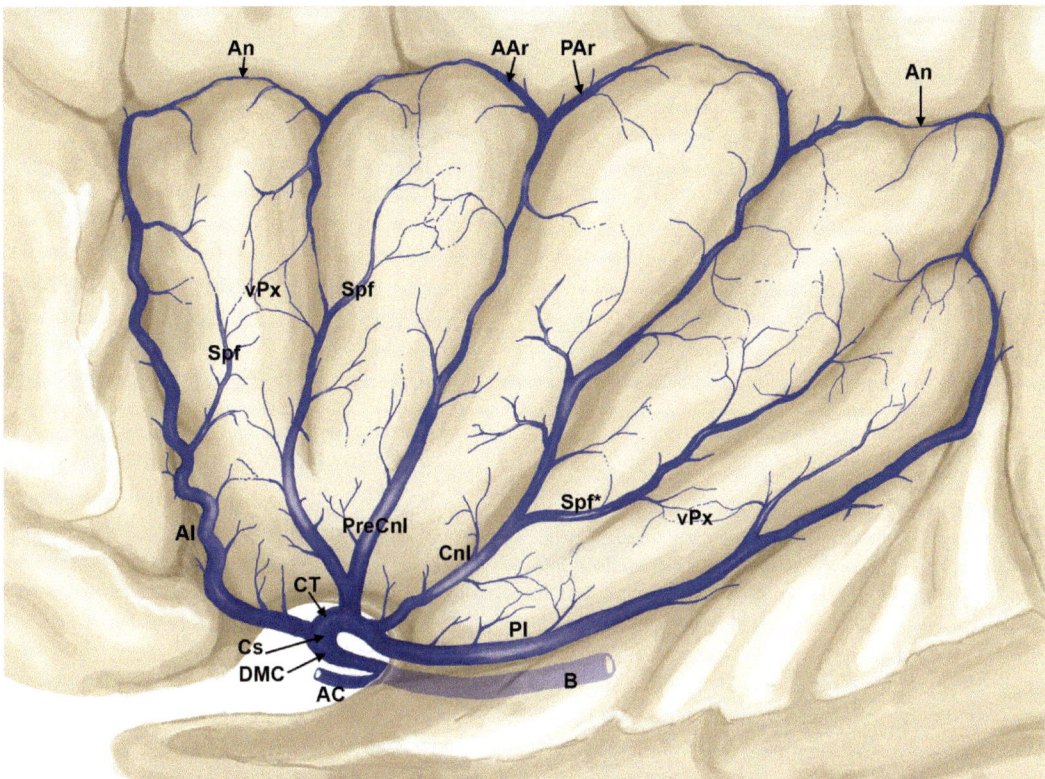

Fig. 5.2 The insular veins are presented in a review picture with operculum retracted. *AAr* anterior arcuate branch of central insular vein, *AC* anterior cerebral vein, *AI* anterior insular vein, *An* anastomosis of arcuate veins, *B* basal vein, *CnI* central insular vein, *CT* common trunk of insular veins, *Cs* confluence of insular veins, *DMC* deep middle cerebral vein, *PAr* posterior arcuate branch of central insular vein, *PI* posterior insular vein, *PreCnI* precentral insular vein, *vPx* pial venous plexus, *Spf* superficial rami of insular vein, *Spf** superficial rami of an insular vein, e.g., the postcentral insular vein

the insula. The deep insular veins come from the inside to the surface of the insula.

5.3.1 The Deep Veins of the Insula

There is little knowledge about the deep veins of the insula, but the general vascular consideration of the brain grey and white matter is also valid for the insula. Microangiographic studies of the cerebral venous system have presented parenchymal veins of the cerebral hemispheres; these veins can be divided into the superficial parenchymal veins, draining superficially, and the deep parenchymal veins, traversing deeply to reach the subependymal veins, which drain into the deep venous system of the cerebrum. The superficial parenchymal veins of the insula consist of the intracortical veins, subcortical veins and superficial medullary veins. The deep medullary veins of the insula consist of the deep parenchymal veins, which create four zones of venous convergence on their way to reach the subependymal veins, especially in the frontoparietal area. These convergence zones consist of some deep insular veins (not yet mentioned) and some intracortical, subcortical and superficial medullary veins, and the anastomoses between the superficial insular veins and the deep venous system of the cerebrum via the deep insular veins, as presented in the Figures in the study of Okudera et al. [8].

Fukusumi [11] presented the CIV as one of the deep veins of the insula. The CIV and the lateral superior striate tributaries of the thalamostriate veins drain the upper and middle portions of the claustrum and the external and internal

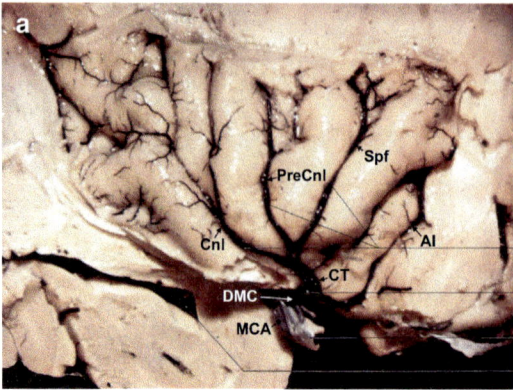
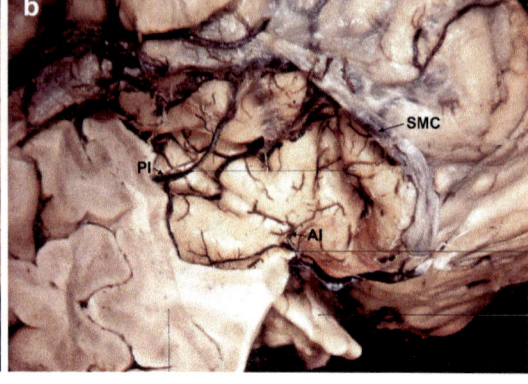

Fig. 5.3 (**a**, **b**) Lateral aspect of right insula and insular veins. (**a**) Insular veins unite to form the deep middle cerebral vein (DMC). Precentral insular vein (PreCnI) and central insular vein (CnI) constitute a common trunk slightly above the limen and the last part of the anterior insular vein (AI), which courses inferior posteriorly to join the common trunk more distally. (**b**) The posterior insular vein (PI) joins the superficial middle cerebral vein (SMC), and the AI is short and drains into the DMC. *AI* anterior insular vein, *CnI* central insular vein, *CT* common trunk of insular veins, *DMC* deep middle cerebral vein (deep Sylvian vein), *MCA* middle cerebral artery, *PI* posterior insular vein, *PreCnI* precentral insular vein, *SMC* superficial middle cerebral vein, *Spf* superficial rami of insular vein *(with permission from Springer Publishing Company, Source: Lang J. Floor and Contents of the Middle Cranial Fossa. In: Lang J, editor (translated by Wilson RR, Winstanley DP). Clinical Anatomy of the Head Neurocranium, Orbit, Craniocervical Regions. Berlin-Heidelberg: Springer-Verlag; 1983. p. 282–3 [13]. "a" in our figure is Fig. 229 in Lang's book and "b" is Fig. 230)*

capsules. The CIV runs superficially in the extreme capsule and joins the superficial insular veins, finally draining into the basal vein of Rosenthal via the insular veins (Fig. 5.4). In some cases, the CIV leaves the insular cortex and joins the opercular vein, continuing with the ascending cortical vein [11].

5.3.2 The Superficial Veins of the Insula

The superficial pial venous plexus presents over the protruding part of the gyri (Figs. 5.2 and 5.6). These plexuses give rise to small veins over the gyri or inside the sulci of the insula, and they join the main superficial insular cortical veins. A middle-sized vein, termed a superficial vein, courses superficially in parallel with the long axis of the gyri and joins the main superficial insular veins (Fig. 5.7). The main superficial insular veins are named for their relationship to the insular sulci and gyri. These veins are the anterior insular vein (AI), precentral insular vein (PreCnI), central insular vein (CnI) and posterior insular vein (PI) (Figs. 5.2, 5.3, 5.5, 5.6 and 5.7) [1, 3, 12].

The anterior insular vein (AI) : The AI was determined in 83.7% of the examined cerebral hemispheres. It courses inferiorly and backwards in the anterior limiting sulcus (ALs) or near the sulcus between the surface of the insula and the frontal operculum (Figs. 5.2 and 5.3). The route of the vein in the sulcus follows a somewhat meandering course, unlike the other insular veins. At the limen insula, the AI runs medially and posteriorly to join the common trunk of the insular veins more distally, forming the DMC (Figs. 5.2 and 5.7a). The AI receives small and middle-sized superficial veins from protruding parts of the gyri. We found that 94.4% of the AI joined the DMC, and the remainder of the AI (5.6%) joined the SMC. In almost half of the examined hemispheres, the AI emptied directly into the DMC, independent of the other insular veins. The drainage areas of the AI were the ALs and the anterior short gyrus (ASg) in the majority of the hemispheres (Fig. 5.8a). Anastomoses of the AI with the tributaries of the SMC, most frequently the frontosylvian veins, have been reported [1, 6, 13, 14].

The precentral insular vein (PreCnI): The PreCnI was present in 90.7% of the examined cerebral hemispheres. The PreCnI courses

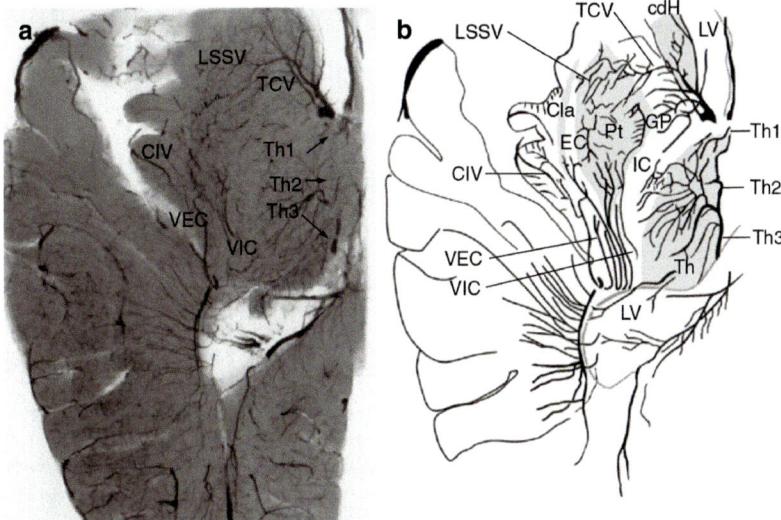

Fig. 5.4 (a) Injected microangiographic view of transaxial slice at the level of the foramen of Monro to the pineal body in a normal adult. (b) Schematic picture of capsulo-insular vein. *CIV* capsulo-insular vein, *Cla* claustrum, *cdH* head of the caudate nucleus, *EC* external capsule, *GP* globus pallidus, *IC* internal capsule, *LSSV* lateral superior striate vein, *LV* lateral ventricle, *Pt* putamen, *TCV* transverse caudate tributaries, *Th* thalamus, *Th1* group (1) thalamic veins that flow superomedially, *Th2* group (2) thalamic veins that flow superiorly, *Th3* group (3) thalamic veins that flow inferiorly, *VIC* vein of the internal capsule, *VEC* vein of the external capsule *(with permission from Springer Publishing Company, Source: Fukusumi A. Normal anatomy of intracranial veins: demonstration with MR angiography, 3D-CT angiography and microangiographic injection study. In: Takahashi S, editor. Neurovascular Imaging, MRI & Microangiography. London: Springer-Verlag 2010. p. 274–276 [5]. "a" in our figure is Fig. 7.20.c in that book and "b" is Fig.7.20.d)*

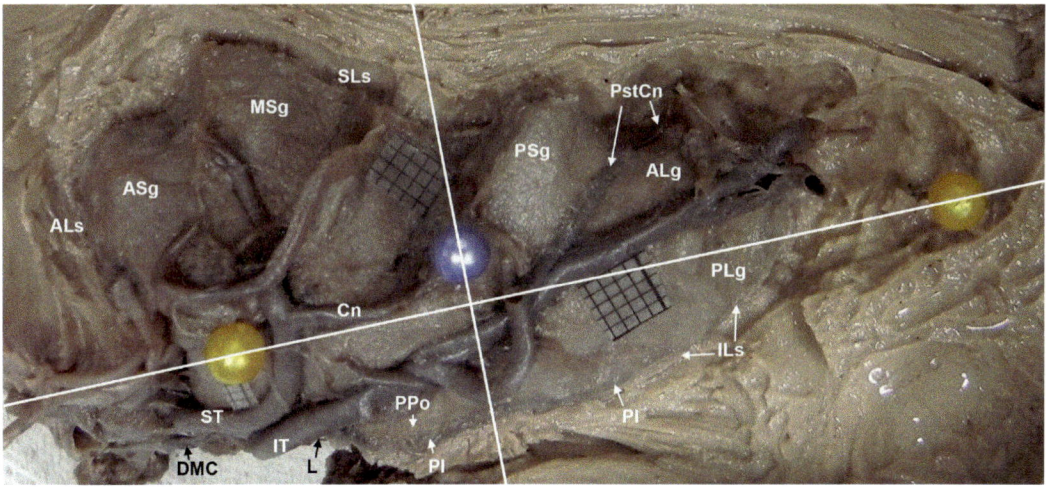

Fig. 5.5 The posterior insular vein (PI) coursing along the inferior limiting sulcus (ILs) of the left hemisphere is presented after removal of the operculum. *The square on the scale bar* is 1 × 1 mm. *ASg* anterior short insular gyrus, *ALs* anterior limiting sulcus, *Cn* central sulcal artery, *DMC* deep middle cerebral vein, *ILs* inferior limiting sulcus, *IT* inferior trunk of the middle cerebral artery, *L* limen of the insula, *MSg* middle short gyrus, *PI* posterior insular vein, *PPo* posterior pole of the insula, *PSg* posterior short gyrus, *PstCns* artery of the postcentral sulcus, *SLs* superior limiting sulcus, *ST* superior trunk of the middle cerebral artery. *Yellow pinheads* anterior and posterior points of the Sylvian fissure pricked before dissection. *Blue pinhead* interventricular foramen (Monro) pricked before dissection. *White lines* Berger-Sanai orientation lines drawn via image software and perpendicular to each other

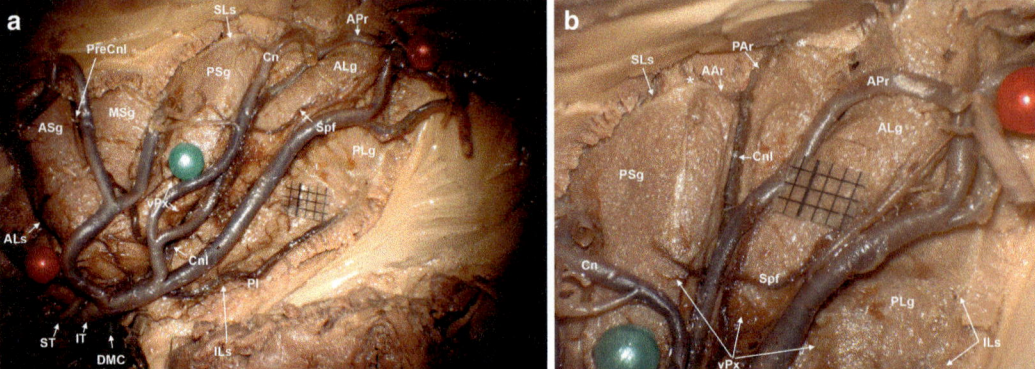

Fig. 5.6 (a, b) Lateral view of vascularity of the left insula in a 10% formalin-fixed cadaver after removal of the operculum. (**a**) The inferior trunk and its branches predominantly supply the insula, and the insular veins course beneath the arteries. (**b**) In close view, the central sulcal artery (Cn) is pulled anteriorly to display the central insular vein (CnI). *The square on the scale bar is 1 × 1 mm.* *AAr* anterior arcuate branch of the central insular vein, *ALs* anterior limiting sulcus, *APr* anterior parietal artery, *ASg* anterior short gyrus, *Cn* central sulcal artery, *CnI* central insular vein, *DMC* deep middle cerebral vein, *IT* inferior trunk of the middle cerebral artery, *ILs* inferior limiting sulcus, *MSg* middle short gyrus, *PAr* posterior arcuate branch of the central insular vein, *PI* posterior insular vein, *PreCnI* precentral insular vein, *vPx* pial venous plexus, *PSg* posterior short gyrus, *SLs* superior limiting sulcus, *Spf* superficial rami of insular vein, *ST* superior trunk of the middle cerebral artery, * tip of the gyrus. Red pinheads anterior and posterior points of Sylvian fissure identified before dissection. Green pinhead interventricular foramen (Monro) identified before dissection

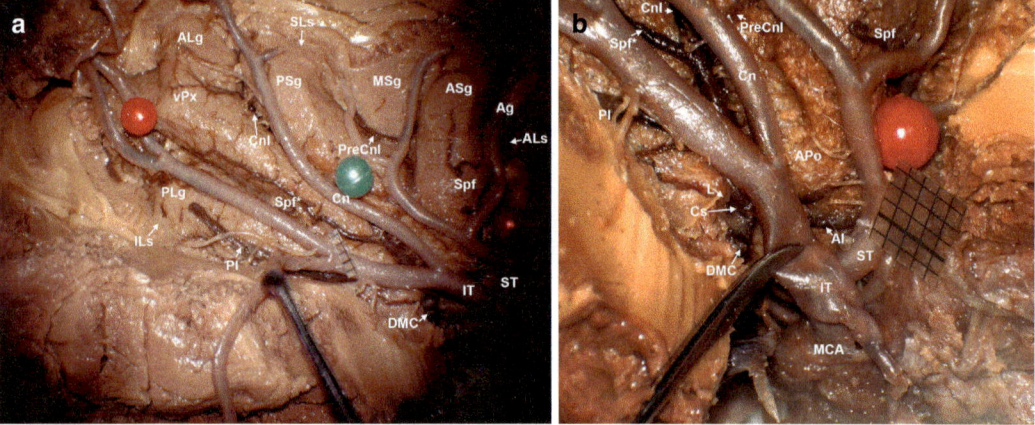

Fig. 5.7 (a, b) Lateral view of vascularity of the right insula in 10% formalin-fixed cadaver after removal of the operculum. (**a**) The course of the superficial rami of the insular vein, e.g., the postcentral insular vein (Spf*) near the postcentral sulcus is presented. (**b**) Close view of the limen, showing the confluence of the insular vein to form the deep middle cerebral vein. *The square on the scale bar is 1 × 1 mm.* *AI* anterior insular vein; *Ag* accessory gyrus; *ALg* anterior long gyrus; *ALs* anterior limiting sulcus; *APo* anterior pole of the insula; *ASg* anterior short gyrus; *Cn* central sulcal artery; *CnI* central insular vein; *Cs* confluence of insular veins; *DMC* deep middle cerebral vein; *ILs* inferior limiting sulcus; *IT* inferior trunk of the middle cerebral artery, inferior terminal branch; *L* limen of the insula; *MCA* middle cerebral artery; *MSg* middle short gyrus; *PI* posterior insular vein; *PLg* posterior long gyrus; *PreCnI* precentral insular vein; *PSg* posterior short gyrus; *SLs* superior limiting sulcus; *Spf* superficial rami of insular vein; *Spf** superficial rami of the insular vein, e.g., the postcentral insular vein; *ST* superior trunk of the middle cerebral artery. Red pinheads anterior and posterior points of Sylvian fissure pricked before dissection. Green pinhead interventricular foramen (Monro) pricked before dissection

5 Anatomy of the Insular Veins

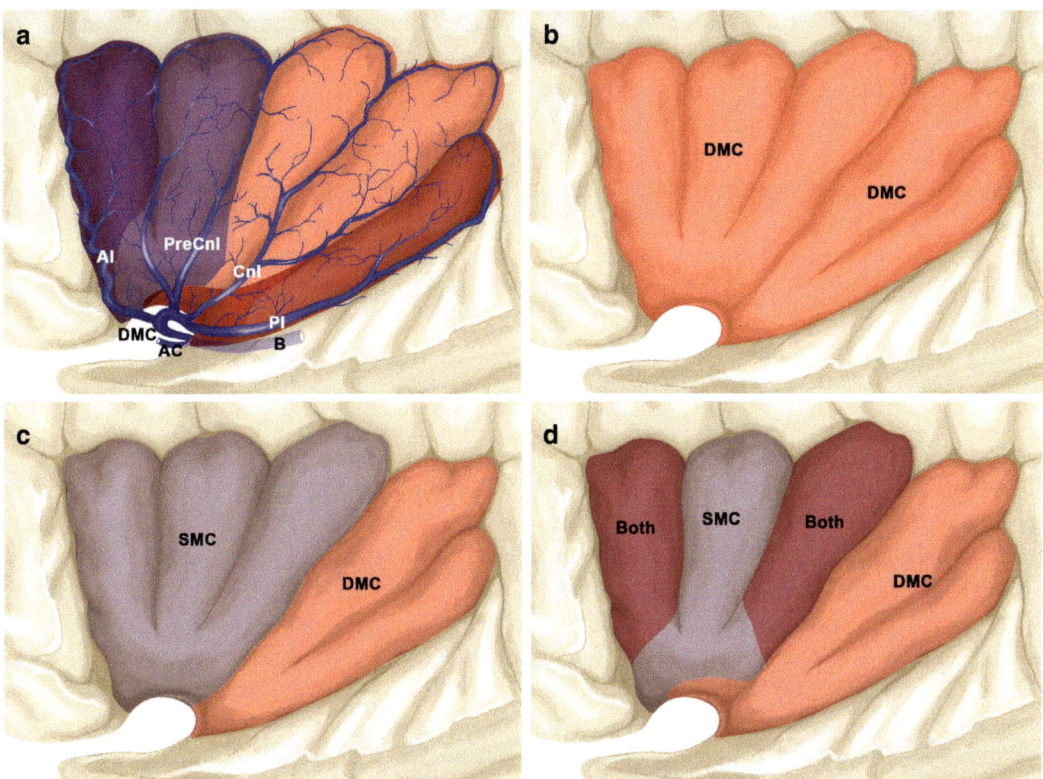

Fig. 5.8 (a–d) Venous drainage of insula. (**a**) Drainage of insular areas of the main superficial veins. (**b**) Historically described drainage, showing entire insula draining to the DMC. (**c**) Drainage to the SMC and DMC. (**d**) Recently described drainage of the insular regions to the SMC, DMC and transitional or mixed zone that drains to both the SMC and DMC. *AC* anterior cerebral vein, *AI* anterior insular vein, *B* basal vein, *Both* transitional or mixed vascular area, *CnI* central insular vein, *DMC* deep middle cerebral vein, *PI* posterior insular vein, *PreCnI* precentral insular vein, *SMC* superficial middle cerebral vein

straight anteroinferiorly, usually on the precentral insular sulcus. It receives small and middle-size superficial veins from protruding parts of the gyri (Figs. 5.2, 5.3a and 5.6a) and in 82% of the hemispheres, the PreCnI joined the CnI and PI to terminate in the DMC. In the remaining hemispheres (18%), the PreCnI coursed superficially to join the SMC. The PreCnI drains the middle short gyrus (MSg) and insular apex more frequently into the SMC than any other insular vein (Fig. 5.8a). The PreCnI anastomosed with the tributaries of the SMC, most commonly the frontosylvian tributaries, in approximately half of the examined hemispheres. At the superior limiting sulcus (SLs), the PreCnI can anastomose with the central insular vein via the arcuate veins (Fig. 5.2) [1, 4, 13].

The central insular vein (CnI): We found that the CnI was present in all of the examined cerebral hemispheres. However, Huang reported that the PI was the most constant and prominent of all the insular veins [12]. The CnI coursed anteroinferiorly along the central insular sulcus (Cns) (Figs. 5.2, 5.3a and 5.6). The CnI received the precentral insular veins in 82% of the examined hemispheres. Also, the long superficial vein that courses near or in the postcentral insular sulcus (PstCns) may join the CnI (Figs. 5.6b and 5.7a). The CnI receives small and middle-sized superficial veins from protruding parts of the gyri. The

CnI emptied into the DMC in 97.7% of the hemispheres. In the remaining 2.3% of the hemispheres, the CnI joined the SMC along the posterior ramus of the Sylvian fissure. In 90% of the hemispheres, the CnI was the unique drainage pathway of the posterior short gyrus (PSg), the ALg and the central insular sulcus (Cns). The CnI, together with the PI, contributed to the drainage of the limen area in almost half of the hemispheres (Fig. 5.8a). In some hemispheres, the CnI anastomosed with the frontosylvian or parietosylvian veins. The CnI can anastomose with the PreCnI via the arcuate vein at the tip of the PSg (Fig. 5.2) [1, 4, 13].

The anterior and posterior long insular gyri are separated by the PstCns. The presence of the PstCns has been reported as not well defined in 56% of hemispheres, as only a small depression in the posterior superior portion of the insula in 18% and as a very shallow indentation in 26% of hemispheres. In our literature search, we met with Lang's term "the postcentral insular vein" [15]. In one of our dissections, we observed long superficial rami of the CnI coursing near the PstCns throughout its length and parallel to it (Fig. 5.7a). It may be consider as a postcentral insular vein due to the previous veins have been named according to course near or in the insular sulci. Also, Tanriover et al. [1] reported similar veins in or near the PstCns, but named these as central veins in their Figures. Very small branches coursing in or near the PstCns were reported in another study, but were named in general terms as insular veins. This uncertainty about, or lack of, the term "postcentral insular vein" may be the result of a shortage of insular venous cadaveric studies, difficulties in preparing these veins for dissection and the variability of the veins [1, 4, 15, 16].

The posterior insular vein (PI): Tanriover et al. [1] reported the PI in all specimens of 43 cadaver brain vessels filled with coloured silicone. The PI commonly courses along the long gyri, usually near or along the inferior limiting sulcus (ILs) (Figs. 5.2, 5.3, 5.4, 5.5, 5.6). It takes a straight anteroinferior course in the lateral view, except at its origin, where it takes a slightly curved course, convex posteroinferiorly. This convexity may lead to an inferoposterior angle (mean 126° ± 2) that is shown on the ILs. The PI is usually the most prominent and constant of all the insular veins. Therefore, it is referred to as the "first portion of the deep middle cerebral vein. The PI most frequently joined with the CnI to form a common stem that ran medially to form the DMC in almost all of the examined hemispheres. The PI can also directly join the SMC (Fig. 5.3b). The PI drained the ILs, posterior long gyrus (PLg) and limen area in almost all of the hemispheres (Fig. 5.8a). Occasionally, it anastomosed with the temporosylvian veins. At the SLs, the PI can anastomose with the central vein via small arcuate veins [1, 5, 13, 14, 17].

Venous arches of the insula: In coloured silicone-injected cadaver specimens, Rhoton et al. reported anterior and posterior small arcuate veins coursing in the junctions of intra-insular and peri-insular sulci (Fig. 5.2). Anterior and posterior arcuate veins anastomosed with each other at the tip of the gyri in the SLs and formed venous arches. Also, arcuate veins joined the main superficial insular veins. Therefore, it could be concluded that the venous insular arches constitute anastomoses between the superficial cortical insular veins. The parallel course of the proximal parts of the insular venous arches located in the SLs differs from the M2 branches of the MCA that course only perpendicular to this sulcus [4, 18].

5.4 Relationships of the Insular Veins

Three to five short and two long insular gyri are located at the base of the Sylvian cistern. The anterior insular, central insular and posterior insular veins outline the ALs, Cns and ILs of the insula. The precentral, central and posterior insular veins are, respectively, in the precentral, central and posterior limiting sulci of the insula. In some cases, these insular veins may lie, not deep within the sulci, but on the surface of the long or short gyri of the insula (Figs. 5.3a and 5.6) [1, 4, 13].

The insular veins lie beneath the insular arteries. The venous network underlying the arteries is generally free of arachnoid trabeculae. Extensions of arachnoid membrane is closer attached to insular veins and their draining venous network than the M2 branches (Figs. 5.5, 5.6, 5.7). The insular veins usually join together at the level of the anterior third of the ILs and travel deep through the stem of the Sylvian fissure to join the DMC just beneath the MCA. Of note, this lower portion of the insular veins, which contains a confluence or convergence, is beneath the MCA and M2 branches at the level of the limen. The upper portion of the insular veins does not lie beneath the arteries at any point throughout their entire course and these veins can be seen over the gyri, near or in the sulci between arteries (Figs. 5.5, 5.6, 5.7) [1, 4, 13, 19].

5.5 Number and Size of Insular Veins

Susceptibility-weighted phase-sensitive magnetic resonance imaging (SW-MRI) enables non-invasive imaging of the cerebral veins, at sub-millimetre resolution, and it makes analysis of the venous architecture possible. The detailed course of the insular veins can be seen on SW-MRI studies, based on differences in the blood oxygen level-dependent induced phase effects between the venous blood and the surrounding brain parenchyma. The diameter of the superficial insular veins was reported to be 0.45 mm in 3 to 7 T MR scans [6, 20].

In insular cortical territories with deep venous communications, four to ten veins coalesced to enter the Sylvian stem, predominantly via the DMC. Four to 11 (mean, 6) small-sized (diameter <1 mm) veins arose from the insular cortex to connect superficially. The central and posterior insular veins have the largest diameter and drainage area of all the insular veins. Very rarely, the diameter of the PI was 1.2 mm, but in all other hemispheres examined, it was less than 1 mm. In the literature, there has been no report of a single dominant vein to which all insular veins were connected, either superficially or deeply [1, 2].

5.6 Confluence of Insular Veins

Typically, the PreCnI, CnI and PI, lying, respectively, in the precentral, central and posterior limiting sulci of the insula, run downwards and forwards towards the pole of the insula, where they unite to form the DMC, which drains into the basal vein (Figs. 5.2, 5.3a and 5.7b). The DMC is formed by the junction of the insular veins at the level of the limen insula. This junction point of the insular veins forming the DMC is called the confluence, convergence or union of insular veins. The confluence of insular veins courses towards the sphenoidal compartment of the Sylvian fissure, underneath the MCA bifurcation [1, 14, 21].

Classically, confluence is described as being formed by the union of the AI, PreCnI, CnI and PI at the limen area to form a common trunk or stem of insular veins that continues as the DMC (Figs. 5.2, 5.3a and 5.7b). The formation of this classic pattern with union of all insular veins was present in 27% of the examined hemispheres. The AI joined the common stem more distally in 42% of the hemispheres. One of the four insular veins, most commonly the PreCnI (seen in 18% of the hemispheres), does not join the common stem, and, instead, may empty into the SMC [1, 2].

5.7 Blood Drainage from the Insula

Dissection and historical angiographic studies have reported that insular drainage occurs primarily via the DMC. Recent studies report that the veins draining the insula emptied predominantly into the DMC in almost all hemispheres; however, the tributaries of the SMC drained some insular areas (Fig. 5.8). There were also anastomoses between the insular veins and the superficial venous system in most hemispheres. Variations in formation, patterns and hypoplasia, as well as the absence of the SMC and DMC, affect draining variations of the insular veins. The variations of the SMC are classified into three categories based on the branching and drainage patterns: superficial, intermediate and basal components. The intermediate category consists of insular veins. The insular veins consist of four main superficial veins of the AIs, precentral sulcus, central sulcus and posterior limiting sulcus, and these unite to form the common stem. Drainage of the common stem of the insular vein can be classified into two types: the classic pattern with a common stem draining into the basal vein, and the nonclassic pattern with a common stem draining into the sphenoparietal sinus. Absence of an anastomosis between the common stem of the insular veins and the SMC means that, primarily, the drainage of the insular vein occurs into the basal vein via the DMC [1, 2, 13, 14, 22].

5.8 Venous Drainage of Insular Regions

Varnavas and Grand [2] and Tanriover et al. [1] reported the numbers of draining veins in the regions of the sulci and gyri in cadaver hemispheres (Table 5.1) [1, 2]. Venous drainage of the insular sulci, gyri and apical and limen areas was as follows:

Anterior limiting sulcus (ALs): Half of the sulci contained three or more veins. In 80% of the hemispheres, drainage was performed by the AI (Fig. 5.8). In 28% of the hemispheres, the frontosylvian veins provided accompanying drainage. However, another study reported that most of the veins (88.7%) from the ALs emptied into the SMC via the frontosylvian veins [2]. As a result, the blood from the ALs emptied into both the DMC and SMC, but these transitional areas emptied three to eight times more frequently into the DMC than into the SMC.

Anterior short gyrus (ASg): In 75.5% of the hemispheres, there were three or more veins over the ASg. This gyrus was drained by the AI, alone or in part, in 77% of the hemispheres (Fig. 5.8, Table 5.1); 94.4% of the AI joined the DMC and the remainder of the AI (5.6%) joined the SMC. In cadaver studies, drainage of the frontosylvian veins that finally ended in the SMC was present in 44% to 96% of the hemispheres. Because of this wide range, the ASg is considered to be a transitional zone empty of blood drainage between the SMC and DMC (Fig. 5.8d) [1].

Middle short gyrus (MSg) and insular apex: In 78.4% of the hemispheres, there were three or more veins over the MSg. This area was drained mainly by the PreCnI (Fig. 5.8, Table 5.1). In most of the hemispheres, the frontosylvian vein participated in the drainage, indicating that the MSg drained into the SMC in most of the hemispheres.

Posterior short gyrus (PSg): In most of the hemispheres (84%), the CnI drained the PSg, and in the remainder of the hemispheres (13%), the PreCnI drained the PSg (Fig. 5.8, Table 5.1). The PSg emptied into the DMC in 82% of cases. Another study reported that this gyrus emptied blood into the SMC in 56.6% of hemispheres and into both the SMC and DMC in 30.2% of hemispheres. Therefore the PSg is considered as another "transitional zone empty of blood drainage" between the SMC and DMC.

Central insular sulcus (Cns) and anterior long gyrus (ALg): The Cns carried one vein in 58.5% of cases, and commonly the CnI drained this area (Fig. 5.8, Table 5.1). The blood of the Cns emptied into the DMC in 68% of hemispheres in a cadaver study and in 88% of cases in another study [1, 2].

Posterior long gyrus (PLg): In most of the hemispheres, the PLg had three or more veins on its surface. The PI drained the PLg in the majority of hemispheres, and in 7% of cases, the CnI contributed to the drainage (Fig. 5.8, Table 5.1). In the majority of cases, the blood from this gyrus drained into the DMC.

Inferior limiting sulcus (ILs): In most of the hemispheres, the ILs harboured only one vein, commonly the PI, which is also considered as the first portion of the DMC. The ILs is drained by the PI. The temporosylvian veins contributed to the drainage of the ILs in 14% of cases (Fig. 5.8, Table 5.1). In the majority of cases, blood from the ILs emptied into the DMC.

Subapical area, limen area: The limen always had three or more draining veins. In almost 50% of the hemispheres drainage occurred via a combination of the CnI and PI. In the remainder of the hemispheres, drainage was via the PI alone. In more than 90% of cases, the limens emptied blood exclusively into the DMC. In the remainder of the hemispheres, the DMC connected to the SMC and drainage was to both the SMC and DMC [1, 2, 13, 22].

The venous connections split the insular cortex into three anatomic zones, with some overlap:

5 Anatomy of the Insular Veins

Table 5.1 Drainage of insular regions. All of the numbers in the studies are shown as percentages (%). Tributaries of the insular veins to the frontosylvian, temporosylvian and parietosylvian veins are given in relation to the superficial middle cerebral vein

Insular region	Number of veins on the region			Draining main superficial insular vein					Empty to (comparison of studies)					
	1v	2v	≥3v	AI	PreCnI	CnI	PI		SMC		DMC		Both	
	V	V	V	T	T	T	T		V	T	V	T	V	T
Anterior limiting sulcus	34	13.2	52.8	84	–	–	–		88.7	21	5.7	72	5.7	7
Anterior short gyrus	17	7.5	75.5	77	9	–	–		96.2	21	1.9	56	1.9	23
Short insular sulcus	18.9	–	81.1	–	–	–	–		96.2	–	1.9	–	1.9	–
Middle short gyrus	17.6	3.9	78.4	2	88	2	–		96.1	21	1.95	16	1.95	60
Precentral sulcus	28.8	3.8	67.3	–	86	5	–		90.4	28	7.7	60	1.9	12
Posterior short gyrus	24.5	20.8	54.7	–	13	84	–		56.6	9	13.2	82	30.2	9
Central sulcus	58.5	20.8	20.8	–	–	100	–		17.1	3	68	88	14.7	9
Anterior long gyrus	15.1	15.1	69.8	–	–	100	–		13.3	3	64.2	90	22.5	7
Postcentral sulcus	53.8	1.9	44.2	–	–	–	–		–	–	–	–	–	–
Posterior long gyrus	19.2	3.8	76.9	–	–	7	90		11.5	2	75	88	13.4	10
Inferior limiting sulcus	88.7	1.9	3.8	–	–	–	100		13.2	0	75.5	86	11.3	14
Apex	–	–	–	5	84	5	–		–	26	–	14	–	60
Subapical/limen area	–	–	100	–	2	58	95		7.6	2	90.6	93	1.9	5

1v one vein, *2v* two veins, *≥3v* three or more veins, *AI* anterior insular vein, *CnI* central insular vein, *DMC* deep middle cerebral vein, *PI* posterior insular vein, *PreCnI* precentral insular vein, *SMC* superficial middle cerebral vein, *T* study of Tanriover et al. (2004), *V* study of Varnavas and Grand (1999)

subapical region (insular pole), anterior lobe and posterior lobe. The bulky anterior lobe connects to the SMC; the posterior lobe and subapical area join and, in part, form the DMC (Fig. 5.8c). The ASg, PSg and ALs are regarded as a "transitional or mixed zone" whose drainage may occur alternately into the DMC or SMC. These transitional areas emptied three to eight times more frequently into the DMC than into the SMC (Fig. 5.8d) [1, 2].

5.9 Variations of Insular Venous Drainage

Variations in the drainage of the insula have been reported in different studies. The insular veins can ascend vertically from the depth of the Sylvian fissure and are three-dimensionally anastomosed to the SMC (Figs. 5.1c–e and 5.3b) [1]. In a cadaver dissection, the PI joined the SMC at the mid-portion of the lateral fissure [15]. The PI, coursing inferior anteriorly, can join the SMC along the greater wing of the sphenoid bone [13]. The SMC may be doubled throughout its entire course, as two trunks, or it can be replaced by groups of different thin veins. The upper groups of these thin veins are connected to the uncal vein and sometimes to an insular vein [4, 23].

Frigeri et al. demonstrated that the frontosylvian, temporosylvian and parietosylvian veins all joined the SMC at the distal Sylvian fissure [24]. These veins between the SMC and insular veins are also called bridging veins [25]. Insular veins can join these veins that finally drain into the SMC (Fig. 5.1) [26].

The insular veins may be replaced by laterally running veins that drain into the SMC. It is common for one of the insular veins (commonly the PI) to drain upwards and backwards into an ascending cortical vein [27].

The prominent PI, CnI and AI unite to form a single trunk in the Sylvian vallecula. This trunk is also joined by a prominent olfactory vein and opens into the cavernous sinus via the uncal vein (or medial vallecular vein) [13]. The drainage of a large insular vein into the uncal vein has been reported [23]. Drainage of the left insular vein into a large basal vein has also been reported; in this case the patient was neurologically intact and had no neuropsychological defect [27].

Variations of the insular veins were barely detectable without indocyanine green videoangiography. The use of indocyanine green videoangiography in microneurosurgery can determine these variable drainings of the insular veins [26]. The SMC and DMC and their tributaries are key structures for achieving a wide opening of the Sylvian fissure.

5.10 Anastomoses of the Insular Veins

Anastomoses of the insular veins are connections with veins other than those in their own main drainage pathways. Insular veins have anastomoses with veins in the deep and superficial venous systems of the brain. The superficial insular veins anastomose with the subependymal veins of the deep venous system via the transcerebral anastomotic medullary veins [8]. The main superficial insular veins anastomose with each other at the tips of the insular gyri [4, 18]. Superficial pial plexuses are connections between insular veins over gyri. The posterior insular vein and superficial cortical veins are connected by a small bridging vein [13]. Anastomoses of insular veins with the uncal vein have been reported [23]. A three-dimensional computed tomography angiographic study of insular veins showed drainage to the mid-portion of the sphenoparietal sinus by a bridging vein. This type of bridging vein limits retraction of the frontal lobe in surgery [28].

We trust that the cadaver dissections and observations outlined here will aid the reader in the clear comprehension of insular vein anatomy.

References

1. Tanriover N, Rhoton AL Jr, Kawashima M, Ulm AJ, Yasuda A. Microsurgical anatomy of the insula and the Sylvian fissure. J Neurosurg. 2004;100(5):891–922. https://doi.org/10.3171/jns.2004.100.5.0891.

2. Varnavas GG, Grand W. The insular cortex: morphological and vascular anatomic characteristics. Neurosurgery. 1999;44(1):127–36. discussion 36-8.
3. Ono M, Rhoton AL Jr, Peace D, Rodriguez RJ. Microsurgical anatomy of the deep venous system of the brain. Neurosurgery. 1984;15:621–57. https://doi.org/10.1097/00006123-198411000-00002.
4. Rhoton AL Jr. The cerebral veins. Neurosurgery, 51. 2002;(4 Suppl):S159–205. https://doi.org/10.1097/00006123-200210001-00005.
5. Standring S. Chapter 19: Vascular supply and drainage of the brain. In: Standring S, editor. Gray's anatomy the anatomical basis of clinical practice, e-book, 2016, Elsevier Health Sciences; 2015. p. 280–290.
6. Fujii S, Kanasaki Y, Matsusue E, Kakite S, Kminou T, Okudera T, et al. Demonstration of deep cerebral venous anatomy on phase-sensitive MR imaging. AJNR Am J Neuroradiol. 2008;18(4):216–23. https://doi.org/10.1007/s00062-008-8027-3.
7. Hassler O. Deep cerebral venous system in man. A microangiographic study on its areas of drainage and its anastomoses with the superficial cerebral veins. Neurology. 1966;16(5):505–11.
8. Okudera T, Huang YP, Fukusumi A, Nakamura Y, Hatazawa J, Uemura K. Micro-angiographical studies of the medullary venous system of the cerebral hemisphere. Neuropathology. 1999;19(1):93–111. https://doi.org/10.1046/j.1440-1789.1999.00215.x.
9. Suzuki Y, Ikeda H, Shimadu M, Ikeda Y, Matsumoto K. Variations of the basal vein: identification using three-dimensional CT angiography. AJNR Am J Neuroradiol. 2001;22(4):670–6.
10. Bradac GB. Cerebral Veins. In: Bradac GB, editor. Cerebral angiography normal anatomy and vascular pathology. 2nd ed. London: Springer; 2014. p. 109–34.
11. Fukusumi A. Normal anatomy of intracranial veins: demonstration with MR angiography, 3D–CT angiography and microangiographic injection study. In: Takahashi S, editor. Neurovascular imaging MRI & microangiography. London: Springer-Verlag; 2010. p. 274–6.
12. Huang YP. Deep cerebral veins. In: Salamon G, Huang YP, editors. Radiologic anatomy of the brain. Berlin, Heidelberg: Springer-Verlag; 1976. p. 210–61.
13. Huang YP. Basal cerebral vein. In: Salamon G, Huang YP, editors. Radiologic anatomy of the brain. Berlin, Heidelberg: Springer-Verlag; 1976. p. 127–72.
14. Yasargil MG, Krisht AF, Ture U, Al-Mefty O, Yasargil DCH. Microsurgery of insular gliomas part II: opening of the sylvian fissure. Contemp Neurosurg. 2002;24(12):1–5.
15. Lang J Floor and contents of the middle cranial fossa. In Lang J editor (translated by Wilson RR, Winstanley DP). Clinical anatomy of the head neurocranium, orbit, craniocervical regions. Berlin, Heidelberg: Springer-Verlag; 1983. p. 282–283.
16. Ture U, Yasargil DC, Al-Mefty O, Yasargil MG. Topographic anatomy of the insular region. J Neurosurg. 1999;90(4):720–33. https://doi.org/10.3171/jns.1999.90.4.0720.
17. Afif A, Mertens P. Description of sulcal organization of the insular cortex. Surg Radiol Anat. 2010;32(5):491–8. https://doi.org/10.1007/s00276-009-0598-4.
18. Wen HT, Rhoton AL Jr, de Oliveira E, LHM C, Figueiredo EG, Teixeira MJ. Microsurgical anatomy of the temporal lobe: part 2-Sylvian fissure region and its clinical application. Neurosurgery. 2009;65(6):1–35. https://doi.org/10.1227/01.Neu.0000336314.20759.85.
19. Yasargil MG. Organization of the cerebral microcirculation. In: Yasargil MG, editor. Microneurosurgery in 4 volumes: III A-AVM of the brain, history, embryology, pathological considerations, hemodynamics, diagnostic studies, microsurgical anatomy. Stuttgart: Georg Thieme Verlag; 1987. p. 284–337.
20. Nowinski WL. Proposition of a new classification of the cerebral veins based on their termination. Surg Radiol Anat. 2012;34(2):107–14. https://doi.org/10.1007/s00276-011-0852-4.
21. Yasargil MG, Krisht AF, Ture U, Al-Mefty O, DCH Y. Microsurgery of insular gliomas part IV: surgical treatment and outcome. Contemp Neurosurg. 2002;24(14):1–5.
22. Kazumata K, Kamiyama H, Ishikawa T, Takizawa K, Maeda T, Makino K, et al. Operative anatomy and classification of the sylvian veins for the distal transsylvian approach. Neurol Med Chir (Tokyo). 2003;43(9):427–33. discussion 34.
23. Galligioni F, Bernardi R, Pellone M, Iraci G. The superficial sylvian vein in normal and pathologic cerebral angiography. Am J Roentgenol Radium Ther Nucl Med. 1969;107(3):565–78.
24. Frigeri T, Paglioli E, de Oliveira E, Rhoton AL Jr. Microsurgical anatomy of the central lobe. J Neurosurg. 2015;122(3):483–98. https://doi.org/10.3171/2014.11.JNS14315.
25. Straus D, Byrne RW, Sani S, Serici A, Moftakhar R. Microsurgical anatomy of the transsylvian translimen insula approach to the mediobasal temporal lobe: Technical considerations and case illustration. Surg Neurol Int. 2013;4:159. https://doi.org/10.4103/2152-7806.123285.
26. Kubota H, Sanada Y, Nagatsuka K, Yoshioka H, Iwakura M, Kato A. Safe and accurate sylvian dissection with the use of indocyanine green videoangiography. Surg Neurol Int. 2016;7(Suppl 14):S427–9. https://doi.org/10.4103/2152-7806.183526.
27. Lasjaunias P, Berenstein A, terBrugge KG, Raybaud C. Intracranial venous system. In: Lasjaunias P, Berenstein A, ter Brugge KG, editors. Surgical neuroangiography volume 1 clinical vascular anatomy and variations. 2nd ed. Berlin: Springer-Verlag; 2001. p. 631–713.
28. Kaminogo M, Hayashi H, Ishimaru H, Morikawa M, Kitagawa N, Matsuo Y, et al. Depicting cerebral veins by three-dimensional CT angiography before surgical clipping of aneurysms. AJNR Am J Neuroradiol. 2001;23(1):85–91.

Middle Longitudinal Fasciculus in the Human Brain from Fiber Dissection

Igor Lima Maldonado, Ilyess Zemmoura, and Christophe Destrieux

6.1 Introduction

Although the middle longitudinal fasciculus (MdLF) is not part of the insular lobe, it penetrates the temporal operculum, which is manipulated or partially removed in surgical approaches to the insula. In this chapter, we present a comprehensive description of that fascicle and its anatomical relationships with neighboring structures that were previously described in Chps. 2 and 3. For the purpose of demonstrating the anatomy, a variant of the Klingler's technique for fiber dissection was used.

This bundle was initially described in the rhesus monkey. First identified by radioisotopic tracing in the 1980s [1, 2], the MdLF was subsequently studied using diffusion imaging techniques in 2007 [3]. In 2008, its trajectory was first traced in humans, using a segmentation technique in MR and tractography [4] and then using fiber dissection techniques [5]. Regarding its trajectory with respect to sulci and gyri, the MdLF extends from the parieto-occipital cortex to anterior portions of the temporal lobe, penetrating the superior temporal gyrus.

6.2 Topography and Trajectory

A stepwise dissection of the temporo-parieto-occipital white matter is presented in Figs. 6.1, 6.2, and 6.3. In those specimens, the arachnoid membrane and the vascular elements were already removed. Before dissection, the first step is the study of the individual morphology of the sulci and gyri on the cerebral surface (Figs. 2.1a and 3.1a).

Although a relatively constant organization of sulci exists, some variants are often noted. The lateral surface of the temporal lobe is generally divided into three gyri, which are parallel to two identifiable sulci (the superior and inferior temporal sulci). A vertical and relatively deep sulcus is often responsible for an apparent discontinuity, particularly of T1 and/or T3. Additionally, cortical folds link T2 to T3 very frequently and less frequently T1 to T2.

In a majority of specimens, the anterior portion of the inferior parietal lobule (IPL) is

I. L. Maldonado, M.D., Ph.D. (✉)
Departamento de Biomorfologia, Instituto de Ciências da Saúde, Universidade Federal da Bahia, Salvador, Brazil

Serviço de Neurocirurgia, Unidade Neuro-Músculo-Esquelética, C.H.U. Prof. Edgard Santos, Salvador, Brazil

UMR 1253, iBrain, Université de Tours, Inserm, Tours, France

LE STUDIUM Loire Valley Institute for Advanced Studies, Orléans, France

I. Zemmoura, M.D., Ph.D. · C. Destrieux, M.D., Ph.D.
UMR 1253, iBrain, Université de Tours, Inserm, Tours, France

CHRU de Tours, Service de Neurochirurgie, Tours, France

© Springer International Publishing AG, part of Springer Nature 2018
M. Turgut et al. (eds.), *Island of Reil (Insula) in the Human Brain*,
https://doi.org/10.1007/978-3-319-75468-0_6

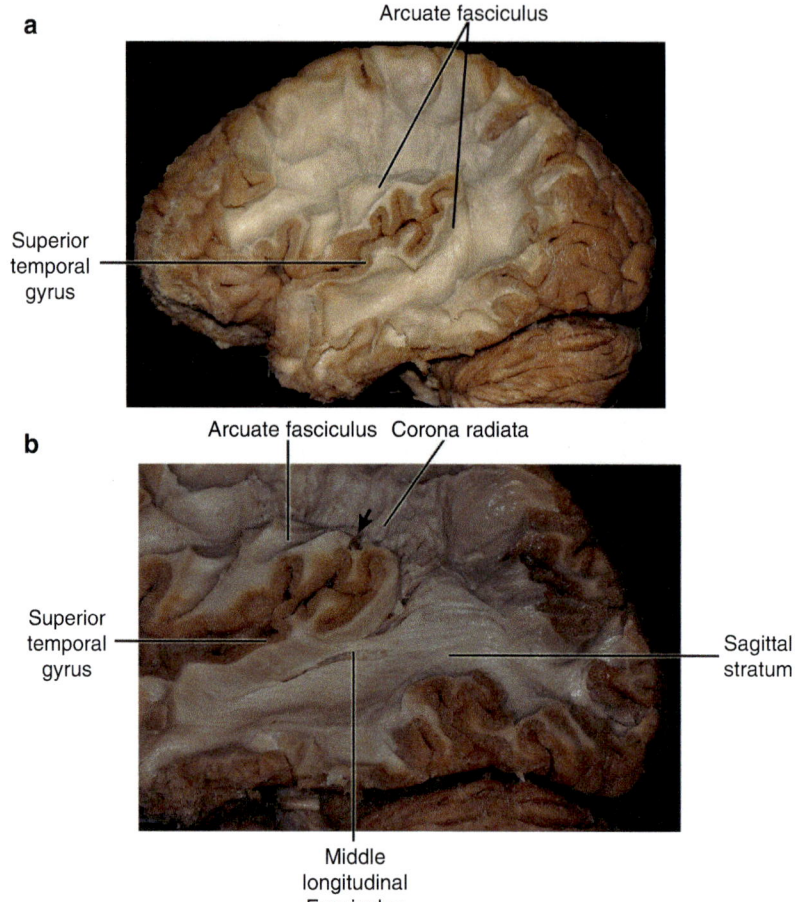

Fig. 6.1 Exposure of the middle longitudinal fasciculus using fiber dissection. The cortex and short association fibers were removed and long association pathways were exposed (**a**). Part of the superior longitudinal fasciculus–arcuate fasciculus complex is seen. To explore the association bundles medially to it, the arcuate fasciculus was sectioned (arrow) and removed (**b**)

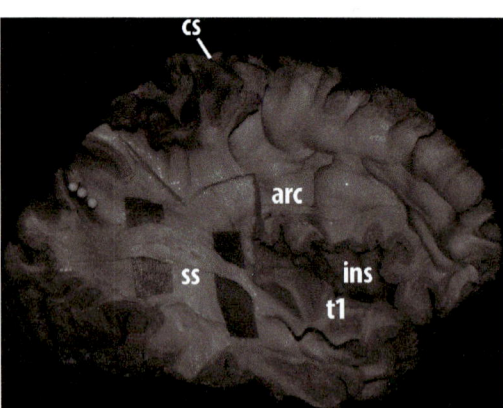

Fig. 6.2 Deep dissection. The MdLF fibers travel in the superficial layer of the sagittal stratum and penetrate the superior temporal gyrus. *Reproduced (with permission) from Maldonado et al. Evidence of a middle longitudinal fasciculus in the human brain from fiber dissection. J Anat 2013; 223(1): 38–45* [5]

limited anteriorly by the postcentral sulcus. The supramarginal gyrus (GSM) begins with a narrow cortical fold that joins to the inferior extremity of the postcentral gyrus. With an irregular trajectory, it contours the posterior extremity (posterior ramus) of the lateral sulcus. A second and frequently interrupted gyrus, the angular gyrus (AG), curves around the posterior extremity of the superior temporal sulcus. The *intermedius primus* (Jensen's) sulcus may be interposed between these two gyri (SMG and AG) with variable depth and length. A pre-AG variant may also potentially be observed in some cases. In this variant, a narrow gyrus is interposed between the AG and the SMG [6]. A pre-SMG variant, comprising a narrow gyrus intercalated between the postcentral gyrus and the SMG, is less frequently observed.

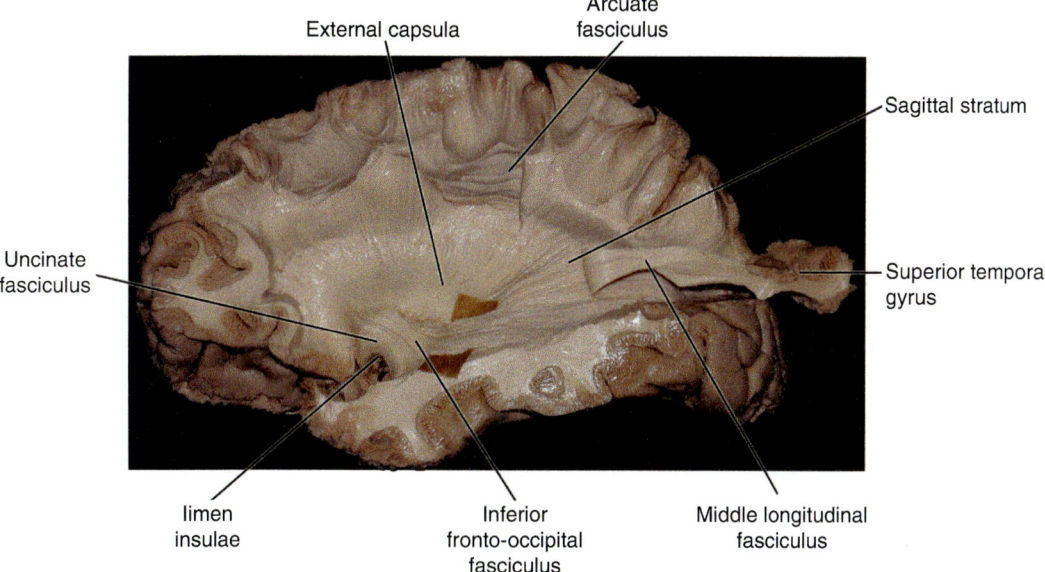

Fig. 6.3 The inferior fronto-occipital fasciculus (IFOF) was exposed, and the middle longitudinal fasciculus (MdLF) was reflected with part of the superior temporal gyrus. The fibers of IFOF penetrate deeper than those of MdLF and reach the anteroinferior portion of the external capsule, posterior to the uncinate fasciculus and the limen of the insula

As expected, the Klingler's technique renders the cortical gray matter brittle and friable, and it may be removed gently and progressively. At the subcortical level, the short U-shaped association fibers may be easily exposed on the lateral side of the temporal lobe and the IPL. This gradual removal allows exposure of the AF in the posterosuperior portion of the temporal lobe and in the depth of the SMG. Note that at this stage, T1 and white matter of AG were still kept in place and partially preserved (Fig. 6.1a).

Once the dissection of the lateral aspect has been completed, the AF may be sectioned in its posterosuperior portion, medial to the parietal operculum, raised inferiorly, and partially removed (Fig. 6.1b). The sagittal stratum is then exposed in contact to the medial aspect of the arcuate fasciculus.

At the level of the most superficial layer of the SS, the middle longitudinal fasciculus is viewed as a group of horizontal fibers joining the white matter of the superior temporal gyrus (Figs. 6.1b and 6.2). They may be followed forward to the caudal portion of the anterior third of T1. Frequently, additional small groups potentially join the same gyrus at adjacent points. Posteriorly, its fibers are partially covered by the thick vertical portion of the SLF/FA complex. Subsequent dissection highlights groups of such fibers continuing beyond the posterior limits of the IPL (Fig. 6.2). There is significant discussion about the posterior terminations of the MdLF, and some variations seem to exist; however the angular gyrus, superior parietal lobule, and occipital lobe (lateral and medial aspects) are most commonly reported [5, 7–10].

6.3 Relationships to Adjacent Fiber Bundles

The MdLF presents a global organization in three segments. The posterior segment is the longest and widest on the sagittal plane. At this level, the fibers are distributed in a thin layer, located medial with respect to SLF/AF. In the middle segment, the bundle leaves the SS to penetrate the posterior and deep portion of T1. Here, its fibers are relatively compacted, and the bundle has a narrower appearance. Finally, the anterior segment is the one in which the MdLF integrates the white matter of the temporal operculum.

The inferior fronto-occipital fasciculus (IFOF) may be identified in the same specimen. In this step, the operator confirms whether the previously dissected fibers are independent from neighboring fasciculi. The deep temporal white matter (deeper to the dissected T1) is explored to demonstrate the IFOF. Additional dissection may expose the uncinate fasciculus and centrifugal fibers of the posterosuperior portion of the external capsule. With this procedure, the IFOF is located deeper than the fibers that penetrate T1. While the IFOF tilt inward to the insular lobe and deeply reach the external capsule, the MdLF does not. Its fibers remain at a relatively superficial level and adopt a much less curved trajectory. Inside the SS, the IFOF is partially covered by the MdLF (Fig. 6.3).

6.4 Surgical and Radiological Considerations

The first observations of an MdLF were obtained in radioisotopic tracing studies in the rhesus monkey [4, 11]. It has only more recently been observed that DTI techniques bring arguments of its existence in humans [4]. Tractography has also been used to distinguish adjacent bundles from the MdLF. Since this fasciculus could provide connections to the posterior part of T1, the hypothesis of a role in language and attention has been formulated.

Despite the fact that some authors suggested that it could be an important pathway for linguistic information in the dominant hemisphere, in patients in which peroperative subcortical electrostimulation was used in the topography of the anterior segment of the fasciculus, no linguistic interference was induced [4, 12]. In addition, no permanent language deficit occurred despite resection of the anterior segment as routinely performed during temporal lobectomy. It may be inferred then that the MdLF potentially plays a role that does not seem to be essential for language processing. As a consequence, at least its anterior segment can be removed safely. In addition, adequate recognition of the elements of the white matter anatomy is important in order to avoid confusion with adjacent bundles, such as the IFOF, while interpreting tractography images or planning surgical approaches to the temporal operculum and insula.

Conclusion

The study of the descriptive and topographic anatomy of the middle longitudinal fasciculus allows an overview of concepts that are important for neuroimaging and neurosurgery. A strong understanding of the white matter anatomy is the basis for precise interpretation of MRI tractography examinations and for surgical planning of parenchymal approaches.

References

1. Schmahmann JD, Pandya DN. Fiber pathways of the brain. Oxford: Oxford University Press, Inc.; 2006.
2. Seltzer B, Pandya DN. Further observations on parieto-temporal connections in the rhesus monkey. Exp Brain Res. 1984;55:301–12.
3. Schmahmann JD, Pandya DN, Wang R, Dai G, D'Arceuil HE, De Crespigny AJ, Wedeen VJ. Association fibre pathways of the brain: parallel observations from diffusion spectrum imaging and autoradiography. Brain. 2007;130:630–53.
4. Makris N, Papadimitriou GM, Kaiser JR, Sorg S, Kennedy DN, Pandya DN. Delineation of the middle longitudinal fascicle in humans: a quantitative, in vivo, DT-MRI study. Cereb Cortex. doi:bhn124 [pii]. 2008. https://doi.org/10.1093/cercor/bhn124.
5. Maldonado IL, de Champfleur NM, Velut S, Destrieux C, Zemmoura I, Duffau H. Evidence of a middle longitudinal fasciculus in the human brain from fiber dissection. J Anat. 2013;223(1):38–45. https://doi.org/10.1111/joa.12055.
6. Kiriyama I, Miki H, Kikuchi K, Ohue S, Matsuda S, Mochizuki T. Topographic analysis of the inferior parietal lobule in high-resolution 3D MR imaging. AJNR Am J Neuroradiol. 2009;30(3):520–4. https://doi.org/10.3174/ajnr.A1417.
7. Makris N, Papadimitriou GM, Kaiser JR, Sorg S, Kennedy DN, Pandya DN. Delineation of the middle longitudinal fascicle in humans: a quantitative, in vivo, DT-MRI study. Cereb Cortex. 2009;19:777–85. doi:bhn124 [pii]. https://doi.org/10.1093/cercor/bhn124.
8. Menjot N, Maldonado IL, Moritz-Gasser S, Le Bars E, Bonafe A, Duffau H. Middle longitudinal fasciculus delineation within language pathways. Eur J Radiol. 2013;82:151–7.

9. Makris N, Preti MG, Wassermann D, Rathi Y, Papadimitriou GM, Yergatian C, Dickerson BC, Shenton ME, Kubicki M. Human middle longitudinal fascicle: segregation and behavioral-clinical implications of two distinct fiber connections linking temporal pole and superior temporal gyrus with the angular gyrus or superior parietal lobule using multi-tensor tractography. Brain Imaging Behav. 2014;7(3):335–52. https://doi.org/10.1007/s11682-013-9235-2.
10. Makris N, Preti MG, Asami T, Pelavin P, Campbell B, Papadimitriou GM, Kaiser J, Baselli G, Westin CF, Shenton ME, Kubicki M. Human middle longitudinal fascicle: variations in patterns of anatomical connections. Brain Struct Funct. 2014;218(4):951–68. https://doi.org/10.1007/s00429-012-0441-2.
11. Ribas GC, Oliveira E. The insula and the central core concept. Arq Neuropsiquiatr. 2007;65(1):92–100. doi: S0004-282X2007000100020 [pii]
12. De Witt Hamer P, Moritz-Gasser S, Gatignol P, Duffau H. Is the human left middle longitudinal fascicle essential for language? A brain electrostimulation study. Hum Brain Mapp. 2010;32(6):962–73.

Structural Connectivity of the Insula

Jimmy Ghaziri and Dang Khoa Nguyen

7.1 Anatomy

The insula is a pyramidal shaped structure hidden underneath the Sylvian fissure and constitutes approximately 1–4% of the total cortical surface [1]. The insular cortex is considered as the fifth lobe of the brain and is one of the last regions to develop along with the frontal lobe. Deemed as a paralimbic region, it is surrounded laterally by the frontal, temporal, and parietal opercula and, medially, by the internal, external, and extreme capsule, as well as the claustrum [2]. More precisely, the median anterior part of the insula is delimited by the internal and external capsule, and the lateral anterior part is adjacent to perisylvian frontal areas. Mid-laterally, the transverse or Heschl's gyrus delimits the posteroinferior insula. The anterior commissure interconnects both anterior insulae, and the posterior commissure, the border of both posterior insulae [3, 4]. The central sulcus divides the insula into an anterior part, formed of three short gyri (anterior, middle, and posterior), and a posterior part, formed of two long gyri (anterior and posterior) which are separated by the postcentral insular sulcus. Cytoarchitectonically, it is divided into three zones: (1) an anterior agranular zone related mostly to processing and integrating autonomic and visceral inputs, (2) a posterior granular zone related to somatomotor systems, and (3) a dysgranular intermediate zone representing a transitional zone making the insular cortex an integrative region of all five senses [3, 5, 6]. The insula is generally composed of pyramidal neurons across its layers and represents Brodmann's area (BA) 13; its ventral anterior part has the particularity of containing von Economo neurons (layer V) thought to be implicated in social cognition [7]. The limen insulae located in the antero-basal region is the entry of the M2 segment of the middle cerebral artery, which forms multiple cortical branches. At the pole of the insula, the orbitofrontal artery, prior to reaching the insular apex, located along the anteroinferior insula, contributes to the supply of the anterior short gyrus. The prefrontal and precentral arteries arise from a common stem artery close to the insular apex, while the prefrontal artery irrigates the middle short gyrus. The precentral artery courses near the precentral insular sulcus, while

J. Ghaziri
Département de Psychologie, Université du Québec à Montréal, Montréal, QC, Canada

Département de Neurosciences, Centre de Recherche du Centre Hospitalier de l'Université de Montréal (CRCHUM), Montréal, QC, Canada

D. K. Nguyen (✉)
Département de Neurosciences, Centre de Recherche du Centre Hospitalier de l'Université de Montréal (CRCHUM), Montréal, QC, Canada

Service de Neurologie, Centre Hospitalier de l'Université de Montréal (CHUM), Montréal, QC, Canada
e-mail: d.nguyen@umontreal.ca

the central artery supplies the central insular sulcus and the posterior short gyrus. Finally, the anterior long gyrus is supplied by a stem artery that gives rise to the angular and the anterior and posterior parietal arteries [8, 9].

The central localization of the insular cortex makes it propitious to a rich array of functions including but not limited to sensorimotor (including pain), visceral and autonomic processes, motor (e.g. speech), language, hearing, olfaction, motivation, emotions such as empathy and disgust, cognitive control, social cognition, awareness, craving, as well as addictions. These will not be reviewed in this chapter, as extensive reviews are available in this book and elsewhere [10–13].

7.2 Connectivity in Nonhuman Primates

Tract-tracing studies in nonhuman primates reported connections with the frontal, temporal, and parietal lobes, as well as the basal ganglia, the amygdala, and the thalamus [14, 15].

Regarding the frontal lobe, the insular cortex has connections with the frontal operculum, olfactory bulb, premotor and supplementary motor areas, the precentral gyrus, the inferior frontal and orbitofrontal gyri, and the anterior cingulate gyrus.

The insula has rich connections with the temporal lobe, including the temporal pole and supratemporal plane; the temporal operculum; the primary and associative auditory cortices; the superior and inferior temporal gyri and the superior temporal sulcus; the prepiriform, piriform, prorhinal, perirhinal, and entorhinal cortices; as well as the parahippocampal gyrus.

Concerning the parietal lobe, the insula has connections with the primary and secondary somatosensory cortices and the parietal operculum.

As for subcortical structures, the insular cortex has been reported to have connections with the amygdala (lateral nucleus, medial amygdaloid area, basolateral part, periamygdaloid), the hippocampus, the thalamus (dorsal, basal, and lentiform nuclei), the hypothalamus, the tail of the caudate, the putamen, and the claustrum.

This extensive pioneering work from Mesulam and Mufson, in parallel with the progress of neuroimaging techniques in the past decades, opened doors to numerous novel and stimulating research on the insular cortex. In the following paragraphs, we will examine recent diffusion tractography-based studies on the connectivity of the human insula.

7.3 Connectivity in Humans

Diffusion tractography is a magnetic resonance imaging technique that estimates the presence of a connection via white matter tracts between two regions [16, 17]. Recent advances in tractography techniques have made the analysis of the structural connectivity of the insula possible.

Because there is a lack of standard methodology in tractography techniques, it is difficult to compare findings from different studies. Therefore, we will describe every region that has been reported based on the four major studies currently available [18–20]. Three of these studies have grossly divided the insula into an anterior, middle, and posterior region. Overall, the insular cortex seems to have a global-like connectivity profile. Figures 7.1 and 7.2 illustrate a sample subject's structural connectivity of the dorsal-anterior, dorsal-posterior, ventral-anterior, and ventral-posterior insular parts.

7.3.1 Frontal Lobe

The superior and inferior frontal gyri are connected with the middle and posterior insular cortices, while the orbitofrontal cortex is connected with the anterior and middle insular cortices. The precentral gyrus is connected with the middle and posterior insular cortices [18–20]. In a recent study, we reported connections with the superior, middle, and inferior frontal gyri, the orbitofrontal cortex including the pars triangularis and pars opercularis, the frontal operculum and the frontal pole with the dorsal and ventral anterior insula, as well as some dorsal and ventral posterior regions of the insula [21].

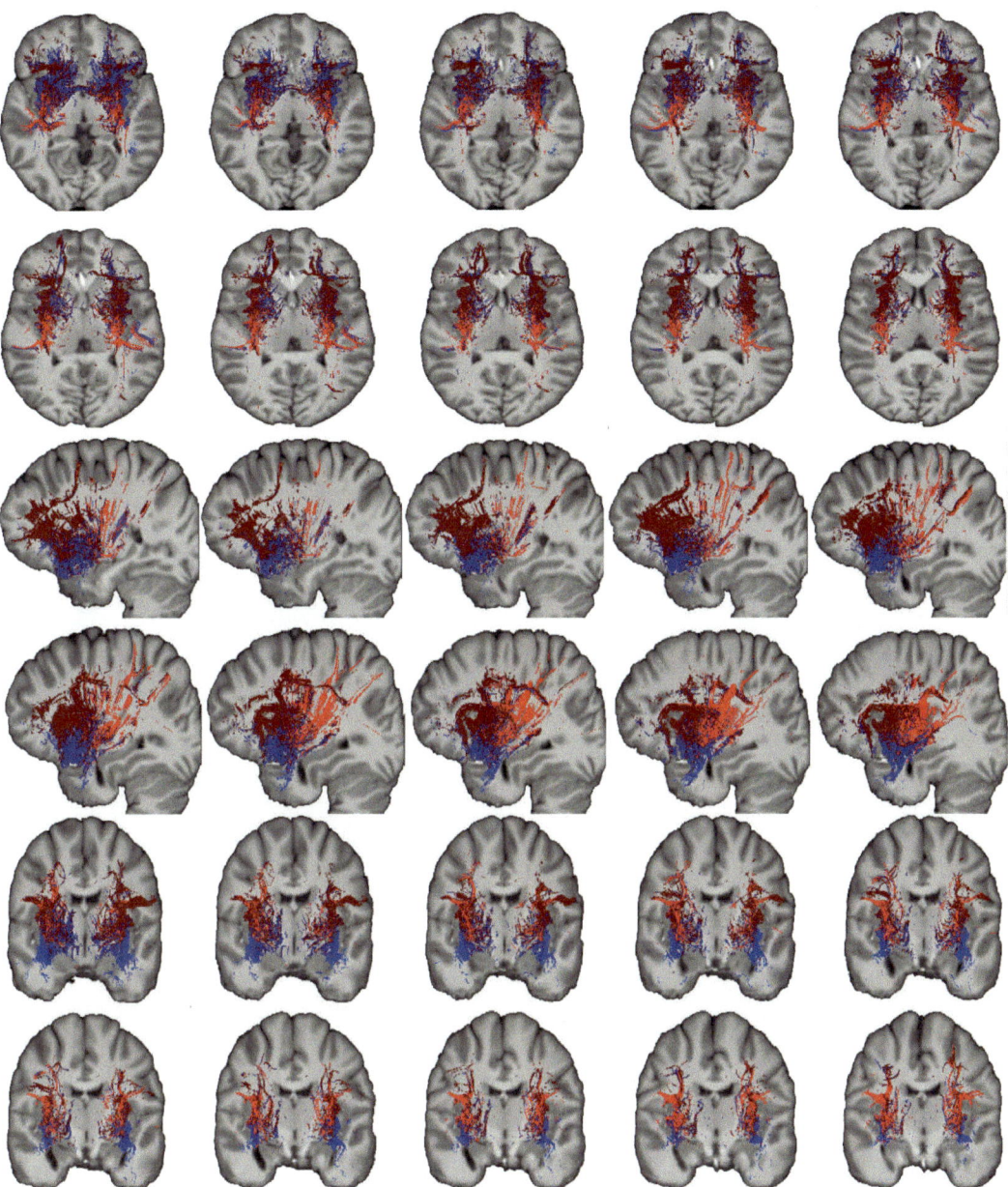

Fig. 7.1 Tracts from a sample subject for illustrative purposes in 2D in axial, sagittal, and coronal view. Light red, dorso-posterior connections; dark red, dorso-anterior connections; dark blue, ventro-posterior connections; light blue, ventro-anterior connections

7.3.2 Temporal Lobe

The superior, middle, and inferior temporal gyri, the superior temporal sulcus, Heschl's gyrus, and the temporal pole are connected to the anterior, middle, and posterior insular cortices [18, 20]. More precisely, the dorsal anterior and posterior insulae are connected to medial regions such as the planum temporale and polare, Heschl's gyrus, and the middle temporal gyrus, whereas more ventral insular regions are connected with the superior and inferior temporal

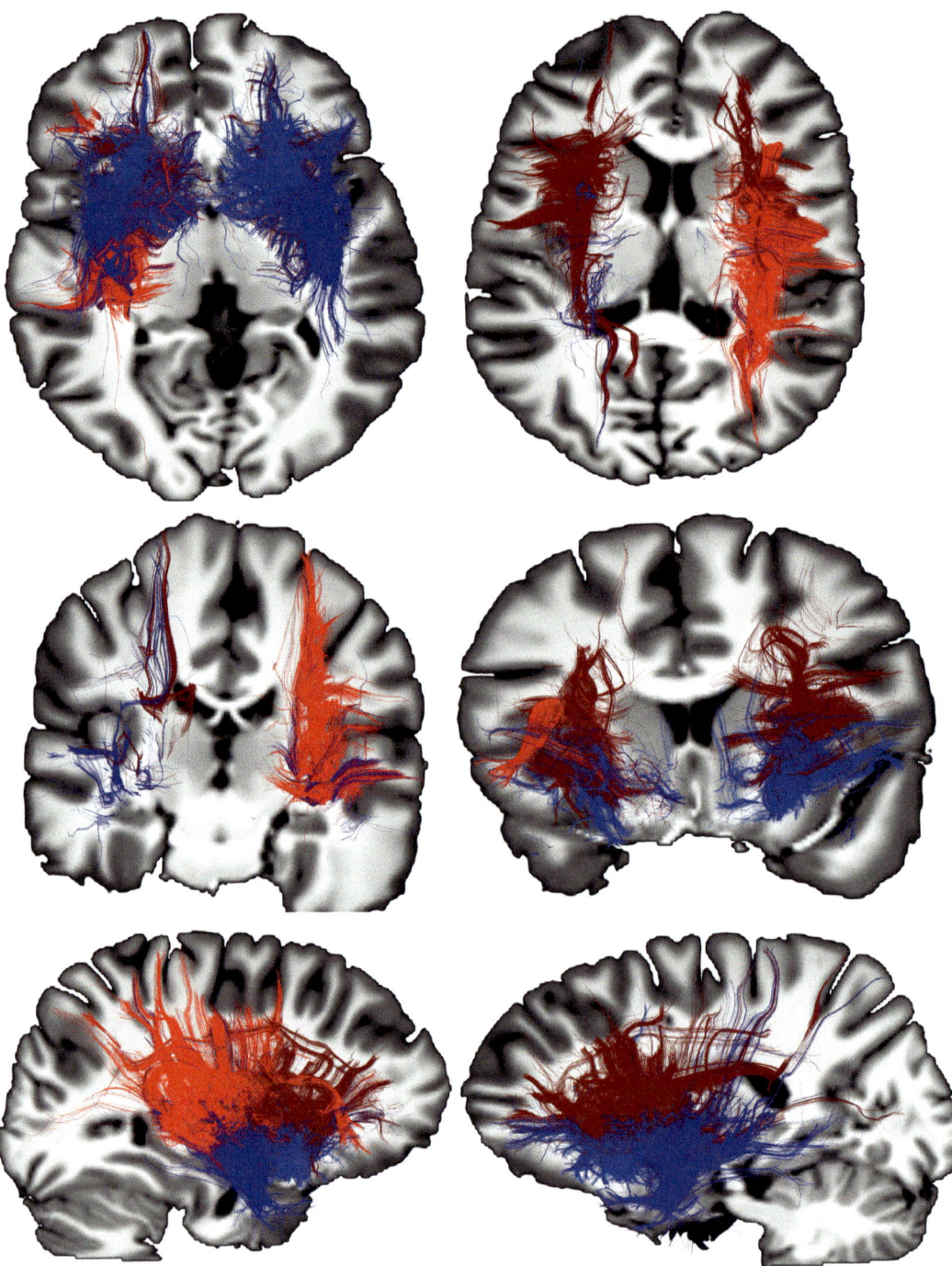

Fig. 7.2 Tracts from a sample subject for illustrative purposes in 3D. From top to bottom, starting on the left: ventral axial view, dorsal axial view, coronal posterior view, coronal anterior view, sagittal right view, sagittal left view. Light red, dorso-posterior connections; dark red, dorso-anterior connections; dark blue, ventro-posterior connections; light blue, ventro-anterior connections

gyri, the temporal pole, and the temporal fusiform gyrus [21].

7.3.3 Parietal Lobe

The supramarginal gyrus has connections with the middle insular cortex, the postcentral gyrus, and the parietal cortex with the middle and posterior insular cortex [18–20]. Moreover, the dorsal anterior and posterior insular cortices have connections with the superior parietal lobule; the postcentral, supramarginal, and angular gyri; as well as the parietal opercula and the precuneus [21].

7.3.4 Occipital Lobe

The anterior and posterior insular cortices seem to have some connections with the occipital cortex [19]. The dorsal posterior part of the insula has connections with the lateral occipital cortex, the cuneus, the occipital fusiform gyrus, and the occipital pole. The dorsal posterior and ventral posterior insular cortices have connections with the lingual and occipital fusiform gyri and the occipital pole [21].

7.3.5 Subcortical Regions

The anterior insular cortex has connections with the thalamus and amygdala [18, 19], while the posterior insular cortex has connections with the thalamus and putamen [19, 20]. More recently, we reported connections with the anterior and posterior parahippocampal gyri and the anterior and posterior cingulate gyri [21]. In our latest submitted work, we identified connections between the insula and the amygdala, thalamus, hippocampus, putamen, caudate nuclei, nucleus accumbens, and globus pallidus. The thalamus is mostly connected to the anterior ventral and rostral and dorsal posterior insula. The putamen is mostly connected to the ventral anterior and posterior insula, as well as its caudal posterior part. The caudate nuclei are mostly connected to the dorsal and ventral posterior part of the insula, as well as the ventral anterior part. The hippocampus, globus pallidus, amygdala, and nucleus accumbens have widespread connections to the insula. It is worthwhile to note that the connectivity profile is quite symmetrical between the two hemispheres.

Conclusion

The extensive connectivity profile of the human insula is now clearly established, but more work is necessary to refine it. Among the challenges is the identification of its specific cytoarchitecture but also its more refined sulco-gyral connectivity as well as its intra-connectivity. This is still quite difficult to achieve as the only available technique to investigate its connectivity in humans is magnetic resonance imaging. Unfortunately, diffusion tractography and functional images are limited by their resolution but also by their inability to differentiate afferent and efferent connections as well as detecting cytoarchitectonic zones [22]. Similarly, methodological limitations in regard to estimating crossing fiber persist but are in constant improvement [23, 24]. Reproducible studies are needed, as well as modern comparison studies between tractography and tract-tracing in humans and nonhuman primates, respectively.

Hopefully, further developments in diffusion tractography techniques will allow us to refine the connectivity of the insula with the hope of improving our understanding of its role in normal and pathological conditions such as epilepsy, Alzheimer, Parkinson, and Huntington diseases [25–27], frontotemporal dementia [28], anxiety disorder [29–31], depression [32], autism [33], and addiction [34, 35].

References

1. Semendeferi K, Damasio H. The brain and its main anatomical subdivisions in living hominoids using magnetic resonance imaging. J Hum Evol. 2000;38:317–32. https://doi.org/10.1006/jhev.1999.0381.
2. Türe U, Yaşargil DC, Al-Mefty O, Yaşargil MG. Topographic anatomy of the insular region. J

Neurosurg. 1999;90:720–33. https://doi.org/10.3171/jns.1999.90.4.0720.
3. Flynn FG. Anatomy of the insula functional and clinical correlates. Aphasiology. 1999;13:55–78. https://doi.org/10.1080/026870399402325.
4. Naidich TP, Kang E, Fatterpekar GM, et al. The insula: anatomic study and MR imaging display at 1.5 T. Am J Neuroradiol. 2004;25:222–32.
5. Kurth F, Eickhoff SB, Schleicher A, et al. Cytoarchitecture and probabilistic maps of the human posterior insular cortex. Cereb Cortex. 2010;20:1448–61. https://doi.org/10.1093/cercor/bhp208.
6. Morel A, Gallay MN, Baechler A, et al. The human insula: architectonic organization and postmortem MRI registration. Neuroscience. 2013;236:117–35. https://doi.org/10.1016/j.neuroscience.2012.12.076.
7. Allman J, Tetreault N. The von Economo neurons in frontoinsular and anterior cingulate cortex in great apes and humans. Brain Struct. 2010;214:495–517. https://doi.org/10.1007/s00429-010-0254-0.
8. Tanriover N, Rhoton AL, Kawashima M, et al. Microsurgical anatomy of the insula and the sylvian fissure. J Neurosurg. 2004;100:891–922. https://doi.org/10.3171/jns.2004.100.5.0891.
9. Türe U, Yaşargil MG, Al-Mefty O, Yaşargil DC. Arteries of the insula. J Neurosurg. 2000;92:676–87. https://doi.org/10.3171/jns.2000.92.4.0676.
10. Uddin LQ, Nomi JS, Hébert-Seropian B, et al. Structure and function of the human insula. J Clin Neurophysiol. 2017;34:300–6. https://doi.org/10.1097/WNP.0000000000000377.
11. Nieuwenhuys R. The insular cortex: a review. Prog Brain Res. 2012;195:123–63. https://doi.org/10.1016/B978-0-444-53860-4.00007-6.
12. Kurth F, Zilles K, Fox PT, et al. A link between the systems: functional differentiation and integration within the human insula revealed by meta-analysis. Brain Struct Funct. 2010:1–16. https://doi.org/10.1007/s00429-010-0255-z.
13. Augustine JR. Circuitry and functional aspects of the insular lobe in primates including humans. Brain Res Rev. 1996;22:229–44. https://doi.org/10.1016/S0165-0173(96)00011-2.
14. Mesulam MM, Mufson EJ. Insula of the old world monkey. III: efferent cortical output and comments on function. J Comp Neurol. 1982;212:38–52. https://doi.org/10.1002/cne.902120104.
15. Mufson EJ, Mesulam MM. Insula of the old world monkey. II: afferent cortical input and comments on the claustrum. J Comp Neurol. 1982;212:23–37. https://doi.org/10.1002/cne.902120103.
16. Mori S, van Zijl PCM. Fiber tracking: principles and strategies—a technical review. NMR Biomed. 2002;15:468–80. https://doi.org/10.1002/nbm.781.
17. Calamante F, Masterton RAJ, Tournier J-D, et al. Track-weighted functional connectivity (TW-FC): a tool for characterizing the structural-functional connections in the brain. Neuroimage. 2013;70:199–210. https://doi.org/10.1016/j.neuroimage.2012.12.054.
18. Cerliani L, Thomas RM, Jbabdi S, et al. Probabilistic tractography recovers a rostrocaudal trajectory of connectivity variability in the human insular cortex. Hum Brain Mapp. 2012;33:2005–34. https://doi.org/10.1002/hbm.21338.
19. Jakab A, Molnár PP, Bogner P, et al. Connectivity-based parcellation reveals interhemispheric differences in the insula. Brain Topogr. 2012;25:264–71. https://doi.org/10.1007/s10548-011-0205-y.
20. Cloutman LL, Binney RJ, Drakesmith M, et al. The variation of function across the human insula mirrors its patterns of structural connectivity: evidence from in vivo probabilistic tractography. Neuroimage. 2012;59:3514–21. https://doi.org/10.1016/j.neuroimage.2011.11.016.
21. Ghaziri J, Tucholka A, Girard G, et al. The corticocortical structural connectivity of the human insula. Cereb Cortex. 2017;27:1216–28. https://doi.org/10.1093/cercor/bhv308.
22. Dell'Acqua F, Catani M. Structural human brain networks: hot topics in diffusion tractography. Curr Opin Neurol. 2012;25:375–83.
23. Tournier J-D, Mori S, Leemans A. Diffusion tensor imaging and beyond. Magn Reson Med. 2011;65:1532–56. https://doi.org/10.1002/mrm.22924.
24. Jbabdi S, Johansen-Berg H. Tractography: where do we go from here? Brain Connect. 2011;1:169–83. https://doi.org/10.1089/brain.2011.0033.
25. Shelley BP, Trimble MR. The insular lobe of Reil—its anatamico-functional, behavioural and neuropsychiatric attributes in humans—a review. World J Biol Psychiatry. 2004;5:176–200. https://doi.org/10.1080/15622970410029933.
26. Namkung H, Kim S-H, Sawa A. The insula: an underestimated brain area in clinical neuroscience, psychiatry, and neurology. Trends Neurosci doi. 2017. https://doi.org/10.1016/j.tins.2017.02.002.
27. Christopher L, Koshimori Y, Lang AE, et al. Uncovering the role of the insula in non-motor symptoms of Parkinson's disease. Brain. 2014;137:2143–54. https://doi.org/10.1093/brain/awu084.
28. Seeley WW. Anterior insula degeneration in frontotemporal dementia. Brain Struct Funct. 2010;214:465–75. https://doi.org/10.1007/s00429-010-0263-z.
29. Nagai M, Kishi K, Kato S. Insular cortex and neuropsychiatric disorders: a review of recent literature. Eur Psychiatry. 2007;22:387–94. https://doi.org/10.1016/j.eurpsy.2007.02.006.
30. Paulus MP, Stein MB. Interoception in anxiety and depression. Brain Struct Funct. 2010;214:451–63. https://doi.org/10.1007/s00429-010-0258-9.
31. Paulus MP, Stein MB. An insular view of anxiety. Biol Psychiatry. 2006;60:383–7. https://doi.org/10.1016/j.biopsych.2006.03.042.
32. Avery JA, Drevets WC, Moseman SE, et al. Major depressive disorder is associated with abnormal interoceptive activity and functional connectivity in the insula. Biol Psychiatry. 2014. https://doi.org/10.1016/j.biopsych.2013.11.027.

33. Uddin LQ, Menon V. The anterior insula in autism: under-connected and under-examined. Neurosci Biobehav Rev. 2009;33:1198–203. https://doi.org/10.1016/j.neubiorev.2009.06.002.
34. Naqvi NH, Rudrauf D, Damasio H, Bechara A. Damage to the insula disrupts addiction to cigarette smoking. Science. 2007;315:531–4. https://doi.org/10.1126/science.1135926.
35. Naqvi NH, Bechara A. The hidden island of addiction: the insula. Trends Neurosci. 2009;32:56–67. https://doi.org/10.1016/j.tins.2008.09.009.

Insular Pharmacology

Hasan Emre Aydın and İsmail Kaya

8.1 Introduction

The insula is a triangular area located deep within the Sylvian fissure covered by frontal, temporal, and parietal lobes. It is a region which covers less than 2% of the cortical surface area and receives sensory stimuli from the thalamus, amygdala, and limbic system and transmits the received input to the premotor cortex and ventral striatum [1]. Tramo et al. compared the whole of the superficial cortex with the insula in structural terms and measured the left and right insula as approximately 17 cm^2 [2]. The insula also plays a role in the regulation of various functions such as pain, speech and social affect [1].

The insular cortex is divided into three equal areas: caudodorsal, granular, rostroventral agranular, and intermediate dysgranular. Due to the rarity of the granules in the intermediate, the complete laminar differentiation is not displayed, and thus the region is called dysgranular [1, 2].

Thalamic nuclei which project to the insula include the ventral posterior superior (VPS) nucleus and the ventral posterior inferior (VPI) nucleus, the ventromedial posterior (VMPo) nucleus, and the parvocellular part of the ventral posteromedial (VPMpc) nucleus. VPS and VPI receive afferents from the vestibular nucleus and 'parietoinsular vestibular cortex' which also includes the posterosuperior part of the insula which reaches the cortical areas of the brain. VMPo which receives the nociceptive and thermoreceptive stimuli is connected to the posterosuperior part of the insular cortex which is called the 'insular nociceptive and thermoreceptive cortex' [1, 3].

Nourishment and smell stimuli activate the central insula. Sensory and motor stimuli activate the mid-posterior part of the insula. The insular somatic association cortex, which is localized in the superior posterior part of the insula, receives somatosensorial signals for the whole cortex [1].

The insular limbic cortex consists of entorhinal (area 28), perirhinal (areas 35, 36), posterior orbitofrontal (areas 13, 14), temporopolar (area 38) and cingulate (areas 23, 24) cortices and the amygdaloid complex and is located in the anterobasal sector of the insula [1–3].

8.2 Intercellular Stimulus Transmission

Intercellular transmission in the cortex is mediated by action potential. In the initial phase there is a rapid flow of sodium (Na) into the cell. Afterwards, Na channels are deactivated and chlorine (Cl) channel opened. This is followed by the opening of slow calcium (Ca) channels in plateau, and the intracellular positive voltage

H. E. Aydın, M.D., Ph.D. (✉) · İ. Kaya, M.D.
Department of Neurosurgery, Dumlupınar University, Medical Faculty, Kutahya, Turkey

increased with Na and Ca. Finally the intracellular positive voltage moves towards extracellular region with the activation of intracellular potassium (K) and K channels, and the voltage returns to its resting membrane potential value.

The medial part of the parvocellular area of the ventral posteromedial (VPMpc) nucleus regulates the stimuli regarding taste. The insular viscerosensory cortex is organized behind the taste area [1].

Many taste neurons in the cortex are activated as a response to taste and somatosensorial stimulation in the mouth cavity and initiate the intercellular transmission. In an experimental study on the parietal and insular cortex connections of neurons related to taste, it was found that the response was multimodal. They demonstrated that caffeine operated via ryanodine receptors and that the application of thapsigargin, which is a calcium channel-blocking agent, prevented the stimuli initiated with electrical stimulation from the parietal cortex to the insular cortex. They showed that calcium channels and ryanodine receptors were effective in the transmission of taste sense on the insular cortex [3].

Several receptors and related mechanisms are effective on the regulation of the signal transmission on different parts of the insula to the cerebral cortex.

8.2.1 Muscarinic Receptors

Muscarinic receptors are clamped together with G protein. They are one of the receptors of cholinergic system. Muscarinic receptors have five subtypes, and M1 subtype of muscarinic receptors is found in the central nervous system. They operate via the Gq subtype of G proteins. Gq phospholipase is activated by protein kinase C enzyme. Diacylglycerol of the second messengers provides activation with protein kinase C stimulus. The other second messenger inositol 3 phosphate increases the intracellular Ca secretion. In a study, saccharine was used as the initial stimulus for taste, and it was observed that in the second encounter, saccharine consumption increased in relation to memory. In the study, scopolamine, which is a muscarinic receptor antagonist, was bilaterally injected into the insular cortex, and saccharine consumption was blocked [4].

In a study examining the effects of the cholinergic neurotransmission related to the M2 receptors in the posterior insular cortex on neuropathic pain, it was shown that central cholinomimetics could be used in the treatment of neuropathic pain. Metabolic changes observed in the posterior insular cortex demonstrated that it is connected to posterior thalamic nucleus and somatosensorial cortex. The finding suggests that the insular cortex may be the central mechanism in the regulation of pain in humans [5].

The relationship between neuropathic pain and the insular cortex has been examined in many studies. The anterior insular cortex plays a role in emphatic pain perception [6]. The most important characteristic of the anterior insular cortex which plays a role in the regulation of limbic system such as anterior cingulate cortex is that it includes pyramidal neurons and fusiform cells in the fifth layer. These different structures are called 'von Economo' neurons. These cells which play a part in emotional functioning, decision-making, social cognitive activity, and awareness, are currently a prominent research subject [1].

8.2.2 NMDA Receptors

NMDA receptors play an important role in the occurrence of spontaneous or stimulus-dependent electrical activation in the insular cortex [7].

NMDA receptors have ion channel characteristics and play the primary part in learning and memory in the central nervous system. There is Na, K and Ca flow through the channel at its activation. Activation of NMDA receptors and muscarinic receptors in the insular cortex play a role in taste-related memory [4, 8, 9]. In a study, saccharine was used as the initial stimulus for taste, and in the second encounter, saccharine consumption increased in relation to memory. In this study which also researched its activity on muscarinic receptors, the NMDA receptor was blocked with AP5 and ketamine, which did not block consumption but removed the dislike of the taste [4].

Fibres rooted in the olfactory piriform cortex are connected to the agranular anterior part of the insula [1]. This region is also connected with gustatory cortices and the primary viscerosensory insular cortex. Moreover, it joins the caudal orbitofrontal cortex and forms the 'orbital network' which provides the integration of the information regarding food [1].

Signal dispersion from the visual region is dependent on both NMDA (N-methyl-D-aspartate) receptors and intracellular calcium secretion [3]. It was observed that when NMDA receptor was blocked, the initial stimuli did not disperse into the insular cortex and parietal cortex [4]. It was shown that during the absence of caffeine, low Mg concentrations are important in NMDA receptor activations and insular dispersion of stimuli [3].

In another study on the relationship between NMDA receptors and taste memory, it was shown that 2B subtype of NMDA receptors modulated the inhibition in the cortex [10].

The fact that the NMDA receptor density is less in the dysgranular cortex, or that the vertical internal connections in this layer of the insular cortex are weak, results in the weakness between parietal and insular connections [3].

In a study examining the electrophysiological relationship between the agranular insular cortex neurons and the NMDA receptors, in order to uncover the mechanism of pain and epileptic diseases, it was seen that the stimulus network was provided in the agranular layer with strong NMDA receptor activation. Activation of the NMDA receptors decrease in the presence of magnesium (Mg). This mechanism which leads to the continuation of action potential is related to the depolarization formed in the synaptic gap, and this hyperexcitability probably accompanies the activation of inhibitor mechanisms related to the GABA-A receptor [7]. In light of these findings, it was shown that it has an important role in the understanding of the mechanisms of pain and epileptic diseases.

In a study examining the effects on cardiac baroreceptors, it was shown that the local NMDA glutamatergic receptors in the insular cortex were effective in parasympathetic activity. Stimulation of the baroreceptors which are sensory nerve endings effective in the regulation of arterial blood pressure provided an important feedback to the central nervous system. It affected the heart rate and vascular resistance in order to regulate the blood pressure and provide the normalization of the pressure [11]. In this context, considering the effects on the heart, it can be suggested that NMDA receptors may have an effect on heart rate increase based on sensory changes.

8.2.3 Dopamine Receptors

Dopamine is a catecholamine commonly found at dopaminergic nerve endings. Although there are many subtypes, functional and pharmacological properties of D1 and D2 receptors have been well defined. All subtypes are G protein-coupled receptors with several segments. It is known that dopamine levels decrease in basal ganglia in Parkinson's disease accompanied by motor coordination failure.

D4 receptors are fundamentally located in the prefrontal region like in the insula. Even though the subject still needs clarification, the insula has an important role in substance and behavioural addiction [12].

Personality is a behavioural pattern, which differs among individuals. In a study examining the mechanisms regarding personality via D2 receptors, it was observed that there was an activity increase in the right insular cortex, whereas there was no increase in other regions. Hence, the right insular cortex was evaluated as the region which regulates the personality changes related to dopamine receptors [13].

The insular cortex is thought to be the upper cortical center to which tactile information from temporal somatosensory cortex is hierarchically transmitted. In the studies performed, it was seen that sensory stimulus directly reaches the posterior ventromedial nucleus of the thalamus. These findings indicate that sensory stimuli can be collected at different centers, but integration is performed through the insular cortex. For example, a stimulus received by mechanoreceptors moves through the thalamus-insular cortex or thalamus-sensory cortex-insular cortex pathways. The area where integration is provided is the insular cortex. Since the unconscious part of sensory transmission is associated with dopamine receptors, newly generated stimuli can be identified [14].

In a study showing the relation of insular cortex to substance dependence, it was shown that the insular cortex ischemia is associated with decreased desire for cigarette smoking. Dopamine receptor systems play an important role in substance dependence. Intense presence of dopamine receptors in the agranular insula has made this region a target for studies on substance abuse. Dopamine projections from ventral tegmental area to nucleus accumbens were also found to be an important pathway for substance dependence. In a study examining the effects of D1 and D2 receptors on the insular cortex, it was shown that the effects on substance dependence are associated with D1 receptors. It was observed that the administration of a D1 antagonist to the insula reduced the need for nicotine by 50% [15].

Furthermore, in a study examining the effects of cocaine dependence on D1 receptors, it was observed that cocaine-dependent stimulation in the agranular insular cortex elevated extracellular dopamine concentration [16].

In a study where D1 receptors were blocked, it was observed that D1 receptors play a role in decision-making in the insular cortex and that agranular cortex plays an important role in decision-making in association with D1 receptors [17].

In association with the development of Parkinson's disease, in addition to striatal dopamine dysfunction, the absence of insular cortex D2 receptors is thought to be associated with mild impairment in cognitive activity in Parkinson's disease [18].

8.2.4 Opioid Receptors

Opioid peptides and opioid analgesics mainly activate mu (MOR), kappa (KOR) and delta (DOR) opioid receptor types and produce their specific effects. Mu receptors are a class of opioid receptors with a high affinity for morphine and beta endorphin. Especially mu1 subtype is found in the insular cortex. Activation of mu receptors leads to inhibition of adenylate cyclase, opening of voltage-dependent calcium channels and thus neuronal inhibition. Delta receptors play a role in formation and regulation of emotional events and cognitive functions related to excitement and affection. Kappa receptors, on the other hand, are effective in spinal analgesia.

Pain is one of the disturbing sensory and emotional components. Morphine and fentanyl are one of the most powerful analgesic agents known, and all opioid receptor subtypes with which they associate, including mu (MOR), kappa (KOR) and delta (DOR), are found in the insular cortex [19–21]. Mu receptors are the main target of analgesics [19].

Opioidergic receptors show their effects in synaptic intervals through G protein. With activation of G protein, a few ion channels go into action. Voltage-dependent calcium channels are inhibited and synaptic transmission is suppressed. Potassium channels are rectified inwardly, and cAMP is reduced to reduce the current that is activated by hyperpolarization [19].

Opioidergic agonists suppress excitatory conduction in synaptic range of the cerebral cortex. Morphine injection to the insular cortex reduces response to harmful thermal stimulation. In a study where responses of cortical areas and areas adjacent to the cortex to dental stimulation via opioid receptors were evaluated, it was revealed that mu receptor agonist suppresses cortical stimulation; delta receptor agonist enhance stimulation and increase its spread to the primary and secondary somatosensory area and insular cortex; and the kappa receptor agonist has minimal effect on cortical stimulation [20].

In another study, excitation potentials of delta receptors were investigated, and the association between delta receptors and the limbic system have been shown to be important in the treatment of chronic pain and affective disorders [22].

The gustatory area of the insular cortex was found to play a role in 'conditioned taste avoidance' induced by morphine in relation to taste [23].

8.2.5 GABA Receptors

GABA plays an important role in the processing of neural information in cerebral cortex [24]. Briefly, it is the main inhibitor neuromediator in the brain. GABA activates receptors located on the synapses, postsynaptic membrane or presynaptic membrane. Thus, it leads to hyperpolarization and associated postsynaptic inhibition.

In studies examining chronic pain, the reduction in GABA activity is thought to be responsible for cortical hyperactivity and hyperalgesia. The reduction in endogenous GABA levels is characterized by an increase in thermal hyperalgesia and mechanical allodynia. These data indicate that excitatory and inhibitory mechanisms act together in cortical structures of the brain such as the insula [23].

GABA-A receptors play a role in electrical stimulation of the brain. They are the receptors inhibited by calcium and chloride channels constitute a certain portion of the receptor. It was observed that the application of bicuculline, a GABA-A receptor antagonist, reduced acute spontaneous epileptiform discharges [24].

GABA-B receptors are metabotropic receptors that manage many mechanisms including presynaptic inhibition of excitatory postsynaptic current [24]. GABA-B receptors are linked via G protein to adenylate cyclase, calcium channels or potassium channels in the neuron membrane. Their effect emerges with different mechanisms.

It was observed that the application of muscimol, a GABAergic agonist, leads to inactivation of the granular insula and reduced need for nicotine in experimental animals [15].

Functionally, the insula is effective in viscerosensory and visceromotor functions; vestibular, somatosensory and somatomotor association; limbic integration; and speech functions [1].

When clinical reflections of the functional activities of a small and hidden part of the cortex which covers less than 2% of the cortex (insula) are considered, the studies showed that the insula is an area effective in the control of senses such as hearing, taste, pain and heat and smell and cognitive activities and in the formation of speech impediments, motor plasticity, Alzheimer's disease, schizophrenia, dementia and substance addiction [1].

When the anterior insula is considered as a whole, it plays a role in the organization of many activities such as attention, vocalization and music, cognitive control, perceptual decision-making, self-recognition, time perception and emotional awareness. In a study conducted by Mutschler et al., it was shown that the back side of the anterior insula is the part firstly activating the hearing and speech functions along with cognitive functions [25]. The front side of the anterior insula is activated by peripheral physiological responses coming from cardiovascular activity and sympathetic skin responses and by sensory stimuli [1]. In a study conducted by the use of radiological methods by Kurth et al., it was shown that amygdala is activated along with these stimuli [1]. It was observed that the anterior insular cortex is an important communication network regarding the regulation of appetite and energy balance in obese patients [26].

References

1. Nieuwenhuys R. The insular cortex: a review. Prog Brain Res. 2012;195:123–63. https://doi.org/10.1016/B978-0-444-53860-4.00007-6.
2. Tramo MJ, Loftus WC, Thomas CE, Green RL, Mott LA, Gazzaniga MS. Surface area of human cerebral cortex and its gross morphological subdivisions: in vivo measurements in monozygotic twins suggest differential hemisphere effects of genetic factors. J Cogn Neurosci. 1995;7(2):292–302. https://doi.org/10.1162/jocn.1995.7.2.292.
3. Yoshimuraa H, Kato N, Honjo M, Sugai T, Segami N, Onoda N. Age-dependent emergence of a parieto-insular corticocortical signal flow in developing rats. Brain Res Dev Brain Res. 2004;149(1):45–51.
4. Parkes SL, De la Cruz V, Bermúdez-Rattoni F, Coutureau E, Ferreira G. Differential role of insular cortex muscarinic and NMDA receptors in one-trial appetitive taste learning. Neurobiol Learn Mem. 2014;116:112–6. https://doi.org/10.1016/j.nlm.2014.09.008.
5. Ferrier J, Bayet-Robert M, Dalmann R, El Guerrab A, Aissouni Y, Graveron-Demilly D, et al. Cholinergic neurotransmission in the posterior insular cortex is altered in preclinical models of neuropathic pain: key role of muscarinic M2 receptors in donepezil-induced antinociception. J Neurosci. 2015;35(50):16438–0.

6. Gu X, Gao Z, Wang X, Liu X, Knight RT, Hof PR, et al. Anterior insular cortex is necessary for empathetic pain perception. Brain. 2012;135:2726–35. https://doi.org/10.1093/brain/aws199.
7. Inaba Y, de Guzman P, Avoli M. NMDA receptor-mediated transmission contributes to network 'hyperexcitability' in the rat insular cortex. Eur J Neurosci. 2006;23(4):1071–6.
8. Rodríguez-Durán LF, Martínez-Moreno A, Escobar ML. Bidirectional modulation of taste aversion extinction by insular cortex LTP and LTD. Neurobiol Learn Mem. 2017;142:85–90. https://doi.org/10.1016/j.nlm.2016.12.014.
9. Escobar ML, Alcocer I, Chao V. The NMDA receptor antagonist CPP impairs conditioned taste aversion and insular cortex long-term potentiation in vivo. Brain Res. 1998;812(1–2):246–51.
10. Rosenblum K, Berman DE, Hazvi S, Lamprecht R, Dudai Y. NMDA receptor and the tyrosine phosphorylation of its 2B subunit in taste learning in the rat insular cortex. J Neurosci. 1997;17(13):5129–35.
11. Alves FH, Crestani CC, Resstel LB, Correa FM. N-methyl-D-aspartate receptors in the insular cortex modulate baroreflex in unanesthetized rats. Auton Neurosci. 2009;147(1–2):56–63. https://doi.org/10.1016/j.autneu.2008.12.015.
12. Cocker PJ, Lin MY, Barrus MM, Le Foll B, Winstanley CA. The agranular and granular insula differentially contribute to gambling-like behavior on a rat slot machine task: effects of inactivation and local infusion of a dopamine D4 agonist on reward expectancy. Psychopharmacology (Berl). 2016;233(17):3135–47. https://doi.org/10.1007/s00213-016-4355-1.
13. Suhara T, Yasuno F, Sudo Y, Yamamoto M, Inoue M, Okubo Y, et al. Dopamine D2 receptors in the insular cortex and the personality trait of novelty seeking. Neuroimage. 2001;13(5):891–5.
14. Chou TS, Bucci LD, Krichmar JL. Learning touch preferences with a tactile robot using dopamine modulated STDP in a model of insular cortex. Front Neurorobot. 2015;9:6. https://doi.org/10.3389/fnbot.2015.00006.
15. Kutlu MG, Burke D, Slade S, Hall BJ, Rose JE, Levin ED. Role of insular cortex D1 and D2 dopamine receptors in nicotine self-administration in rats. Behav Brain Res. 2013;256:273–8. https://doi.org/10.1016/j.bbr.2013.08.005.
16. Di Pietro NC, Mashhoon Y, Heaney C, Yager LM, Kantak KM. Role of dopamine D1 receptors in the prefrontal dorsal agranular insular cortex in mediating cocaine self-administration in rats. Psychopharmacology (Berl). 2008;200(1):81–91. https://doi.org/10.1007/s00213-008-1149-0.
17. Pattij T, Schetters D, Schoffelmeer AN. Dopaminergic modulation of impulsive decision making in the rat insular cortex. Behav Brain Res. 2014;270:118–24. https://doi.org/10.1016/j.bbr.2014.05.010.
18. Christopher L, Marras C, Duff-Canning S, Koshimori Y, Chen R, Boileau I, et al. Combined insular and striatal dopamine dysfunction are associated with executive deficits in Parkinson's disease with mild cognitive impairment. Brain. 2014;137(Pt 2):565–75. https://doi.org/10.1093/brain/awt337.
19. Yokota E, Koyanagi Y, Yamamoto K, Oi Y, Koshikawa N, Kobayashi M. Opioid subtype- and cell-type-dependent regulation of inhıbitory synaptc transmission in the rat insular cortex. Neuroscience. 2016;339:478–90. https://doi.org/10.1016/j.neuroscience.2016.10.004.
20. Yokota E, Koyanagi Y, Nakamura H, Horİnuki E, Oi Y, Kobayashi M. Opposite effects of mu and delta opioid receptor agonists on excitatory propagation induced in rat somatosensory and insular cortices by dental pulp stimulation. Neurosci Lett. 2016;628:52–8. https://doi.org/10.1016/j.neulet.2016.05.065.
21. Burkey AR, Carstens E, Wenniger JJ, Tang J, Jasmin L. An opioidergic corticalantinociception triggering site in the agranular insular cortex of the rat that contributes to morphine antinociception. J Neurosci. 1996;16(20):6612–23.
22. Chu Sin Chung P, Kieffer BL. Delta opioid receptors in brain function anddiseases. Pharmacol Ther. 2013;140(1):112–20. https://doi.org/10.1016/j.pharmthera.2013.06.003.
23. Watson CJ. Insular balance of glutamatergic and GABAergic signaling modulates pain processing. Pain. 2016;157(10):2194–207. https://doi.org/10.1097/j.pain.0000000000000615.
24. Fujita S, Koshikawa N, Kobayashi M. GABA(B) receptors accentuate neural excitation contrast in rat insular cortex. Neuroscience. 2011;199:259–71. https://doi.org/10.1016/j.neuroscience.2011.09.043.
25. Mutschler I, Wieckhorst B, Kowalevski S, Derix J, Wentlandt J, Schulze-Bonhage A, et al. Functional organization of the human anterior insular cortex. Neurosci Lett. 2009;457(2):66–70. https://doi.org/10.1016/j.neulet.2009.03.101.
26. Frank S, Kullmann S, Veit R. Food related processes in the insular cortex. Front Hum Neurosci. 2013;7:499. https://doi.org/10.3389/fnhum.2013.00499.

Neuroimaging Techniques for Investigation of the Insula

9

Ersen Ertekin, Özüm Tunçyürek, Mehmet Turgut, and Yelda Özsunar

Abbreviations

ADC	Apparent diffusion coefficient
BOLD	Blood oxygen level-dependent
CBF	Cerebral blood flow
CBV	Cerebral blood volume
CSF	Cerebrospinal fluid
CT	Computed tomography
DSA	Digital subtraction angiography
DWI	Diffusion-weighted MRI
FLAIR	Fluid attenuation inversion recovery
fMRI	Functional MRI
MCA	Middle cerebral artery
MRI	Magnetic resonance imaging
MTT	Mean transit times
NAA	*N*-Acetyl aspartate
SWI	Susceptibility-weighted MRI

9.1 Introduction

The insular cortex, which constitutes 2% of the cortical surfaces of the brain, is located deep in the lateral sulcus. Frontal, parietal, and temporal lobe parenchyma surrounding the insula are called the opercula. Although some authors consider it a separate lobe of the telencephalon, most authors regard it as a distinct area that does not belong to neighboring lobes. It looks like an irregular pyramid because its apex lies laterally in the Sylvian fissure. The central sulcus of the insula, which lies obliquely from the posterosuperior to the anteroinferior, divides the insula into two lobules. The anterior lobule usually consists of three gyri including the anterior, middle, and posterior short insular gyri, whereas posterior lobule consists of two gyri including anterior and posterior long insular gyri. The insula has connection pathways to the frontal-temporal-parietal-occipital lobes, corpus callosum, and limbic system [1, 2]. The insula is usually fed through perforating branches that separate from the M2 segment of the middle cerebral artery (MCA) [3].

Recent research, especially functional magnetic resonance imaging (fMRI) studies, have revealed quite useful information about the functions of the insula. Studies have shown that the primary gustatory cortex is located in the insula [4]. Furthermore, it also plays a significant role in speech planning and coordination [5–7], vestibular integration [8, 9], visual-vestibular inhibition [10], cardiopulmonary regulation [11, 12], emotion [13], empathy, interoception [14], pain [15–17], somatosensation, visceral sensation, thermosensation, and memory functions [18, 19].

There are also publications that indicate specific pathologies caused by injury or volume loss

E. Ertekin (✉) · Ö. Tunçyürek · Y. Özsunar
Department of Radiology, Adnan Menderes University School of Medicine, Aydın, Turkey

M. Turgut
Department of Neurosurgery, Adnan Menderes University School of Medicine, Aydın, Turkey

occurring in the insula, for example, cocaine and smoking addiction [20, 21], emotional decision-making problems [22], abnormalities in the perception of justice and equality and the urge for revenge [16, 23], and neuropsychiatric diseases, such as borderline personality disorder, anxiety, and psychosis [24–26].

9.2 Imaging of Insula

9.2.1 Imaging Methods

9.2.1.1 Radiography
Radiography is a radiological imaging method that provides two-dimensional imaging using X-rays. Although it has the highest spatial resolution, its uses are limited due to overlapping of the tissues in two-dimensional images and low-contrast resolution (capacity of separating the different density tissues). There is no contribution to brain imaging except for the evaluation of skull bone pathologies and cerebral calcific lesions.

9.2.1.2 Computed Tomography
Computed tomography (CT) is the radiological method that composes a cross-sectional image of the region to be examined by using X-ray. As the X-ray source in the gantry rotates around the patient, the examination region is exposed to X-ray. The X-rays passing through the patient are collected by detectors in the gantry. The data obtained by detectors are converted into an optical image with the aid of a computer. The contrast ratio, which is significantly higher than that of the radiograph, provides significant diagnostic superiority to CT. With the development of multi-row detector systems, the shooting times are considerably shortened and the slice thicknesses are rather thin. Decrease in the slice thickness allows multiplanar imaging. In recent years, there has been considerable progress in reducing the amount of radiation dosages given to patients with dose-reducing software or exposures. Despite all these developments, radiation still remains a disadvantage. In the cranial examination, it is often the first referenced imaging method because of the fact that easy access and fast shooting times allow for early diagnosis.

9.2.1.3 Magnetic Resonance Imaging
Magnetic resonance imaging (MRI) is the imaging method with the highest soft tissue resolution. It uses the natural structures found in our bodies to make imaging. For this purpose hydrogen atoms are used, which are the most common in the body and have a single proton. Signals obtained from H atoms in the region of interest, which are stimulated by radio-frequency waves in a strong magnetic field, are processed by computer programs to generate images. MRI devices used for diagnostic purposes have 0.2–3 tesla magnetic force. There is no significant side effect of MRI on humans. However, due to the strong magnetic field, MRI is contraindicated in patients with medical devices such as cardiac pacemakers, metallic prostheses, and intubation devices. For patients with claustrophobia, low-dose open MRI devices can be used. Another disadvantage of MRI is the length of time of the shooting, despite all technological developments. The ability to perform vascular imaging without the use of contrast medium (time of flight MR angiography and phase contrast MR angiography) is a significant advantage of MRI. In addition to conventional MRI sequences that provide anatomopathologic information, advanced MRI methods such as diffusion-weighted imaging (DWI), susceptibility-weighted imaging (SWI), perfusion MRI, and magnetic resonance spectroscopy (MRS) allow qualitative and quantitative information about tissue physiology in the brain. Furthermore, functional imaging of the brain can be performed with functional MRI (fMRI).

9.2.1.4 Angiography
Angiography is the gold standard imaging method in vascular imaging. It is based on the visualization of the contrast agent in the vessel while using X-ray. In this way, as the vascular luminal imaging is provided, endovascular treatment in cases of aneurysm, dissection, or thromboembolism can also be performed. In the cranial examination, it can be used to evaluate the loca-

tion of aneurysms and evaluate the rupture presence in subarachnoid hemorrhage (SAH) cases, evaluate the affected vessels in stroke cases, evaluate the tumor vascularization, investigate the postoperative recurrence, and evaluate the treatment response after radio-chemotherapy. Because of improvements in CT and MRI technologies and as it is an invasive examination, its usage decreases gradually and tends to be limited to patients to be treated endovascularly.

9.2.2 Imaging Findings of Diseases

Both CT and MRI are used for imaging of the insula. While CT is superior to MRI in acute hemorrhagic cases and the detection of calcification, soft tissue contrast is insufficient to show many pathologies. Perfusion CT can show tissue perfusion, but ionizing radiation is an important disadvantage. MRI is the standard imaging method in brain imaging, because it provides high-resolution images by giving better contrast between different tissues. In addition to standard MRI that provides useful anatomopathologic information, it is possible to evaluate the displacement of protons in the tissues with diffusion MRI, tissue blood flow by perfusion MRI, and neural connection pathways between different regions of the brain by MR tractography. Functional imaging of the insula can be performed with fMRI, which provides functional analysis by showing the activity in the tissues (Fig. 9.1). Technically, fMRI is accomplished by detecting increased signal activity in the brain region that is stimulated by various stimuli. There are two reasons for signal increase in fMRI: the first is that oxygen-rich and oxygen-poor blood show different magnetic properties and the latter is that although the cause is not completely known, neural activity is more likely to affect blood flow than oxygen metabolism. As a result, the increased activity increases the oxygenated blood in the respective regions. Blood oxygen level-dependent (BOLD) technique, that shows the oxygenated blood level, is the basis of fMRI [27]. Normal CT and MRI images of the insula are shown in Figs. 9.2 and 9.3.

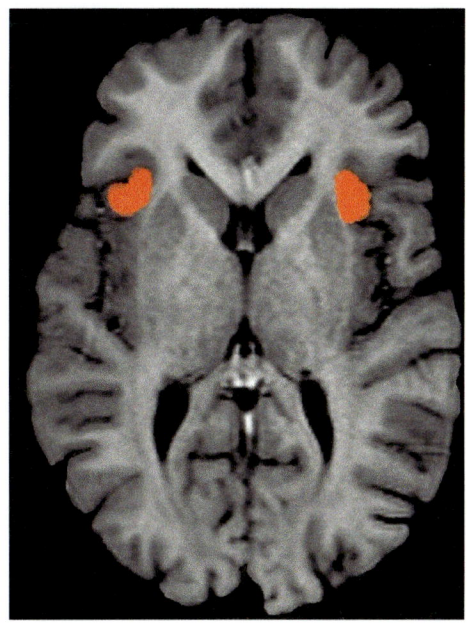

Fig. 9.1 fMRI image; there is an increased activity during emotional stimulus

9.2.2.1 Stroke

The insula is prone to early cytotoxic edema due to its poor collateral network. In the MCA acute stroke, hypodensity occurs due to the edema on the insular cortex. Because of hypodensity, the separation of the insular gray-white matter becomes unclear, which is called "loss of insular ribbon sign" (Fig. 9.4) [28, 29]. Herpes encephalitis, which may cause similar appearance, should be kept in mind in differential diagnosis. Diffusion restriction on DWI helps to diagnose acute stroke. In the acute infarct area, marked hyperintensity is seen on DWI, while ADC is hypointense (Fig. 9.4). Perfusion MRI shows reduced cerebral blood flow (CBF) and cerebral blood volume (CBV) and prolonged mean transit times (MTT) in the infarcted area.

9.2.2.2 Infections

Herpes encephalitis, usually from herpes simplex type 1, is one of the common cranial infections in childhood and adulthood, particularly affecting the temporal lobe, the inferior frontal gyrus, and the insular cortex. It can be seen unilaterally or bilaterally. Cerebrospinal fluid (CSF)

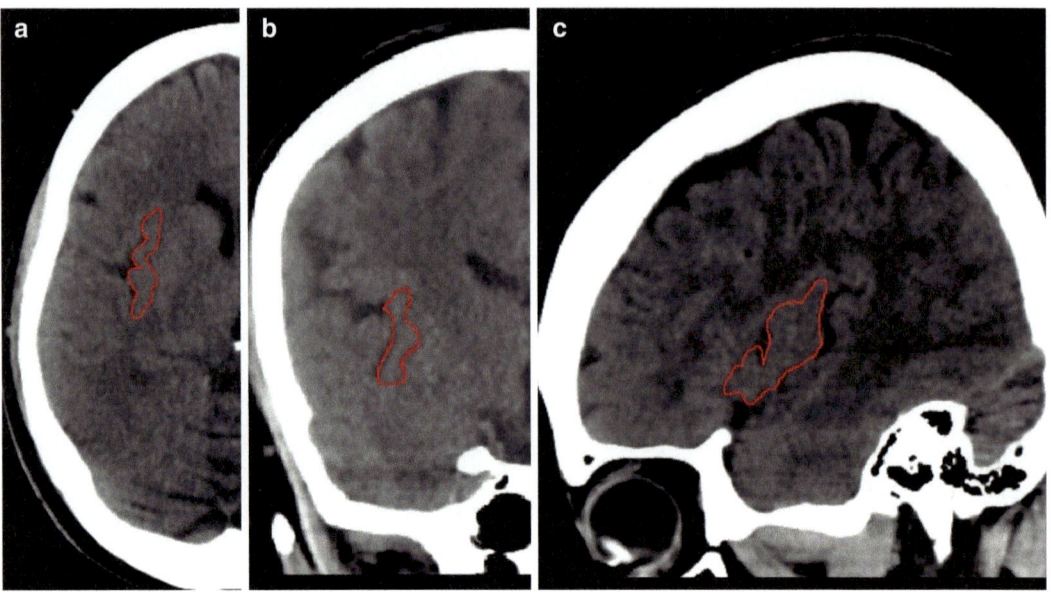

Fig. 9.2 Normal CT images of the insula; (**a**) axial, (**b**) coronal, (**c**) sagittal images

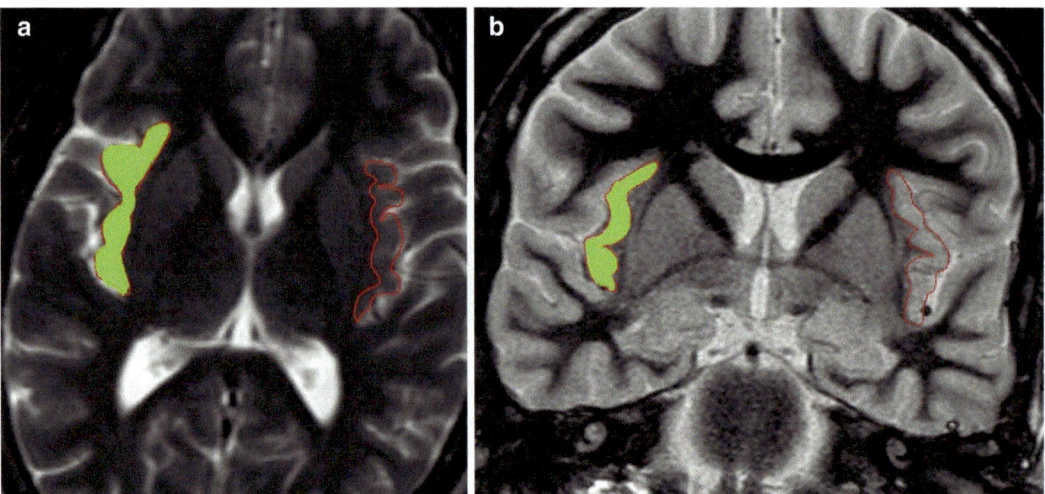

Fig. 9.3 Normal MRI images of the insula; (**a**) axial T2, (**b**) coronal T2, (**c** and **d**) sagittal T2 images. *A* Anterior lobule of insula, *P* Posterior lobule of insula

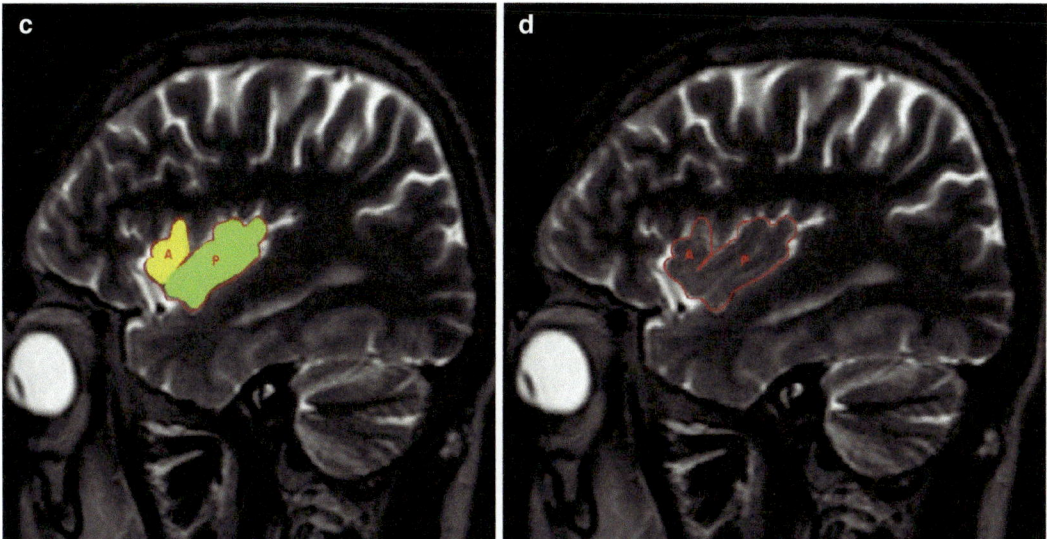

Fig. 9.3 (continued)

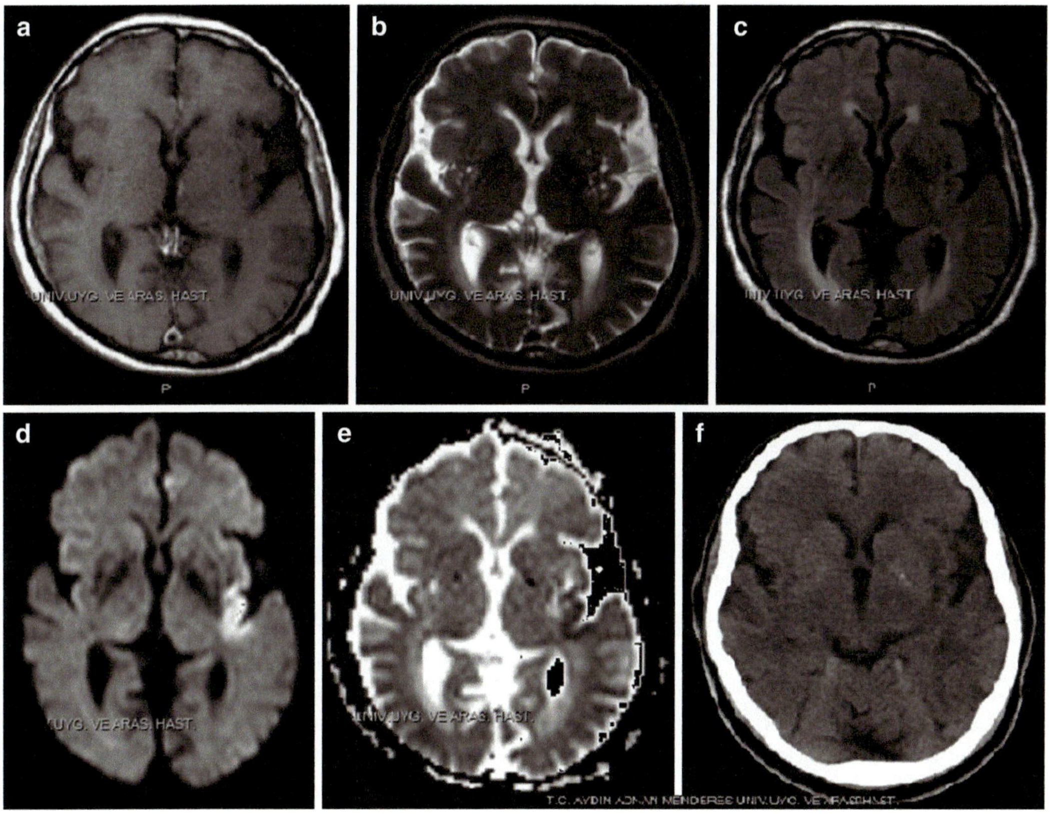

Fig. 9.4 Left MCA acute infarct; T1W (**a**), T2W (**b**), FLAIR (**c**), diffusion (**d**), and ADC (**e**) images on MRI and non-contrast CT (**f**): "loss of insular ribbon sign"

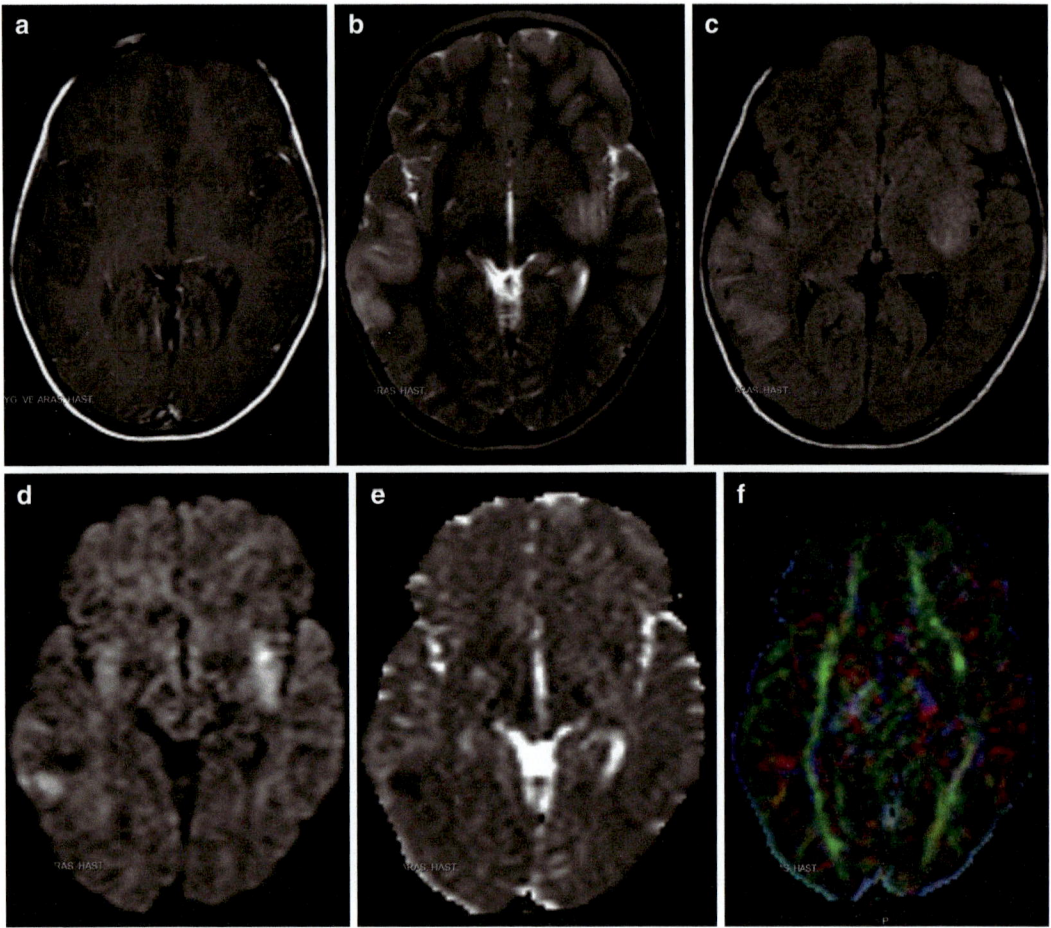

Fig. 9.5 Herpes encephalitis of a 16-year-old boy. Right temporal and occipital lobes, left frontal lobe, and insular involvement are seen on T1W (**a**), T2W (**b**), FLAIR (**c**), diffusion (**d**), ADC (**e**), and tractography images (**f**) on MRI

examination and MRI play an important role in diagnosis. MRI findings are more evident after the third day and include hypointensity on T1-weighted images and hyperintensity on T2-weighted and FLAIR images (Fig. 9.5). Diffusion restriction due to cytotoxic edema may be seen in DWI, but less hyperintense than acute infarction. Unlike infarcts, protection of the basal ganglia usually occurs with encephalitis. Hemorrhagic areas that may be seen due to microhemorrhages can be shown as hypointensities on the SWI images [30, 31].

9.2.2.3 Hemorrhages

Hemorrhage in the insular region is a form of SAH secondary to rupture of aneurysms in MCA and its branches due to its proximity. The insula can be affected from traumatic intracerebral hemorrhage and rarely from hypertensive hemorrhage. Non-contrast CT is the first choice for imaging cerebral hemorrhage, based on fast shooting times and relatively easy access. Early-stage hemorrhages are seen as hyperdense in CT. On MRI, gradient-weighted T2 images and SWI can detect early-stage hemorrhage with as much sensitivity as non-contrast CT (Fig. 9.6) [32]. CT and MR angiography are imaging methods used as an alternative to digital subtraction angiography (DSA), in order to demonstrate aneurysmatic dilatation causing the bleeding [33].

9.2.2.4 Neoplasies

Tumoral formation in insular cortex is relatively low. The majority of these are glial tumors such

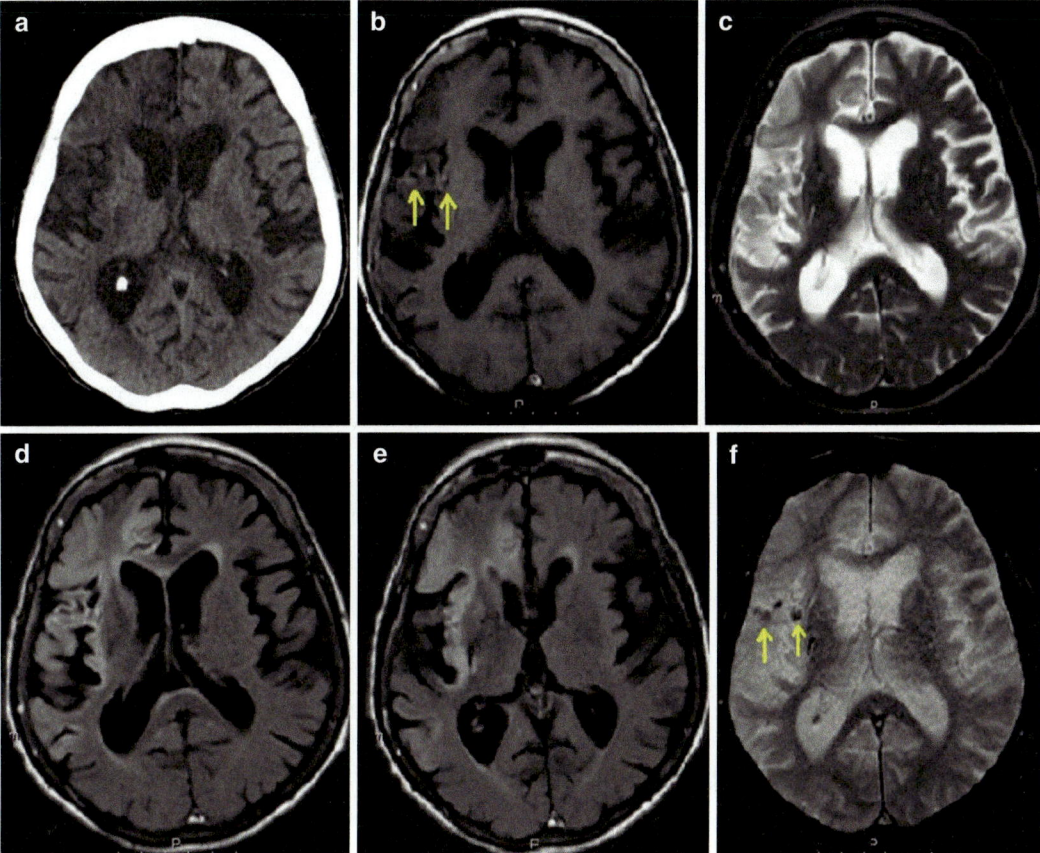

Fig. 9.6 Right insular and perisylvian subarachnoid hemorrhage; Non-contrast CT (**a**) and MRI (**b–f**) images, (**b**) T1W, (**c**) T2W, (**d, e**) FLAIR, (**f**) T2* images

as astrocytomas and oligodendrogliomas. Other common tumors in the insula are metastases and meningiomas. While astrocytomas are located in white matter, oligodendrogliomas are prone to occur in cortical or subcortical areas. Calcification, cystic-necrotic areas, and hemorrhage are more frequent within oligodendrogliomas. Lung, kidney, breast, and colorectal cancers and malignant melanoma constitute 80% of brain metastases. Meningiomas are extra-axial tumors originating from the meninges [34].

The first preferred method for brain tumor imaging is MRI. In MRI, glial tumors are iso-hypointense on T1W images and hyperintense on T2W images (Fig. 9.7). While low-grade tumors do not show postcontrast enhancement, heterogeneous enhancement can be observed in high-grade tumors. Hemorrhages and calcifications can better be visualized in gradient-weighted images and SWI. In diffusion-weighted MRI, low ADC values are in favor of oligodendroglioma [35]. Increased CBV and CBF values are observed in MR perfusion and are more pronounced with high-grade tumors. On MR spectroscopy increased choline and lipid peaks and decreased N-acetyl aspartate (NAA) peak support the diagnosis of the tumor [36].

Metastatic tumors are generally iso-hypodense on CT, iso-hypointense on T1W MR images, hyperintense on T2W MR images, except melanoma that is seen hyperdense on CT, and hyperintense on T1W MR images. In postcontrast imaging, ring enhancement is an important marker for metastasis. However, uniform or punctate contrast enhancement can also be observed. Vasogenic edema is seen in varying proportions around the metastatic mass. Increased choline and decreased NAA levels on MR spectroscopy

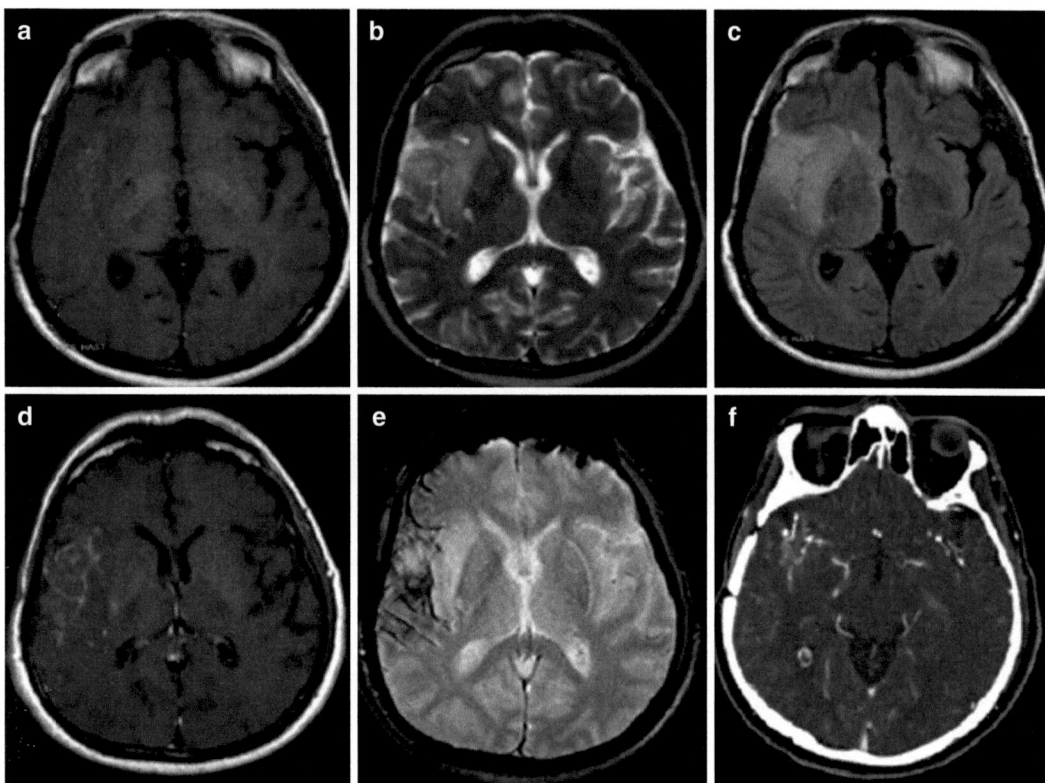

Fig. 9.7 Right temporal lobe-insular anaplastic astrocytoma; T1W (**a**), T2W (**b**), FLAIR (**c**), postcontrast T1W (**d**), and gradient sequence (**e**) MRI and postcontrast CT images (**f**)

and increased CBV values on perfusion MRI can be observed with metastatic tumors [37].

A meningioma is seen as a mass that is isodense with gray matter on non-contrast CT. On MRI, they are isointense on T1W and iso-hyperintense on T2W images with gray matter. Diffuse homogeneous enhancement occurs in the postcontrast examination both on CT and MRI. Heterogeneous appearance and heterogeneous enhancement should suggest atypical or malign formations. Diffusion MRI and perfusion MRI are not useful for the diagnosis of meningiomas. In the spectroscopic examination, while the NAA pike cannot be observed due to the absence of neurons, the increased alanine and increased glutamine/glutamate peaks help to diagnose meningioma [38].

Conclusion

Imaging of the insula is important to demonstrate the involvement of it in both intrinsic and extrinsic pathologies. Additionally, it is also possible to identify brain regions to protect the social and vital functions in the patients planned for operation. New developments in brain imaging will allow us to have more information about the insula.

References

1. Dennis EL, Jahanshad N, McMahon KL, de Zubicaray GI, Martin NG, Hickie IB, et al. Development of insula connectivity between ages 12 and 30 revealed by high angular resolution diffusion imaging. Hum Brain Mapp. 2014;35(4):1790–800. https://doi.org/10.1002/hbm.22292.
2. Cauda F, D'Agata F, Sacco K, Duca S, Geminiani G, Vercelli A. Functional connectivity of the insula in the resting brain. Neuroimage. 2011;55(1):8–23. https://doi.org/10.1016/j.neuroimage.2010.11.049.
3. Naidich TP, Kang E, Fatterpekar GM, Delman BN, Gultekin SH, Wolfe D, et al. The insula: anatomic

study and MR imaging display at 1.5 T. AJNR Am J Neuroradiol. 2004;25(2):222–32.
4. Ogawa H. Gustatory cortex of primates: anatomy and physiology. Neurosci Res. 1994;20(1):1–13.
5. Dronkers NF. A new brain region for coordinating speech articulation. Nature. 1996;384(6605):159–61.
6. Wise RJ, Greene J, Büchel C, Scott SK. Brain regions involved in articulation. Lancet. 1999;353(9158):1057–61.
7. Nagao M, Takeda K, Komori T, Isozaki E, Hirai S. Apraxia of speech associated with an infarct in the precentral gyrus of the insula. Neuroradiology. 1999;41(5):356–7.
8. Fasold O, von Brevern M, Kuhberg M, Ploner CJ, Villringer A, Lempert T, et al. Human vestibular cortex as identified with caloric stimulation in functional magnetic resonance imaging. Neuroimage. 2002;17(3):1384–93.
9. Dieterich M, Brandt T. Vestibular system: anatomy and functional magnetic resonance imaging. Neuroimaging Clin N Am. 2001;11(2):263–73.
10. Brandt T, Bartenstein P, Janek A, Dieterich M. Reciprocal inhibitory visual-vestibular interaction. Visual motion stimulation deactivates the parieto-insular vestibular cortex. Brain. 1998;121(9):1749–58.
11. Oppenheimer SM, Kedem G, Martin WM. Left-insular cortex lesions perturb cardiac autonomic tone in humans. Clin Auton Res. 1996;6(3):131–40.
12. Cheung RT, Hachinski V. The insula and cerebrogenic sudden death. Arch Neurol. 2000;57(12):1685–8.
13. Spiegel DR, Pattison A, Lyons A, Ansari U, Mccroskey AL, Luehrs E, et al. The role and treatment implications of peripheral and central processing of pain, pruritus, and nausea in heightened somatic awareness: a review. Innov Clin Neurosci. 2017;14(5–6):11–20.
14. Gu X, Hof PR, Friston KJ, Fan J. Anterior insular cortex and emotional awareness. J Comp Neurol. 2013;521(15):3371–88. https://doi.org/10.1002/cne.23368.Review.
15. Burkey AR, Carstens E, Jasmin L. Dopamine reuptake inhibition in the rostral agranular insular cortex produces antinociception. J Neurosci. 1999;19(10):4169–79.
16. Corradi-Dell'Acqua C, Tusche A, Vuilleumier P, Singer T. Cross-modal representations of first-hand and vicarious pain, disgust and fairness in insular and cingulate cortex. Nat Commun. 2016;18(7):10904. https://doi.org/10.1038/ncomms10904.
17. Karshikoff B, Jensen KB, Kosek E, Kalpouzos G, Soop A, Ingvar M, et al. Why sickness hurts: a central mechanism for pain induced by peripheral inflammation. Brain Behav Immun. 2016;57:38–46. https://doi.org/10.1016/j.bbi.2016.04.001.
18. Stephani C, Fernandez-Baca Vaca G, Maciunas R, Koubeissi M, Lüders HO. Functional neuroanatomy of the insular lobe. Brain Struct Funct. 2011;216(2):137–49. https://doi.org/10.1007/s00429-010-0296-3.
19. Kurth F, Zilles K, Fox PT, Laird AR, Eickhoff SB. A link between the systems: functional differentiation and integration within the human insula revealed by meta-analysis. Brain Struct Funct. 2010;214(5–6):519–34. https://doi.org/10.1007/s00429-010-0255-z.
20. Risinger RC, Salmeron BJ, Ross TJ, Amen SL, Sanfilipo M, Hoffmann RG, et al. Neural correlates of high and craving during cocaine self-administration using BOLD fMRI. Neuroimage. 2005;26(4):1097–108.
21. Goudriaan AE, de Ruiter MB, van den Brink W, Oosterlaan J, Veltman DJ. Brain activation patterns associated with cue reactivity and craving in abstinent problem gamblers, heavy smokers and healthy controls: an fMRI study. Addict Biol. 2010;15(4):491–503. https://doi.org/10.1111/j.1369-1600.2010.00242.x.
22. Clark L, Bechara A, Damasio H, Aitken MR, Sahakian BJ, Robbins TW. Differential effects of insular and ventromedial prefrontal cortex lesions on risky decision-making. Brain. 2008;131(5):1311–22. https://doi.org/10.1093/brain/awn066.
23. Emmerling F, Schuhmann T, Lobbestael J, Arntz A, Brugman S, Sack AT. The role of the insular cortex in retaliation. PLoS One. 2016;11(4):e0152000. https://doi.org/10.1371/journal.pone.0152000.
24. Zhou Q, Zhong M, Yao S, Jin X, Liu Y, Tan C, et al. Hemispheric asymmetry of the frontolimbic cortex in young adults with borderline personality disorder. Acta Psychiatr Scand. 2017;136(6):637–47. https://doi.org/10.1111/acps.12823.
25. Li H, Chen L, Li P, Wang X, Zhai H. Insular muscarinic signaling regulates anxiety-like behaviors in rats on the elevated plus-maze. Behav Brain Res. 2014;15(270):256–60. https://doi.org/10.1016/j.bbr.2014.05.017.
26. Hatton SN, Lagopoulos J, Hermens DF, Hickie IB, Scott E, Bennett MR. Short association fibres of the insula-temporoparietal junction in early psychosis: a diffusion tensor imaging study. PLoS One. 2014;9(11):e112842. https://doi.org/10.1371/journal.pone.0112842.
27. Gore JC. Principles and practice of functional MRI of the human brain. J Clin Invest. 2003;112(1):4–9.
28. Koga M, Saku Y, Toyoda K, Takaba H, Ibayashi S, Iida M. Reappraisal of early CT signs to predict the arterial occlusion site in acute embolic stroke. J Neurol Neurosurg Psychiatry. 2003;74(5):649–53.
29. Sarikaya B, Provenzale J. Frequency of various brain parenchymal findings of early cerebral ischemia on unenhanced CT scans. Emerg Radiol. 2010;17(5):381–90. https://doi.org/10.1007/s10140-010-0870-2.
30. Eran A, Hodes A, Izbudak I. Bilateral temporal lobe disease: looking beyond herpes encephalitis. Insights Imaging. 2016;7(2):265–74. https://doi.org/10.1007/s13244-016-0481-x. Epub 2016 Feb 24.
31. Misra UK, Kalita J, Phadke RV, Wadwekar V, Boruah DK, Srivastava A, et al. Usefulness of various MRI sequences in the diagnosis of viral encephalitis. Acta Trop. 2010;116(2):206–11. https://doi.org/10.1016/j.actatropica.2010.08.007.
32. Long B, Koyfman A, Runyon MS. Subarachnoid hemorrhage: updates in diagnosis and management.

Emerg Med Clin North Am. 2017;35(4):803–24. https://doi.org/10.1016/j.emc.2017.07.001.
33. HaiFeng L, YongSheng X, YangQin X, Yu D, ShuaiWen W, XingRu L, et al. Diagnostic value of 3D time-of-flight magnetic resonance angiography for detecting intracranial aneurysm: a meta-analysis. Neuroradiology. 2017;59(11):1083–92. https://doi.org/10.1007/s00234-017-1905-0.
34. Brant-Zawadzki M, Badami JP, Mills CM, Norman D, Newton TH. Primary intracranial tumor imaging: a comparison of magnetic resonance and CT. Radiology. 1984;150(2):435–40.
35. Tozer DJ, Jäger HR, Danchaivijitr N, Benton CE, Tofts PS, Rees JH, et al. Apparent diffusion coefficient histograms may predict low-grade glioma subtype. NMR Biomed. 2007;20(1):49–57.
36. Sibtain NA, Howe FA, Saunders DE. The clinical value of proton magnetic resonance spectroscopy in adult brain tumours. Clin Radiol. 2007;62(2):109–19.
37. Chiang IC, Kuo YT, Lu CY, Yeung KW, Lin WC, Sheu FO, et al. Distinction between high-grade gliomas and solitary metastases using peritumoral 3-T magnetic resonance spectroscopy, diffusion, and perfusion imagings. Neuroradiology. 2004;46(8):619–27.
38. Demir MK, Iplikcioglu AC, Dincer A, Arslan M, Sav A. Single voxel proton MR spectroscopy findings of typical and atypical intracranial meningiomas. Eur J Radiol. 2006;60(1):48–55.

Measurements of the Insula Volume Using MRI

10

Niyazi Acer and Mehmet Turgut

Abbreviations

BD	Bipolar disorder
CT	Computed tomography
DARTEL	Diffeomorphic Anatomical Registration Through Exponentiated Lie Algebra
DTI	Diffusion tensor imaging
fMRI	Functional MRI
GM	Gray matter
IBASPM	Individual Brain Atlas using Statistical Parametric Mapping
MDD	Major depressive disorder
MNI	Montreal Neurological Institute
MRI	Magnetic resonance imaging
NTS	Nucleus tractus solitarius
OCD	Obsessive-compulsive disorder
PDD	Pervasive developmental disorders
PET	Positron emission tomography
PTSD	Post-traumatic stress disorder
rCBF	Regional cerebral blood flow
ROI	Region of interest
SPM	Statistical Parametric Mapping
VBM	Voxel-based morphometry
VPM	Ventral posteromedial thalamic nucleus
WM	White matter

N. Acer, Ph.D. (✉)
Department of Anatomy, Erciyes University School of Medicine, Kayseri, Turkey

M. Turgut
Department of Neurosurgery, Adnan Menderes University School of Medicine, Aydın, Turkey

10.1 Introduction

The insular cortex, island of Reil, or Brodmann areas 13–16, was first defined by J.C. Reil in 1809 [1]. In 1955, Penfield et al. [2] researched the effect of the insular cortex in temporal lobe epilepsy. They studied the symptoms that occurred during temporal lobe seizures and their similarity to the ones evoked by insular cortex stimulation [3]. According to the results of Penfield et al. [2], the insular cortex played a role in gustatory (taste) and visceral sensations and visceral motor responses. The insular cortex mainly takes part in the processing of attention, pain, emotion, vestibular function, visceral sensory, motor, verbal, musical information, gustatory, visual, olfactory, auditory, and tactile data [4].

The insular cortex is not visible on an exterior view of the brain, as it is fully covered laterally by the opercula of the parietal, frontal, and temporal lobes. The insular cortex is situated on the lateral aspect of the cerebral hemispheres. It overlies the claustrum and is interposed between the motor, sensory, orbital, auditory, and piriform (primary olfactory) cortices [5]. Most primates including humans have larger temporal and parietal opercula. In humans, the insular cortex is not visible within the depths of the lateral sulcus of the brain. Directly medial to the insula are the extreme capsule and the claustrum which also cover the external capsule, putamen, and globus pallidus. The central sulcus (Rolandic fissure) of the insula is the most inferior extension of the central sulcus, which separates the frontal and parietal lobes (Fig. 10.1) [5, 6].

The insular cortex is anatomically separated into two main sections by the central sulcus of the insula: anterior lobule and posterior lobule [5] (Fig. 10.1). The central sulcus of the insula designates the prominent sulcus, which angles obliquely across the insula and, parts it into the anterior lobule and posterior lobule [7, 8]. The insula consists of five to seven sulci in humans, and a larger left than right insula by adults [5, 7]. There are three or four short gyri within an anterior lobule of the insula, while there are two long gyri in a posterior lobule. The three short gyri in the anterior lobule are the anterior, middle, and posterior short gyri. On the ventral margin of the anterior part of the insula, there is also accessory gyrus. The anterior gyri converge at the ventral-oriented apex. The posterior part has two long gyri—an anterior and a posterior long gyrus [5]. These two groups of gyri are separated at the level of the central sulcus of the insula [5, 9] (Fig. 10.2).

According to Singer et al. [10], the insular cortex consists of the ventroanterior, dorsoanterior, and posterior insula subregions. It is a connection of self-relevant feelings, empathy, and uncertainty prediction in the brain [11]. The ventroanterior subregion is related to socioemotional processing [10, 12]; the dorsoanterior network region is involved with cognitive processing [13, 14]; and the posterior insula subregion is related to somatovisceral sensations [15] and auditory information processing [16]. It is possible that feelings of worthlessness are connected with the ventroanterior subregion because it is related to socioemotional processing [6].

In the human brain, the insula has many connections with the rostral and dorsolateral prefrontal cortex, anterior cingulate cortex, entorhinal cortex, amygdala, parietal and temporal lobes,

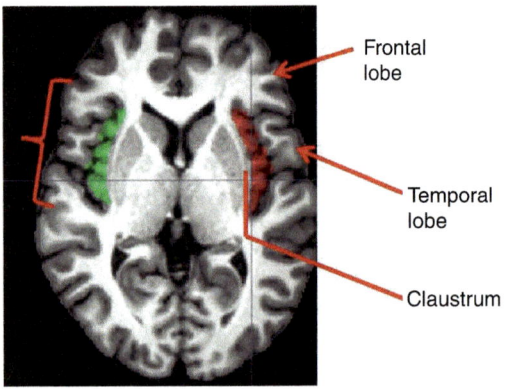

Fig. 10.1 The insula is illustrated on an axial MRI. Red, left; green, right

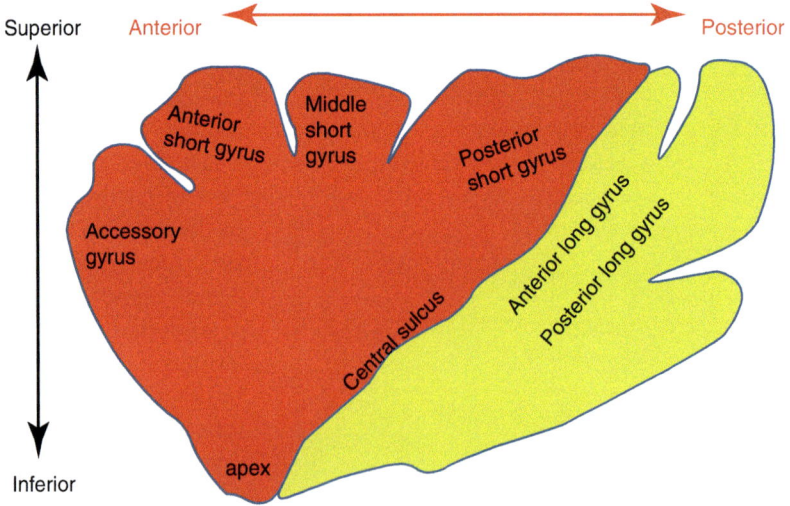

Fig. 10.2 Simplified sagittal drawing of the insula depicts the major gyri

thalamus, and hypothalamus. The insula plays a role in visceromotor, viscerosensory, and somatosensory functions [17]. The functional neuroimaging studies report insular activation depending on an individual's emotional state. This indicates an internal representation of the body and provides a possible basis for subjective awareness in the social interactions [11, 18].

The insular cortex, as a part of the limbic region, has a critical function in integration of perceptual experiences, thus producing a balanced behavior [1]. The insular cortex is hidden by the frontal, temporal, and parietal lobes [1]. All of the following information is processed by the insula: pain, temperature, visceral sensory, visceral motor, vestibular, visual, auditory, language, and tactile [1]. The insula is a complex structure with increased complexity throughout primate evolution that is characterized by a marked heterogeneity in architecture, connectivity with other parts of the brain [19].

The anterior insula has connections with the orbitofrontal, piriform, parahippocampal, and temporopolar regions; these structures provide the control of autonomic regulation. The insula is anatomically related to behavioral functions and emotional reactions [20]. In primates, the anterior insular cortex has some connections with the ventral posteromedial thalamic nucleus (VPM); therefore, it is considered part of the primary gustatory cortex [21]. It is thought to be a crucial site in swallowing because of its connections to the VPM thalamic nucleus and the nucleus tractus solitarius (NTS) as well as the primary and supplementary motor cortex [22]. The posterior insular lobule has connections with the temporal regions; somatosensory, auditory, lower precentral, and postcentral gyri; and motor areas [20].

Imaging studies have shown that the ventral anterior insula has an activation during cognitive tasks and emotion, thus providing the regulation of interpersonal interactions [23].

Recent studies revealed that the insular cortex plays a role in the processing of visceral sensory, visceral motor, emotion, vestibular, pain, temperature, and language inputs, in addition to visual, auditory, tactile, olfactory, and gustatory information [1, 4, 19]. In particular, the human insula has important roles in self-recognition, time perception, empathy, awareness of emotions, decision-making, and the processing of music and language [24].

10.2 Volume of the Insular Cortex and Its Clinical Importance

In 2007, Nagai et al. [25] suggested that the insular cortex is important in schizophrenia, mood disorder, eating disorders, obsessive-compulsive disorder, and panic disorder.

There is a functional difference between the left and right anterior insula. Activation of the "right" anterior insula occurs predominantly with feelings of negative emotional valence and sympathetic activation. On the other hand, positive affect and parasympathetic function are a result of activation of the "left" anterior insula [26].

Both right and left human insula, just beneath the white matter (WM) and including cortex volume was a total of 17.4 cm^3 based on measurements of MRIs [27]. The left insula is larger than the right in humans [26, 28, 29].

Recently, many studies have revealed that the insular cortex has many connections with the other regions of the brain including the somatosensory areas (primary and secondary), motor cortex, prefrontal cortex, frontal and parietal operculum, superior temporal gyrus and pole, orbitofrontal cortex, anterior cingulate cortex, amygdaloid body, primary auditory cortex and auditory association cortex, visual association cortex, olfactory bulbs, entorhinal cortex, and hippocampus. As a result of these connections, the neurons located within the insular cortex have numerous interconnections [4].

Regional cerebral blood flow (rCBF) in the brains of healthy adults during transient sadness was studied by George et al. [30]. Increased rCBF was shown in the insular cortex in the left side in both males and females and among females during sadness.

Neuroimaging studies have reported both local and global morphological changes in the brain in psychiatric diseases such as schizophrenia, obsessive-compulsive disorder, bipolar disorder, and mood disorder. Most studies have shown that the volume of the insula in these psychiatric diseases was smaller than in healthy people [20, 31, 32].

A recent region of interest (ROI)-based volumetric MRI study demonstrated a reduction in the anterior insular volume in the left side in patients with schizophrenia [20]. On the other hand, Yamada et al. [20] found a reduction in posterior insular volume in the right side using a voxel-based morphometry (VBM)-based volumetric study. In 2007, Saze et al. [33] reported that patients with schizophrenia have a prominent decrease in insular gray matter (GM) volumes compared to those of healthy subjects. Crespo-Facorro et al. [34] explored the morphology of the insular cortex in 25 healthy male volunteers and 25 male patients with schizophrenia using MRI [35]. According to Crespo-Facorro et al. [34], patients with schizophrenia revealed a significant reduction in both the cortical surface area and GM volumes in the insular cortex in the left side.

In the last decade, several investigators reported significant reductions in the volume of GM using VBM in patients with schizophrenia [36–38]. Furthermore, some authors found that GM concentrations in the insular cortex are significantly reduced in patients with paranoid schizophrenia [39, 40]. Recently, Morgan et al. [41] found a significant reduction of GM in patients with affective psychosis. In 2009, Takahashi et al. [42] revealed that the volume of the anterior and posterior insular cortices is reduced in both the right side and left side of patients with schizophrenia in relation to the controls.

Insular volume changes have been reported about obsessive-compulsive disorder (OCD), indicating a minor difference in size between OCD and healthy controls according to VBM studies [43, 44]. In 2011, Nishida et al. [45] found that the patients with OCD had a significantly increased GM volume of the anterior part of the insular cortex in both the right side and left side, compared to the normal controls.

In 2010, Kosaka et al. [47] studied regional GM volumes which were compared in infants with pervasive developmental disorders (PDD) and control groups using VBM with the Diffeomorphic Anatomical Registration Through Exponentiated Lie Algebra (DARTEL). They reported that the volume of GM in the insula in the right side is significantly less in the high-functioning PDD group than that of control group [47].

In the current literature, there are several papers regarding insular cortex activation using neuroradiological studies: the positron emission tomography (PET) study [46] and functional MRI studies [33, 48] for "higher" insular cortex activation, while functional MRI (fMRI) studies [49] for "lower" insular cortex activation. Furthermore, it has been reported that OCD patients increased fractional anisotropy in the diffusion tensor MRI study [50].

In 2010, Takahashi et al. [51] stated that patients with major depressive disorder (MDD) showed a significant reduction in the volume of the anterior insular cortex in the left side in comparison with the healthy controls using an ROI approach. Peng et al. [27] found a GM volume reduction in various bilateral anterior insular cortices in MDD using VBM. It has been reported that patients with BD have a reduced volume in the ventral anterior insula in both the right side and left side using automated VBM methods [34, 51, 52].

Recently, Tang et al. [8] found that volumes of GM in the dorsal anterior insula are larger in the left side of patients with bipolar disorder (BD). They also noted that volumes of GM in the left ventral anterior insula are decreased in patients with BD [52]. When the close integration between the ventrolateral prefrontal cortex and the dorsal anterior insular cortex is considered, it is possible that the increased volumes of GM in the dorsal anterior insula in the left side are related to existence of an increased effort for inhibition of subcortical brain activity [30]. In 2004, Lochhead et al. [29] observed that the volume of the dorsal anterior insular region in the left side was increased in patients with BD, who had been medication-free for 14 days before scanning. More recently, Pompei et al. [52] found that euthymic BD patients have a growth in both the insula and the ventrolateral prefrontal cortex.

Most of the studies found decreased volume of the insula in patients with schizophrenia [20, 31, 39]. A meta-analysis revealed that insular volume in patients with schizophrenia is reduced compared to control subjects, and the difference is more prominent in the anterior insula than in the posterior one [20, 31, 39].

10.3 Volumetric Methods for Insular Cortex Volume Calculation

Nowadays, there are several recent software tools that are developed to automatically obtain volumetric measurements for volume of the insula using different strategies such as the Statistical Parametric Mapping (SPM), FreeSurfer, FSL, etc. Among these, SPM is the widely used tool for the analysis of alterations in global GM or WM. VBM toolbox, an extension of SPM, is also used for the evaluation of local GM and WM atrophy [2, 25, 52]. Now, we can obtain the volume of the insular cortex using MRIcroGL (Fig. 10.3).

10.3.1 Voxel-Based Morphometry (VBM)

Technically, VBM is a fully automated whole-brain measurement technique that maps the statistical association between cognition and regional tissue volume or density [25]. Using the latest version of SPM12, MRIs were segmented into GM, WM, and cerebrospinal fluid images via a tissue segmentation procedure following an image-intensity nonuniformity correction. Afterward, with the help of DARTEL (Wellcome Department of Imaging Neuroscience), the segmented GM and WM images were normalized to the customized template in the standardized anatomic space [2]. Using the Jacobean determinants derived from the normalization by DARTEL, the GM and WM volumes within each voxel were preserved by modulating the images, and lastly these were smoothed with a 12 mm FWHM Gaussian kernel. For the correlation analyses with global cognitive performance and memory scores, multiple regression analysis and an uncorrected threshold of $p < 0.001$ with a cluster size of >50 were used.

At present, there are several recent atlas and web-based tools, which are developed to automatically produce these volumetric measurements of the insula, such as IBA, SPM, and MriCloud.

10.3.2 IBASPM (Individual Brain Atlas Using Statistical Parametric Mapping)

IBASPM (Cuban Neuroscience Center) software, an atlas-based method for automatic segmentation of brain structures, is available as a freeware toolbox for the SPM package [53, 54].

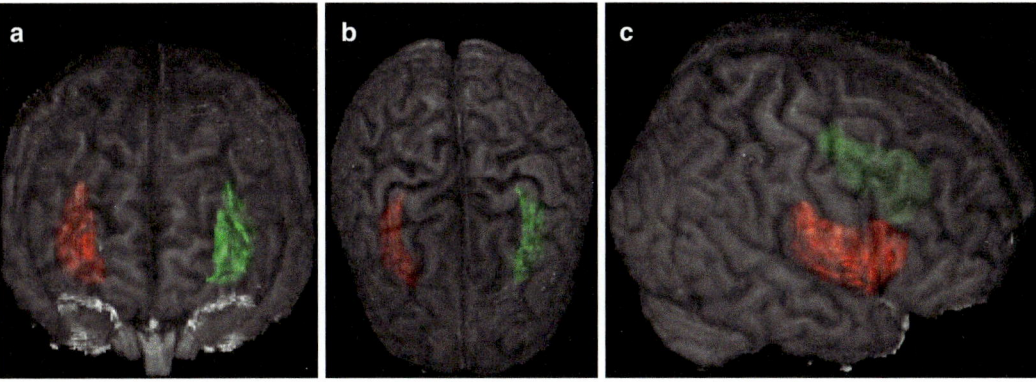

Fig. 10.3 Insular cortex rendering on a T1 image: (**a**) coronal view; (**b**) axial view; and (**c**) sagittal view showing the insula. Red, left; green, right

Technically, it is based on the registration of the brain image to an anatomical template labeled with an anatomical atlas [53, 54]. This toolkit allows selecting both the anatomical template for registration and the atlas template for segmentation [54]. Furthermore, customized atlases may be produced following the IBASPM procedure, upon a group of labeled brains with estimations of the likelihood of each structure [54] (Fig. 10.1).

The Cuban Neuroscience Center validated its segmentation with real datasets of healthy brains, based on volumetric measurements and using default parameters: ICBM 152 T1 registration template in Montreal Neurological Institute (MNI) and its corresponding labeled atlas MNI AAL [53, 54].

Mri_convert or MRIcron is used to convert scans into the ANALYZE format (.img) to ensure compatibility with IBASPM [54]. The resulting images are visually inspected to verify proper orientation and the absence of warping [54]. Afterward, images are processed using the IBASPM toolbox (Cuban Neuroscience Center, Havana, Cuba) functions of SPM8 (Wellcome Department of Cognitive Neurology, University College, London, UK), implemented in MATLAB 10a (MathWorks, Natick, MA) [54]. Using the parameters of this study, IBASPM (Individual Brain Atlases using Statistical Parametric Mapping) completed the following five sequential processes, the last three of which work outside SPM8 (segmentation, normalization, labeling, atlasing) [54]. The IBASPM, SPM8, and SPM12 software packages and their documentation are available at http://www.thomaskoenig.ch/Lester/ibaspm.htm and http://www.fil.ion.ucl.ac.uk/spm/, respectively, without any fee. We overlaid the insula on MRI T1 images for the axial, coronal, and sagittal sections (Fig. 10.4a–c). Thus, you can calculate insula volume automatically using IBASPM without any difficulty.

10.3.3 MriCloud

Susumu Mori developed image analysis tools for brain MRI more than 15 years of experience and in sharing the tools with research communities at Johns Hopkins University. In 2001, they produced DtiStudio as an executable program; this program can be downloaded from their website to do tensor calculation of diffusion tensor imaging (DTI) and/or 3D WM tract reconstruction [53, 55]. In 2006, two new programs (i.e., RoiEditor and DiffeoMap) were added to the family called MriStudio. In fact, these programs were designed to perform ROI-based image quantification for any type of brain MRI data. The ROI may be defined manually. However, DiffeoMap introduced our first capability for automated brain segmentation. Now, research studies are based on a single-subject atlas with at least 100 certain brain regions that are automatically deformed to image data, and thus transfer the predefined ROIs in the atlas to produce automated brain segmentation of the target [41, 56].

The T1-weighted images are segmented using our online resource MriCloud [55, 57] (www.mricloud.org), through a fully automated T1 image segmentation pipeline. A number of studies have demonstrated improved segmentation accuracy with multi-atlas fusion as opposed to single-atlas approaches [42, 58, 59]. The whole brain was segmented into 289 structures, which could be grouped at five levels of granularity based on their ontological relationships [57]. Structural volumes were obtained based on the T1 segmentation at this level. The brain parcellation maps are illustrated in Fig. 10.5 [55].

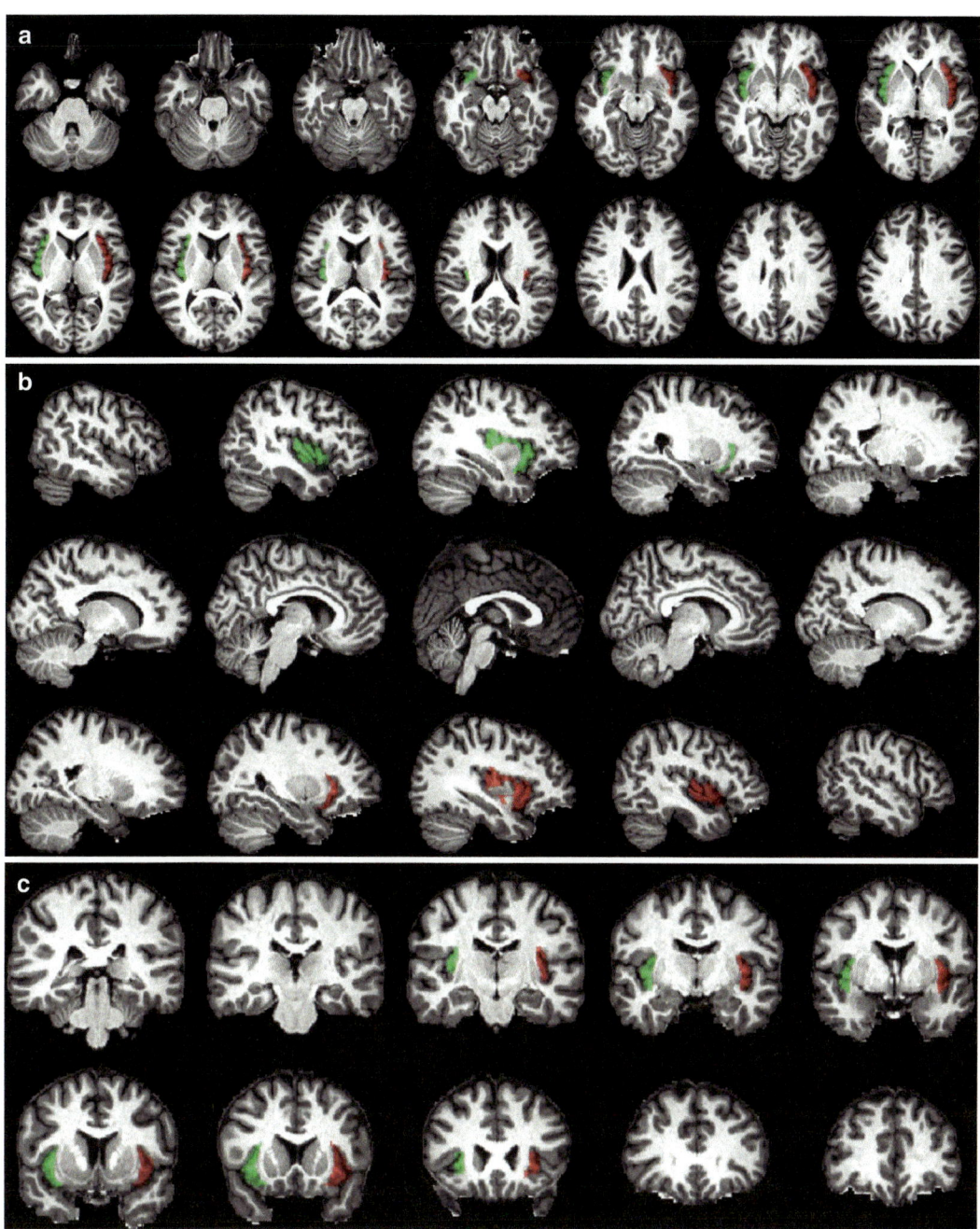

Fig. 10.4 Insular cortex displayed on a T1 image: (**a**) axial section; (**b**) sagittal section; and (**c**) a coronal section showing the insula

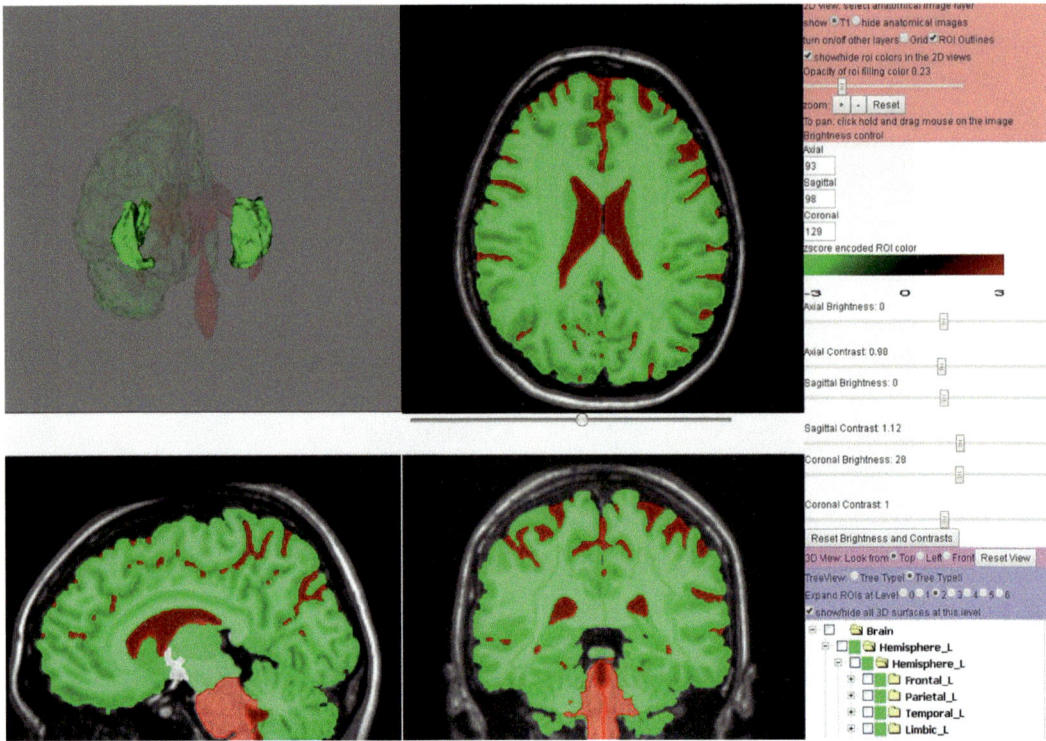

Fig. 10.5 The brain parcellation maps are illustrated using MriCloud

Conclusion

Based on the close relationship between the ventral anterior insula and the limbic regions, it is concluded that the reduction of volumes of GM in the ventral anterior insula in the left side may suggest over activity of the ventral limbic region. The decreased insular volume is not a specific illness marker, but it may be a general marker for the regarding psychopathology. However, the diagnostic specificity of insular cortex abnormalities among various neuropsychiatric disorders will be a topic for new studies in the future. It is considered that a reduction in volume in the insular cortex may be a possible explanation for the impairment in self-consciousness observed in some neuropsychiatric disorders. Considering the abovementioned literature data, we strongly suggest that the insular cortex may play a key function in human mental processing. Recent neuroimaging studies, including structural and functional MRI and voxel- and ROI-based morphometry, demonstrated that volumetric change of the insular cortex was involved in many neuropsychiatric diseases such as PTSD, OCD, schizophrenia, BD, eating disorders, and mood and panic disorders.

References

1. Augustine JR. Circuitry and functional aspects of the insular lobe in primates including humans. Brain Res Brain Res Rev. 1996;22(3):229–44.
2. Ashburner J. A fast diffeomorphic image registration algorithm. Neuroimage. 2007;38(1):95–113. Epub 2007 Jul 18.
3. Penfield W, Faulk ME Jr. The insula; further observations on its function. Brain. 1955;78(4):445–70.
4. Nagai M, Kishi K, Kato S. Insular cortex and neuropsychiatric disorders: a review of recent literature. Eur Psychiatry. 2007;22(6):387–94. Epub 2007 Apr 9.
5. Ture U, Yasargil DC, Al-Mefty O, Yasargil MG. Topographic anatomy of the insular region. J Neurosurg. 1999;90(4):720–33.
6. Cauda F, D'Agata F, Sacco K, Duca S, Geminiani G, Vercelli A. Functional connectivity of the insula in the resting brain. Neuroimage. 2011;55(1):8–23. https://doi.org/10.1016/j.neuroimage.2010.11.049. Epub 2010 Nov 24.

7. Guenot M, Isnard J, Sindou M. Surgical anatomy of the insula. Adv Tech Stand Neurosurg. 2004;29:265–88.
8. Stephani C, Fernandez-Baca Vaca G, Maciunas R, Koubeissi M, Luders HO. Functional neuroanatomy of the insular lobe. Brain Struct Funct. 2011;216(2):137–49. https://doi.org/10.1007/s00429-010-0296-3. Epub 2010 Dec 14.
9. Varnavas GG, Grand W. The insular cortex: morphological and vascular anatomic characteristics. Neurosurgery. 1999;44(1):127–36. discussion 136-8.
10. Chang LJ, Smith A, Dufwenberg M, Sanfey AG. Triangulating the neural, psychological, and economic bases of guilt aversion. Neuron. 2011;70(3):560–72. https://doi.org/10.1016/j.neuron.2011.02.056.
11. Singer T, Critchley HD, Preuschoff K. A common role of insula in feelings, empathy and uncertainty. Trends Cogn Sci. 2009;13(8):334–40. https://doi.org/10.1016/j.tics.2009.05.001. Epub 2009 Jul 28.
12. Sanfey AG, Rilling JK, Aronson JA, Nystrom LE, Cohen JD. The neural basis of economic decision-making in the Ultimatum Game. Science. 2003;300(5626):1755–8.
13. Dosenbach NU, Visscher KM, Palmer ED, Miezin FM, Wenger KK, Kang HC, Burgund ED, Grimes AL, Schlaggar BL, Petersen SE. A core system for the implementation of task sets. Neuron. 2006;50(5):799–812.
14. Eckert MA, Menon V, Walczak A, Ahlstrom J, Denslow S, Horwitz A, Dubno JR. At the heart of the ventral attention system: the right anterior insula. Hum Brain Mapp. 2009;30(8):2530–41. https://doi.org/10.1002/hbm.20688.
15. Craig AD. Interoception: the sense of the physiological condition of the body. Curr Opin Neurobiol. 2003;13(4):500–5.
16. Bamiou DE, Musiek FE, Luxon LM. The insula (Island of Reil) and its role in auditory processing. Literature review. Brain Res Brain Res Rev. 2003;42(2):143–54.
17. Critchley HD. Neural mechanisms of autonomic, affective, and cognitive integration. J Comp Neurol. 2005;493(1):154–66.
18. Craig AD. How do you feel—now? The anterior insula and human awareness. Nat Rev Neurosci. 2009;10(1):59–70. https://doi.org/10.1038/nrn2555.
19. Mesulam MM, Mufson EJ. Insula of the old world monkey. III: efferent cortical output and comments on function. J Comp Neurol. 1982;212(1):38–52.
20. Makris N, Goldstein JM, Kennedy D, Hodge SM, Caviness VS, Faraone SV, Tsuang MT, Seidman LJ. Decreased volume of left and total anterior insular lobule in schizophrenia. Schizophr Res. 2006;83(2–3):155–71. Epub 2006 Jan 31.
21. Rolls ET. Information processing in the taste system of primates. J Exp Biol. 1989;146:141–64.
22. Daniels SK, Foundas AL. The role of the insular cortex in dysphagia. Dysphagia. 1997;12(3):146–56.
23. Mutschler I, Wieckhorst B, Kowalevski S, Derix J, Wentlandt J, Schulze-Bonhage A, Ball T. Functional organization of the human anterior insular cortex. Neurosci Lett. 2009;457(2):66–70. https://doi.org/10.1016/j.neulet.2009.03.101. Epub 2009 Apr 5.
24. Bauernfeind AL, de Sousa AA, Avasthi T, Dobson SD, Raghanti MA, Lewandowski AH, Zilles K, Semendeferi K, Allman JM, Craig AD, Hof PR, Sherwood CC. A volumetric comparison of the insular cortex and its subregions in primates. J Hum Evol. 2013;64(4):263–79.
25. Ashburner J, Friston KJ. Voxel-based morphometry—the methods. Neuroimage. 2000;11(6 Pt 1):805–21.
26. Craig AD. Forebrain emotional asymmetry: a neuroanatomical basis? Trends Cogn Sci. 2005;9(12):566–71. Epub 2005 Nov 4.
27. Semendeferi K, Damasio H. The brain and its main anatomical subdivisions in living hominoids using magnetic resonance imaging. J Hum Evol. 2000;38(2):317–32.
28. Takahashi T, Wood SJ, Soulsby B, Tanino R, Wong MT, McGorry PD, Suzuki M, Velakoulis D, Pantelis C. Diagnostic specificity of the insular cortex abnormalities in first-episode psychotic disorders. Prog Neuropsychopharmacol Biol Psychiatry. 2009;33(4):651–7. https://doi.org/10.1016/j.pnpbp.2009.03.005. Epub 2009 Mar 17.
29. Takahashi T, Wood SJ, Yung AR, Phillips LJ, Soulsby B, McGorry PD, Tanino R, Zhou SY, Suzuki M, Velakoulis D, Pantelis C. Insular cortex gray matter changes in individuals at ultra-high-risk of developing psychosis. Schizophr Res. 2009;111(1–3):94–102. https://doi.org/10.1016/j.schres.2009.03.024. Epub 2009 Apr 5.
30. Jiao Q, Ding J, Lu G, Su L, Zhang Z, Wang Z, Zhong Y, Li K, Ding M, Liu Y. Increased activity imbalance in fronto-subcortical circuits in adolescents with major depression. PLoS One. 2011;6(9):e25159. https://doi.org/10.1371/journal.pone.0025159. Epub 2011 Sep 16.
31. Saze T, Hirao K, Namiki C, Fukuyama H, Hayashi T, Murai T. Insular volume reduction in schizophrenia. Eur Arch Psychiatry Clin Neurosci. 2007;257(8):473–9. Epub 2007 Sep 27.
32. Yamada M, Hirao K, Namiki C, Hanakawa T, Fukuyama H, Hayashi T, Murai T. Social cognition and frontal lobe pathology in schizophrenia: a voxel-based morphometric study. Neuroimage. 2007;35(1):292–8. Epub 2007 Jan 19.
33. Mataix-Cols D, Wooderson S, Lawrence N, Brammer MJ, Speckens A, Phillips ML. Distinct neural correlates of washing, checking, and hoarding symptom dimensions in obsessive-compulsive disorder. Arch Gen Psychiatry. 2004;61(6):564–76.
34. Matsuo K, Kopecek M, Nicoletti MA, Hatch JP, Watanabe Y, Nery FG, Zunta-Soares G, Soares JC. New structural brain imaging endophenotype in bipolar disorder. Mol Psychiatry. 2012;17(4):412–20. https://doi.org/10.1038/mp.2011.3. Epub 2011 Feb 15.

35. Crespo-Facorro B, Kim J, Andreasen NC, O'Leary DS, Bockholt HJ, Magnotta V. Insular cortex abnormalities in schizophrenia: a structural magnetic resonance imaging study of first-episode patients. Schizophr Res. 2000;46(1):35–43.
36. Goldstein JM, Goodman JM, Seidman LJ, Kennedy DN, Makris N, Lee H, Tourville J, Caviness VS Jr, Faraone SV, Tsuang MT. Cortical abnormalities in schizophrenia identified by structural magnetic resonance imaging. Arch Gen Psychiatry. 1999;56(6):537–47.
37. Sigmundsson T, Suckling J, Maier M, Williams S, Bullmore E, Greenwood K, Fukuda R, Ron M, Toone B. Structural abnormalities in frontal, temporal, and limbic regions and interconnecting white matter tracts in schizophrenic patients with prominent negative symptoms. Am J Psychiatry. 2001;158(2):234–43.
38. Wright IC, Ellison ZR, Sharma T, Friston KJ, Murray RM, McGuire PK. Mapping of grey matter changes in schizophrenia. Schizophr Res. 1999;35(1):1–14.
39. Ha TH, Youn T, Ha KS, Rho KS, Lee JM, Kim IY, Kim SI, Kwon JS. Gray matter abnormalities in paranoid schizophrenia and their clinical correlations. Psychiatry Res. 2004;132(3):251–60.
40. Hulshoff Pol HE, Schnack HG, Mandl RC, van Haren NE, Koning H, Collins DL, Evans AC, Kahn RS. Focal gray matter density changes in schizophrenia. Arch Gen Psychiatry. 2001;58(12):1118–25.
41. Oishi K, Faria A, Jiang H, Li X, Akhter K, Zhang J, Hsu JT, Miller MI, van Zijl PC, Albert M, Lyketsos CG, Woods R, Toga AW, Pike GB, Rosa-Neto P, Evans A, Mazziotta J, Mori S. Atlas-based whole brain white matter analysis using large deformation diffeomorphic metric mapping: application to normal elderly and Alzheimer's disease participants. Neuroimage. 2009;46(2):486–99.
42. Langerak TR, van der Heide UA, Kotte AN, Viergever MA, van Vulpen M, Pluim JP. Label fusion in atlas-based segmentation using a selective and iterative method for performance level estimation (SIMPLE). IEEE Trans Med Imaging. 2010;29(12):2000–8. https://doi.org/10.1109/TMI.2010.2057442. Epub 2010 Jul 26.
43. Pujol J, Soriano-Mas C, Alonso P, Cardoner N, Menchon JM, Deus J, Vallejo J. Mapping structural brain alterations in obsessive-compulsive disorder. Arch Gen Psychiatry. 2004;61(7):720–30.
44. Yoo SY, Roh MS, Choi JS, Kang DH, Ha TH, Lee JM, Kim IY, Kim SI, Kwon JS. Voxel-based morphometry study of gray matter abnormalities in obsessive-compulsive disorder. J Korean Med Sci. 2008;23(1):24–30. https://doi.org/10.3346/jkms.2008.23.1.24.
45. Nishida S, Narumoto J, Sakai Y, Matsuoka T, Nakamae T, Yamada K, Nishimura T, Fukui K. Anterior insular volume is larger in patients with obsessive-compulsive disorder. Prog Neuropsychopharmacol Biol Psychiatry. 2011;35(4):997–1001. https://doi.org/10.1016/j.pnpbp.2011.01.022. Epub 2011 Feb 17.
46. Kosaka H, Omori M, Munesue T, Ishitobi M, Matsumura Y, Takahashi T, Narita K, Murata T, Saito DN, Uchiyama H, Morita J, Kikuchi M, Mizukami K, Okazawa H, Sadato N, Wada Y. Smaller insula and inferior frontal volumes in young adults with pervasive developmental disorders. Neuroimage. 2010;50(4):1357–63. https://doi.org/10.1016/j.neuroimage.2010.01.085. Epub 2010 Feb 1.
47. Rauch SL, Savage CR, Alpert NM, Fischman AJ, Jenike MA. The functional neuroanatomy of anxiety: a study of three disorders using positron emission tomography and symptom provocation. Biol Psychiatry. 1997;42(6):446–52.
48. Schienle A, Schafer A, Stark R, Walter B, Vaitl D. Neural responses of OCD patients towards disorder-relevant, generally disgust-inducing and fear-inducing pictures. Int J Psychophysiol. 2005;57(1):69–77. Epub 2005 Apr 22.
49. Gilbert AR, Akkal D, Almeida JR, Mataix-Cols D, Kalas C, Devlin B, Birmaher B, Phillips ML. Neural correlates of symptom dimensions in pediatric obsessive-compulsive disorder: a functional magnetic resonance imaging study. J Am Acad Child Adolesc Psychiatry. 2009;48(9):936–44. https://doi.org/10.1097/CHI.0b013e3181b2163c.
50. Nakamae T, Narumoto J, Shibata K, Matsumoto R, Kitabayashi Y, Yoshida T, Yamada K, Nishimura T, Fukui K. Alteration of fractional anisotropy and apparent diffusion coefficient in obsessive-compulsive disorder: a diffusion tensor imaging study. Prog Neuropsychopharmacol Biol Psychiatry. 2008;32(5):1221–6. https://doi.org/10.1016/j.pnpbp.2008.03.010. Epub 2008 Mar 25.
51. Selvaraj S, Arnone D, Job D, Stanfield A, Farrow TF, Nugent AC, Scherk H, Gruber O, Chen X, Sachdev PS, Dickstein DP, Malhi GS, Ha TH, Ha K, Phillips ML, McIntosh AM. Grey matter differences in bipolar disorder: a meta-analysis of voxel-based morphometry studies. Bipolar Disord. 2012;14(2):135–45. https://doi.org/10.1111/j.1399-5618.2012.01000.x.
52. Tang LR, Liu CH, Jing B, Ma X, Li HY, Zhang Y, Li F, Wang YP, Yang Z, Wang CY. Voxel-based morphometry study of the insular cortex in bipolar depression. Psychiatry Res. 2014;224(2):89–95. https://doi.org/10.1016/j.pscychresns.2014.08.004. Epub 2014 Aug 28.
53. Rolls ET, Joliot M, Tzourio-Mazoyer N. Implementation of a new parcellation of the orbitofrontal cortex in the automated anatomical labeling atlas. Neuroimage. 2015;122:1–5. https://doi.org/10.1016/j.neuroimage.2015.07.075. Epub 2015 Aug 1.
54. Tzourio-Mazoyer N, Landeau B, Papathanassiou D, Crivello F, Etard O, Delcroix N, Mazoyer B, Joliot M. Automated anatomical labeling of activations in SPM using a macroscopic anatomical parcellation of the MNI MRI single-subject brain. Neuroimage. 2002;15(1):273–89.

55. Ma J, Ma HT, Li H, Ye C, Wu D, Tang X, Miller M, Mori S. A fast atlas pre-selection procedure for multi-atlas based brain segmentation. Conf Proc IEEE Eng Med Biol Soc. 2015;2015:3053–6. https://doi.org/10.1109/EMBC.2015.7319036.
56. Ceritoglu C, Oishi K, Li X, Chou MC, Younes L, Albert M, Lyketsos C, van Zijl PC, Miller MI, Mori S. Multi-contrast large deformation diffeomorphic metric mapping for diffusion tensor imaging. Neuroimage. 2009;47(2):618–27. https://doi.org/10.1016/j.neuroimage.2009.04.057. Epub 2009 May 3.
57. Djamanakova A, Tang X, Li X, Faria AV, Ceritoglu C, Oishi K, Hillis AE, Albert M, Lyketsos C, Miller MI, Mori S. Tools for multiple granularity analysis of brain MRI data for individualized image analysis. Neuroimage. 2014;101:168–76. https://doi.org/10.1016/j.neuroimage.2014.06.046. Epub 2014 Jun 27.
58. Artaechevarria X, Munoz-Barrutia A, Ortiz-de-Solorzano C. Combination strategies in multi-atlas image segmentation: application to brain MR data. IEEE Trans Med Imaging. 2009;28(8):1266–77. https://doi.org/10.1109/TMI.2009.2014372. Epub 2009 Feb 18.
59. Warfield SK, Zou KH, Wells WM. Simultaneous truth and performance level estimation (STAPLE): an algorithm for the validation of image segmentation. IEEE Trans Med Imaging. 2004;23(7):903–21.

11

Neurocognitive Mechanisms of Social Anxiety and Interoception

Yuri Terasawa and Satoshi Umeda

11.1 Introduction

The relationship between subjective emotion and associated somatic responses has long been a subject of research. In 1884, William James proposed that emotions result from the perception of specific and unique patterns in the somatovisceral response [1]. James' proposition provided a framework within which to understand the mechanisms of emotion.

Amusing fantasy makes us smile and blush; however, painful imagination makes us choke up and feel our face stiffen. In one study, participants colored in a drawing of a body to express enhanced or attenuated bodily responses to emotions such as anger, happiness, or envy [2]. The participants presented similar body images for each emotion across nationality and race. In particular, anxiety was expressed as enhanced activity in the chest and throat and reduced activity in the hands and feet.

Anxiety disorders present both psychological and physiological symptoms, such as disconcertedness and restlessness, palpitations, shallow breathing, nausea, coldness in the limbs, and tightness in chest. These symptoms are more or less associated with autonomic nervous functions. Interestingly, patients with autonomic hyper-reactive syndrome such as postural tachycardia syndrome show high frequency of anxiety disorders [3]. These physiological symptoms are consistent with the results of Nummenmaa et al. [2], indicating that anxiety is generated by complex and integrated processes of the mind and body. In this chapter, we introduce some studies examining the cognitive neuroscientific mechanism of anxiety. The studies were based on the concept of "interoception"—the perception of afferent information arising from within the body [4, 5].

11.2 Self-Referential Thought and Interoception in SAD

The DSM-V states that anxiety disorders present with excessive fear and anxiety, as well as related behavioral disturbances. Anxiety disorders are categorized as follows: generalized anxiety disorder, specific phobia, social anxiety disorder (SAD—also known as social phobia), panic disorder, and agoraphobia. This chapter focuses on SAD, which is characterized by fear or anxiety about social situations. Individuals with SAD report a strong fear and desire to avoid social situations like giving a speech or performing in front of others. These psychological symptoms are often associated with somatic manifestations.

Individuals with SAD show enhanced idiosyncratic negative belief and self-referential

Y. Terasawa · S. Umeda (✉)
Department of Psychology, Keio University, Tokyo, Japan
e-mail: umeda@flet.keio.ac.jp

thought [6]. In social situations, they tend to focus intense attention on themselves and on how they are seen and perceived by others, not on the contents of interaction itself. Furthermore, they may falsely believe that others perceive them in a negative light. For example, when a person narrows his/her gaze in response to a bright light, or when he/she ask the same thing several times in a noisy room, individuals with SAD may conclude that they have offended the person or that they said something nonsensical. These biased interpretations lead to excess self-referential processing and consequent somatic symptoms, such as strong anxiety, palpitations, shallow breath, and feelings of nausea.

Neuroimaging studies have indicated that excessive self-referential processing in SAD correlates with enhanced insular cortex activation in a resting state, and with processing of negative emotions. In this chapter, we discuss insular activation in SAD within the framework of self-referential processing, which refers to attention oriented toward the self.

Neuroimaging studies have revealed that the following brain areas are involved in self-referential processing: the medial prefrontal cortex (MPFC), posterior cingulate cortex (PCC), inferior parietal lobule, and insular cortex [7]. We believe that the most fundamental factor in self-referential processing is attention to interoception. Interestingly, attention to the body is the first step of mindfulness training. Interoception is important for perceiving internal bodily conditions, among other functions. Both perceptual processing through sensory organs and evaluative processing are essential to understanding the value of the objects within various environments. In other words, we can integrate exteroceptive and interoceptive information to decide whether an object is subjectively good or bad for us. In addition, one recent theoretical paper suggested that this integrative process is the foundation for recognizing the present, which is continuously changing, and for emotional experiences [8].

There are two approaches to researching the neural substrates of interoception: (1) modulation of internal bodily states using respiratory load, rectal distention, or similar challenge and (2) reorientation of attention from the outside toward the inside of the body, without stimulating internal organs. Studies applying these approaches have shown that the brainstem, thalamus, anterior insular cortex, and anterior cingulate cortex are important neural substrates of interoception [9]. In particular, the insular cortex is considered essential in the generation of subjective emotions from internal bodily signals [10–13]. Highly aroused subjective feelings and interoception activate the anterior insular cortex [14, 15]. Interestingly the activation level in the insular cortex corresponds to the emotional value of successive events and to the certainty that the events happened [16]. Neuroimaging studies have implied that this brain area integrates exteroceptive and interoceptive signals, helping us make appropriate decisions [16–18].

Conversely, in studies that have attempted to define the neural substrate of anxiety, threatening images, sounds, or electric shocks have been associated with some figures or response options during neuroimaging sessions. Furthermore, such interventions lead to associative learning; thus, neural responses to the conditioned stimuli—both figures and options—are considered anxiety responses. This procedure has identified the amygdala, insular cortex, anterior cingulate cortex (ACC), and thalamus as the base for anxiety, and many studies have reported enhanced activity or functional connectivity in these areas [19].

Which brain functions are modified in individuals with SAD? Reviews about the neural basis of SAD have suggested that the fear circuit is hyper-activated in the condition [19–21]. Thus, enhanced activity in the ACC, MPFC, PCC, precuneus, and cuneus, as well as in the amygdala and insular cortex, are found in SAD. Furthermore, neuroimaging studies have consistently reported enhanced insular activation [21] in individuals with SAD and that this activity subsided when SAD was treated using cognitive behavioral therapy or cognitive reappraisal. It follows that the pathological mechanisms of SAD involve insular functioning. Generally, the insula is more active when the dynamics of neural networks, such as the salient network and the default mode network

(DMN), are changed [22]. However, insular activity is involved in the DMN of highly anxious individuals [23], suggesting that they have a default state of high arousal.

The neural substrates of interoception and anxiety appear to have many commonalities and shared mechanisms that support mental processing. This suggests that enhanced self-referential thought in SAD evokes vivid sensation of somatic responses, such as palpitations and blushing. These sensations subsequently enhance anxiety, as the individual may think "I am very anxious because my heart is beating fast!" In fact, several studies have tried to elucidate the mechanism of anxiety disorders by focusing on the neural substrates they share with anxiety and interoception. These studies indicated that insular activation during emotion-laden tasks is a useful biomarker for anxiety disorders [24–26]. According to Paulus and Stein [24], hypersensitivity to interoception allows a subject to detect subtle bodily changes when predicting follow-up events, and it generates anxiety as a result. Specific phobias could be understood as a form of anticipatory anxiety. Thus, there is probably a close connection between phobias and anterior insular activation. However, this connection would not occur in generalized anxiety, where the target of the anxiety is unclear.

11.3 fMRI Studies About Social Anxiety and Interoception

Activation in the insular cortex and ACC are important in the neurocognitive mechanism of anxiety through interoception, as described in the previous section. Interestingly, these activations are positively correlated with individuals' anxiety level. However, no previous studies have yet disentangled whether it is enhanced autonomic activity or attention to one's internal body itself that elevates anxiety level. Therefore, we conducted an fMRI study to examine how SAD is related to insular activation caused by attention to interoception [13]. Participants were required to evaluate their own emotional and bodily state during an fMRI scan. Since we did not elicit the participants' emotions in any way, we could focus on neural responses that were specific to the participants' awareness of their emotions or bodily state. We showed that the right anterior insular cortex and left ventromedial prefrontal cortex (VMPFC) are important for monitoring emotion and bodily states. Previous studies have reported that the insula and VMPFC are pivotal to interoceptive awareness and activation, and that they are associated with the evaluation of subjective emotions. Our own results indicated that the integration of interoception and contextual or environmental information is the foundation for subjective feelings of emotion.

Next, we examined activation in brain regions related to interoception while participants evaluated their own emotional and bodily states, as well as their level of social anxiety [27]. We used the Social Anxiety Disorder Scale questionnaire, which is used in Japan to assess SAD. The scale takes into account four factors: fear of social situations (social fear), avoidance of social situations, somatic symptoms related to anxiety, and interference in daily life. We found that the level of social fear was positively correlated with activation in the right anterior insular cortex, middle/superior prefrontal cortex, cuneus, parahippocampal gyrus, and inferior frontal gyrus. In addition, activation in the right anterior cortex was strongly associated with personality. That is, neuroticism was positively correlated, and agreeableness, extraversion, and openness to experience were negatively correlated with activation in this area.

In this study, we sought to clarify how interoceptive sensibility modulates anxiety level. To disentangle this relationship, we focused on how the insular cortex modulates the association between social anxiety and activity in the neural substrates of interoceptive sensibility. During this experiment, we also asked the subjects about their emotional state. We then employed mediation analysis to examine activity in regions associated with interoceptive sensibility; in this way, we identified which regions indicate the extent of interoceptive processing.

In the first step of the mediation analysis—regression analysis—we examined the neural regions in which activity predicted the level of social anxiety. Activation in the left thalamus was significantly associated with levels of social fear. We had hypothesized that the insular cortex is essential in the association between interoception and social anxiety. For this reason, we focused on whether activation in the anterior insular cortex was related to that in the left thalamus. The results revealed a statistically significant association between these two regions.

Next, we assessed whether activation in the right anterior insular cortex mediates this relationship between activation in the left thalamus and the level of social fear. As shown in Fig. 11.1, the analysis revealed that the left thalamus has a significant indirect effect on the level of social fear and that this effect is mediated via the right anterior insula. That is, higher activation in the left thalamus was associated with higher activation in the right anterior insula, and higher activation in the right anterior insula was associated with higher levels of social fear. Importantly, the direct correlation between the left thalamus and social fear became nonsignificant when the right anterior insula mediated these two factors, indicating that this association was fully mediated by the right anterior insula.

Our study indicated that anterior insular cortex was closely related to anxiety trait scores and that its activation mediated social anxiety and activation in the left thalamus, which is related to somatic perception. The brain regions most closely related to somatic perception in the left thalamus were the ventral posterior lateral and mediodorsal areas. Relay nuclei are concentrated in the ventral posterior lateral region, and these relay information from the cranial nerve and medulla to the primary sensory and somatosensory areas [28]. In addition, these areas are involved in the pathogenic mechanisms of intractable pain, and stimulation of this brain region induces visceral pain. Furthermore, the onset of myocardial ischemia without subjective symptoms enhances activation in this area [29]. Taken together, these results suggest that these regions mediate interoception by monitoring the internal bodily state [4]. The thalamic nuclei of the ventral posterior region convey visceral afferent information that is transferred from the brainstem to the insular cortex [30]. In addition, enhanced functional connectivity occurs in the thalamocortical circuits, including the insula and the medial prefrontal cortex, in social anxiety disorders [19, 31]. Taken together, these results indicate that attention to interoceptive information can impact subjective feelings of anxiety, rather than the perception of bodily state per se. In support of this conclusion, Lovero et al. [32] revealed that the anterior insula was activated during the anticipation of touch, but not during the actual tactile sensation, and that the intensity of anticipated touch was predicted by its activity.

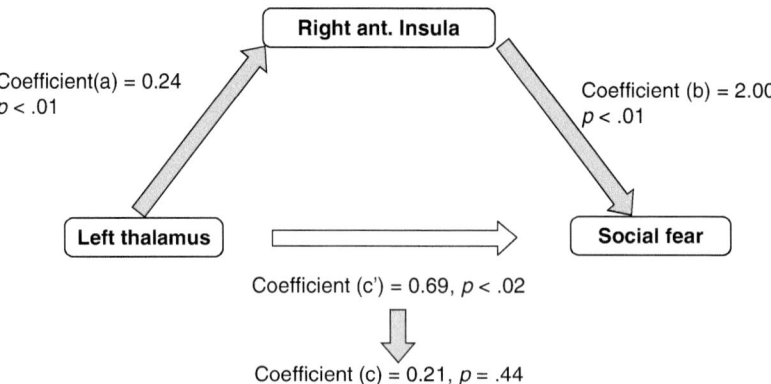

Fig. 11.1 Outline of the mediation analysis, which focused on whether the insula modulates the association between social anxiety levels and activity in the left thalamus (activity was correlated with somatic sensibility level). The analysis revealed that activity in the right anterior insular cortex mediated both activity in the left thalamus and the level of social fear ($p < 0.05$)

These findings support our hypothesis that activity in the anterior insular cortex represents attention to interoceptive information, not only to bodily perception, and that interoceptive processing is important in the subjective experience of emotion.

Patients with SAD often report disproportionate pain responses during anxious situations. Our research implies that enhanced attention toward the body causes patients to detect homeostatic fluctuations in visceral activities and that these detected sensations lead to anxiety.

11.4 Subcategories of Interoception and Anxiety Disorders

In the previous section, we suggested that exaggerated interoceptive sensibility elevates anxiety level. However, many researchers have reported inconsistent results regarding interoception in highly anxious persons. Some have reported that such individuals show enhanced interoception, while others have stated that they show attenuated interoception [33]. To disentangle this situation, researchers must conduct detailed examinations of each anxiety disorder (e.g., SAD, generalized anxiety disorder, and panic disorder). Thus, we must focus on the methods used to measure interoceptive sensitivity.

Several methods are used to measure interoceptive sensitivity. Some are questionnaires (e.g., Modified Somatic Perception Questionnaire [34], Body Perception Questionnaire [35]; Mehling, Gopisetty [36] [review]). Others are experimental methods, such as the heartbeat perception task, which asks participants to evaluate the timing or numbers of their heartbeats [37, 38]. In this regard, researchers must be careful about what exactly each method measures, and consider whether these are conceptually identical or not. For example, using different methods, "interoceptive sensitivity" may denote either perceptual sensitivity (thresholds) or an individual's propensity for attention toward the internal body, which is an aspect of personality.

According to Khalsa and Lapidus [39], the number of studies involving interoception had increased rapidly in the previous 5 years. Many studies involving psychiatric or psychosomatic disorders have tried to explain emotional processing problems in terms of interoceptive processing. However, these studies have been inconsistent in their results. For example, Scholz et al. [40] and Barsky [41] reported enhanced interoceptive sensitivity in patients with somatoform disorder, while Brown [42] and Pollatos et al. [43] reported attenuated interoceptive sensitivity in this disorder. This discrepancy may have arisen from differences in the methods used to measure interoceptive sensitivity. Garfinkel and Critchley [44] reported that interoceptive sensitivity differs depending on what methods are used to measure it. They categorized three types of interoception on the basis of the measuring methods: *interoceptive awareness, interoceptive sensitivity/accuracy, and interoceptive sensibility* (Garfinkel and Critchley, 2013; Garfinkel et al., 2016). They defined (1) interoceptive awareness as a metacognitive awareness involving interoceptive accuracy of the self, (2) interoceptive accuracy as performance accuracy that can be measured using experimental methods, and (3) interoceptive sensibility as an attention propensity toward the internal state or self, which can be measured using questionnaires. These definitions provide a good framework within which to consider inconsistency among previous studies. For example, some individuals with anxiety disorders or psychosomatic diseases report that they are intensely aware of their own internal bodily state. Thus, their interoceptive sensibility was high when they answered the questionnaires. However, the same individuals often showed low performance in tasks that measured interoceptive accuracy, such as the heartbeat perception task [38]. These results seem to conflict. However, a large discrepancy between self-report and performance in experimental tasks may indicate that subjects have lowered interoceptive awareness, which may be important in the pathological mechanism of anxiety disorders. In fact, patients who have a strong attention propensity for the self will probably report that they feel internal bodily changes vividly, even though these changes were just homeostatic fluctuations.

Yoris et al. [45] found enhanced interoceptive sensibility but normal interoceptive accuracy in patients with panic disorder. Thus, lowered interoceptive awareness may impact the severity of panic disorder. Garfinkel et al. [46] reported that individuals with autism spectrum disorder (ASD) showed similar patterns to those with panic disorder. These results suggest that predictions about somatic sensations are inflated in ASD and that these predictions prevent the patient from feeling bodily responses in context. In this study, a large discrepancy between interoceptive sensibility and accuracy ("interoceptive awareness") was associated with difficulties in emotion recognition and anxiety symptoms, suggesting that appropriate interoceptive awareness is essential for smooth and appropriate emotion processing.

11.5 Modified Functional Connectivity in SAD

The study introduced in Sect. 3 showed that the right anterior insular cortex mediates between interoceptive sensibility and social anxiety, since the employed task in fMRI corresponded to the questionnaires. This indicates that the anterior insular cortex is a neural substrate for interoceptive sensibility, not accuracy, and that activation in this area represents enhanced self-referential thought, as observed in SAD. Excessive interoceptive sensibility, in contrast with interoceptive accuracy, generates false beliefs about internal bodily condition.

Bruhl et al. [21] reviewed neuroimaging studies involving SAD and devised a cognitive neuroscientific model of SAD that was based on the modified functional connectivity found in that condition. According to this model, functional connectivity from the frontal cortex and limbic area to the PCC and precuneus is significantly decreased in patients with SAD. The PCC is pivotal in the DMN; it ensures appropriate energy distribution throughout the body, as well as attention switching via dynamic connections with other brain networks. Lowered connectivity in this area may indicate that regulation system of psychological and physiological functions for responding environmental demands is attenuated in SAD. In the context of self-referential thought, the area is important for integrating multimodal sensory information and controlling consciousness levels. Hyper- and isolated activation in this area is correlated with enhanced self-referential thought in SAD, and these symptoms manifest as a dysfunctional attention switching and persistent heightened alertness. Future studies should involve a detailed examination of connectivity between the subregions of the insular cortex and DMN that is based on the model mentioned. Such a study would help researchers understand SAD in the context of interoception.

References

1. James W. What is an emotion? Mind. 1884;19:188–205.
2. Nummenmaa L, Glerean E, Hari R, Hietanen JK. Bodily maps of emotions. Proc Natl Acad Sci U S A. 2014;111(2):646–51.
3. Umeda S, Harrison NA, Gray MA, Mathias CJ, Critchley HD. Structural brain abnormalities in postural tachycardia syndrome: a VBM-DARTEL study. Front Neurosci. 2015;9:34.
4. Cameron OG. Interoception: the inside story—a model for psychosomatic processes. Psychosom Med. 2001;63(5):697–710.
5. Sherrington CS. The integrative action of the nervous system. New Haven: Yale University Press; 1906.
6. Wells A, Papageorgiou C. Social phobic interoception: effects of bodily information on anxiety, beliefs and self-processing. Behav Res Ther. 2001;39(1):1–11.
7. Farb NA, Segal ZV, Mayberg H, Bean J, McKeon D, Fatima Z, et al. Attending to the present: mindfulness meditation reveals distinct neural modes of self-reference. Soc Cogn Affect Neurosci. 2007;2(4):313–22.
8. Harshaw C. Interoceptive dysfunction: toward an integrated framework for understanding somatic and affective disturbance in depression. Psychol Bull. 2015;141(2):311–63.
9. Critchley HD, Harrison NA. Visceral influences on brain and behavior. Neuron. 2013;77(4):624–38.
10. Craig AD. How do you feel—now? The anterior insula and human awareness. Nat Rev Neurosci. 2009;10(1):59–70.
11. Critchley HD, Wiens S, Rotshtein P, Ohman A, Dolan RJ. Neural systems supporting interoceptive awareness. Nat Neurosci. 2004;7(2):189–95.

12. Lee KH, Siegle GJ. Common and distinct brain networks underlying explicit emotional evaluation: a meta-analytic study. Soc Cogn Affect Neurosci. 2012;7(5):521–34.
13. Terasawa Y, Fukushima H, Umeda S. How does interoceptive awareness interact with the subjective experience of emotion? An fMRI study. Hum Brain Mapp. 2013;34(3):598–612.
14. Phan KL, Taylor SF, Welsh RC, Ho SH, Britton JC, Liberzon I. Neural correlates of individual ratings of emotional salience: a trial-related fMRI study. Neuroimage. 2004;21(2):768–80.
15. Iaria G, Committeri G, Pastorelli C, Pizzamiglio L, Watkins KE, Carota A. Neural activity of the anterior insula in emotional processing depends on the individuals' emotional susceptibility. Hum Brain Mapp. 2008;29(3):363–73.
16. Singer T, Critchley HD, Preuschoff K. A common role of insula in feelings, empathy and uncertainty. Trends Cogn Sci. 2009;13(8):334–40.
17. Huettel SA, Song AW, McCarthy G. Decisions under uncertainty: probabilistic context influences activation of prefrontal and parietal cortices. J Neurosci. 2005;25(13):3304–11.
18. Feinstein JS, Stein MB, Paulus MP. Anterior insula reactivity during certain decisions is associated with neuroticism. Soc Cogn Affect Neurosci. 2006;1(2):136–42.
19. Etkin A, Wager TD. Functional neuroimaging of anxiety: a meta-analysis of emotional processing in PTSD, social anxiety disorder, and specific phobia. Am J Psychiatry. 2007;164(10):1476–88.
20. Freitas-Ferrari MC, Hallak JE, Trzesniak C, Filho AS, Machado-de-Sousa JP, Chagas MH, et al. Neuroimaging in social anxiety disorder: a systematic review of the literature. Prog Neuropsychopharmacol Biol Psychiatry. 2010;34(4):565–80.
21. Bruhl AB, Delsignore A, Komossa K, Weidt S. Neuroimaging in social anxiety disorder-a meta-analytic review resulting in a new neurofunctional model. Neurosci Biobehav Rev. 2014;47:260–80.
22. Menon V, Uddin LQ. Saliency, switching, attention and control: a network model of insula function. Brain Struct Funct. 2010;214(5–6):655–67.
23. Dennis EL, Gotlib IH, Thompson PM, Thomason ME. Anxiety modulates insula recruitment in resting-state functional magnetic resonance imaging in youth and adults. Brain Connect. 2011;1(3):245–54.
24. Paulus MP, Stein MB. An insular view of anxiety. Biol Psychiatry. 2006;60(4):383–7.
25. Simmons AN, Stein MB, Strigo IA, Arce E, Hitchcock C, Paulus MP. Anxiety positive subjects show altered processing in the anterior insula during anticipation of negative stimuli. Hum Brain Mapp. 2011;32(11):1836–46.
26. Paulus MP. The breathing conundrum-interoceptive sensitivity and anxiety. Depress Anxiety. 2013;30(4):315–20.
27. Terasawa Y, Shibata M, Moriguchi Y, Umeda S. Anterior insular cortex mediates bodily sensibility and social anxiety. Soc Cogn Affect Neurosci. 2013;8(3):259–66.
28. Blumenfeld H. Neuroanatomy through clinical cases. Sunderland, MA: Sinauer; 2002.
29. Rosen SD, Paulesu E, Nihoyannopoulos P, Tousoulis D, Frackowiak RS, Frith CD, et al. Silent ischemia as a central problem: regional brain activation compared in silent and painful myocardial ischemia. Ann Intern Med. 1996;124(11):939–49.
30. Cameron OG. Visceral sensory neuroscience, interoception. New York, NY: Oxford University Press; 2002.
31. Gimenez M, Pujol J, Ortiz H, Soriano-Mas C, Lopez-Sola M, Farre M, et al. Altered brain functional connectivity in relation to perception of scrutiny in social anxiety disorder. Psychiatry Res. 2012;202(3):214–23.
32. Lovero KL, Simmons AN, Aron JL, Paulus MP. Anterior insular cortex anticipates impending stimulus significance. Neuroimage. 2009;45(3):976–83.
33. Domschke K, Stevens S, Pfleiderer B, Gerlach AL. Interoceptive sensitivity in anxiety and anxiety disorders: an overview and integration of neurobiological findings. Clin Psychol Rev. 2010;30(1):1–11.
34. Main CJ. The modified somatic perception questionnaire (MSPQ). J Psychosom Res. 1983;27(6):503–14.
35. Porges S. Body perception questionnaire: laboratory of developmental assessment: University of Maryland; 1993.
36. Mehling WE, Gopisetty V, Daubenmier J, Price CJ, Hecht FM, Stewart A. Body awareness: construct and self-report measures. PLoS One. 2009;4(5):e5614.
37. Brener J, Jones JM. Interoceptive discrimination in intact humans: detection of cardiac activity. Physiol Behav. 1974;13(6):763–7.
38. Schandry R. Heart beat perception and emotional experience. Psychophysiology. 1981;18(4):483–8.
39. Khalsa SS, Lapidus RC. Can interoception improve the pragmatic search for biomarkers in psychiatry? Front Psych. 2016;7:121.
40. Scholz OB, Ott R, Sarnoch H. Proprioception in somatoform disorders. Behav Res Ther. 2001;39(12):1429–38.
41. Barsky AJ. Amplification, somatization, and the somatoform disorders. Psychosomatics. 1992;33(1):28–34.
42. Brown RJ. Psychological mechanisms of medically unexplained symptoms: an integrative conceptual model. Psychol Bull. 2004;130(5):793–812.
43. Pollatos O, Herbert BM, Wankner S, Dietel A, Wachsmuth C, Henningsen P, et al. Autonomic imbalance is associated with reduced facial recognition in somatoform disorders. J Psychosom Res. 2011;71(4):232–9.
44. Garfinkel SN, Critchley HD. Interoception, emotion and brain: new insights link internal physiology to social behaviour. Commentary on:: "Anterior insular cortex mediates bodily sensibility and social anxiety" by Terasawa et al. (2012). Soc Cogn Affect Neurosci. 2013;8(3):231–4.

45. Yoris A, Esteves S, Couto B, Melloni M, Kichic R, Cetkovich M, et al. The roles of interoceptive sensitivity and metacognitive interoception in panic. Behav Brain Funct. 2015;11:14.

46. Garfinkel SN, Tiley C, O'Keeffe S, Harrison NA, Seth AK, Critchley HD. Discrepancies between dimensions of interoception in autism: Implications for emotion and anxiety. Biol Psychol. 2016;114:117–26.

Part II

Functions of the Human Insula

Participation of the Insula in Language

Alfredo Ardila

Unfortunately, the problem of insular aphasia, which would be so important for our considerations, has not so far been clarified by clinical observation. Meynert, de Boyer, Wernicke himself and others maintain that the insula belongs to the speech area, while Bernard and others, following Charcot, emphatically deny such a relation. Nothing decisive concerning this problem emerged from Naunyn's survey. Although it seems highly probable that lesions of the insula cause speech disorders (not only because of anatomical contiguity to the so-called center), it is nevertheless impossible to state whether the speech disorder is of a specific type and if so of what type. (Freud [1], pp. 12–13).

12.1 Introduction

The controversies regarding the participation of the insula in language have extended for over 100 years. During the nineteenth century, it was assumed that the insula had an evident participation in language. During the early twentieth century, Dejerine [2] proposed the concept of a "brain language area," including three cortical regions: (a) left frontal (posterior part of the foot of F3, the frontal operculum, and the immediate surrounding zone, including the foot of F2, and *probably* extending to the anterior insula), (b) left temporal (encompassing the posterior first and second temporal gyri), and (c) parietal (the angular gyrus) areas. This proposal was rapidly accepted by virtually all researchers in the aphasia area (e.g., [3–7]). It was assumed that the brain language zone corresponded in consequence to the perisylvian area of the left hemisphere; and the insula was forgotten. Until the end of the twentieth century, the insula virtually disappeared as a language area. However, during the last couple of decades, the interest in the potential participation of the insula in language has reemerged.

The insula indeed is located in the center of the language area. The anterior segment of the insula extends to and interfaces with Broca's area, while its posterior elements adjoin Wernicke's area. Due to its location, in the epicenter of the human language area, it is understandable that the insula may play a significant role in language, as it was assumed during the nineteenth century.

12.2 Involvement of the Insula in Aphasia

Brain pathology in cases of aphasia frequently includes an insular extension [8]. Ever since Wernicke [9], an extensive body of clinical research has supported that the insula is frequently involved in so-called "major" or perisylvian apha-

A. Ardila
Department of Communication Sciences and Disorders, Florida International University,
Miami, FL, USA
e-mail: ardilaa@fiu.edu

sic syndromes: Broca's aphasia, conduction aphasia, and Wernicke's aphasia. Wernicke himself directly related insula damage with conduction aphasia; noteworthy, the first reported case of conduction aphasia presented an insular pathology. Involvement of the anterior part of the insula in Broca's aphasia was reported many years ago, since the very beginning of the twentieth century [1, 10]. The word-deafness component of Wernicke's aphasia, on the other hand, has been associated with posterior insula damage [11]. Thus, the insula is frequently involved in these three major aphasia syndromes.

Pathology involving only the insula, nonetheless, is rarely reported. Alexander et al. [12] presented two cases with CT evidence of pathology limited to the left insula and subjacent extreme-external capsules. An aphasia with mildly paraphasic production and agraphia was noted in both patients. Nielsen and Friedman [13] reported several cases from the literature with autopsies suggesting that the left insular damage was associated with mild aphasia. Starkstein et al. [14] observed crossed aphemia associated with a right insular lesion. Fifer [15] reported a patient presenting a lesion involving the right insula as well as the adjacent white matter; this patient presented with a unilateral disorder in processing auditory information when speech materials were presented to the left ear. Habib et al. [16] found a case of bilateral insula pathology, extending to a small part of the striatum on the left side and to the temporal pole on the right hemisphere. It was observed mutism lasting several weeks; this patient did not respond to any auditory stimulation and made no effort to communicate. Baratelli et al. [17] reported a herpes simplex encephalitis patient with a left hemisphere lesion limited to the left insula and to the left anterior parahippocampal region. The aphasia type and language recovery were analyzed. The language deficit was of the fluent type, without phonological impairment, and showed a good but incomplete recovery after four months.

In an extensive study, Lemieux et al. [18] reviewed the consecutive stroke database from two centers to identify patients' insular ischemic strokes during a four-year period. Clinicoradiological correlation was performed by distinguishing insular ischemic strokes involving the anterior or posterior insular cortex. The authors identified seven patients from their institutions and 16 previously published cases. Infarcts were limited to the anterior ($n = 4$) or posterior ($n = 12$) insula or affected both ($n = 7$). The five most frequent symptoms were somatosensory deficits ($n = 10$), aphasia ($n = 10$), dysarthria ($n = 10$), a vestibular-like syndrome ($n = 8$), and motor deficits ($n = 6$). So, aphasia is without a doubt a frequent clinical manifestation of insular ischemic stroke.

Mutism has been sometimes reported in cases of insular pathology. Transient mutism may be found in cases of left inferior motor cortex damage extending to the insula [19, 20]; lasting mutism, on the other hand, appears to be associated with bilateral lesions of the frontal operculum and anterior insula [21–24]. A transient mutism and persistent auditory agnosia, due to two successive ischemic infarcts mainly involving the insular cortex on both hemispheres, were reported by Habib et al. [16]. Shuren [25] observed a patient who presented language initiation defects as a result of a left anterior insular infarct. The author proposed that dominant hemisphere anterior insular lesions impair the speech initiation loop. Alexander et al. [20] suggested that left cortical and subcortical opercular lesions frequently result in a total speech loss associated with a right hemiparesis. Right hemiparesis has a good recovery, but the oral apraxia responsible for the mutism improves only slowly.

Although most of the reports of insular aphasia refer to vascular aphasia, other insular pathologies, such as insular tumor, can also be associated with aphasia [26].

12.3 Apraxia of Speech

The interest in the potential involvement of the insula in language reappeared during the late twentieth century; this renewed interest in the insula was significantly associated with the publication of Dronkers' paper "A new brain region for coordinating speech articulation" [27]. In this paper, it is reported that the left precentral gyrus of the insula is involved in motor planning of speech. Twenty-five stroke patients with a disorder in motor planning of articulatory movements, which Dronkers labeled as "apraxia of speech," were compared with 19 individuals without such deficit. A robust double dissociation was observed. All patients with articulatory planning impairments presented lesions including the anterior insula. This area was completely spared in all patients without these articulatory defects. The conclusion was obvious: anterior insula represents the crucial brain area in motor planning and organization of speech. Verbal apraxia and agrammatism represent the two cardinal characteristics of Broca's aphasia, and hence, the anterior insula is directly involved in Broca's aphasia. Several reports have confirmed the involvement of the anterior insula in speech praxis. During the following years, several papers were published emphasizing the participation of the insula in language [28, 29]. For instance, Nagao et al. [30] reported a patient with apraxia of speech who showed an acute infarct limited to the precentral gyrus of the insula.

In a critical study, 33 left hemisphere stroke patients with varying degrees of speech impairment were asked to perform multiple repetitions of single words; these words varied along three separate dimensions: number of syllables, degree of articulatory travel (i.e., change between places of articulation for consonants), and presence/absence of an initial consonant cluster [31]. The authors found that precentral gyrus of the insula was critical for performance on the articulation task across all three conditions. The authors concluded that the left precentral gyrus of the insula is a critical area for intra- and inter-syllabic coordination of complex articulatory movements, prior to end-stage execution of speech commands.

12.4 Neuroimaging Studies

Contemporary neuroimaging technique research has supported the active involvement of the insula in linguistic processes. Activation of the insula has been demonstrated in a diversity of linguistic tasks, including during word generation task performance [32, 33] and naming [34]. Lexical knowledge and word retrieval evidently represent central linguistic processes frequently impaired in aphasia. Activation of the insula has also been reported during phonological decision tasks [35]. The insula has also been found to be involved in the motor aspects of speech production [36] and naming [37].

Using fMRI, it has been found that the insular cortex is sensitive to phonological processing, particularly sublexical spelling-sound translation [38]. It has been also reported that processing of semantic violations relied primarily on the mid-portion of the superior temporal region bilaterally and the insular cortex bilaterally [39].

Rousseaux et al. [40] correlated linguistic impairment with cerebral blood flow and related verbal comprehension, naming, and paraphasias to an asymmetry index of the insula and the lenticular nucleus. The insula would accordingly not participate in some unique and isolated language dimension, but rather would be active in different linguistic aspects. It may therefore be concluded that contemporary neuroimaging studies support the assumption of a significant participation of the insula in language. The insula would be involved not in a single linguistic activity, but in several verbal processes simultaneously. The anterior portion of the insula would be involved in the organization and planning of language articulation, while the middle and posterior portions would be involved with lexical knowledge, word retrieval, language understanding, and phonological discrimination. By the same token, functional imaging studies demonstrate that the insula participates in several key auditory processes, such as allocating auditory attention and tuning in to novel auditory stimuli, temporal processing, phonological processing, and visual-auditory integration [41]. Zaccarella and Friederici [42] using fMRI attempted to specify the involvement of the insula during visual word processing using a subregional parcellation approach. They found that word compared to letter string processing was strongly subregional sensitive within the anterior-dorsal cluster only and was left-lateralized.

Using graph theoretical analysis of functional MRI (fMRI) data in healthy subjects, Fuertinger et al. [43] quantified the speech network topology by constructing functional brain networks of increasing hierarchy. The authors identified a segregated network of highly connected local neural communities (hubs) in the primary sensorimotor and parietal regions. Compared to other tasks, speech production was characterized by the formation of six distinct neural communities including the prefrontal cortex, insula, putamen, and thalamus, which collectively forged the formation of the functional speech connectome.

12.5 Contemporary Meta-analytic Studies

Recently, two major papers were published presenting meta-analyses of insula fMRI activation, assessed using the activation likelihood estimation (ALE) method [44, 45]. In the first meta-analysis, 42 fMRI studies including 639 healthy adults were performed, comparing insula activation during performance of language (expressive and receptive) and speech (production and perception) tasks. Both tasks activated bilateral anterior insula. The authors found that speech perception tasks preferentially activated the left dorsal mid-insula, whereas expressive language tasks activated left ventral mid-insula.

The second meta-analytic study used as well as the activation likelihood estimation (ALE) technique was developed. Search conditions were (a) studies reporting insula activation; (b) studies using fMRI; (c) context, normal subjects; (d) activations, activation only; (e) handedness, right-handed subjects; (f) age 20–60 years; and (g) domain, cognition, subtype, language. By means of the BrainMap functional and database, 26 papers corresponding to 39 paradigms including 522 participants were selected. Thirteen different activation clusters were found. Insula connections included not only the classical areas involved in language production (Broca's area) and language understanding (Wernicke's area) but also areas involved in language repetition (supramarginal gyrus) and other linguistic functions such as the left prefrontal lobe involved in language control and the Brodmann area (BA) 37 involved in lexico-semantic associations. The first cluster includes the left claustrum (the insular subcortical gray matter). The second cluster includes the anterior cingulate gyrus (BA24) (involved in motor organization—motor preparation/planning and cognitive/motor inhibition— and language initiative) and left BA6 (medial frontal gyrus). BA6 includes the supplementary motor area (SMA). Cluster 3 included the right insula and the insular subcortical gray matter (claustrum) and indicates an integrated activity of the left and right insula. Cluster 4 referred to the left BA7 (superior parietal lobe). Therefore, the insula would also be part of the language system related to some contextual and motor learning aspects of speech. The following cluster involved BA37 (posterior inferior temporal gyrus, middle temporal gyrus, fusiform gyrus), and the cerebellar culmen. The following two clusters (6 and 7) referred to two areas traditionally involved in language: the left BA22 (superior temporal gyrus—part of Wernicke's area) and the left BA40 (supramarginal gyrus). Clusters 8 and 9 may contribute to the motor aspects of speech and the attention control of language. Cluster 10 on the other hand is similar to Cluster 7 and includes the left BA40 (inferior parietal lobe).

It was concluded that the insula represents a core area in language processing, and it is related not just with language production functions but also with language understanding processing. It could consequently be conjectured that the insula is a core hub for language. Its strategic location between the anterior and posterior language areas would be crucial to play a language coordinating function.

12.6 Language Motivation and Affect

Several reports have suggested that the insula may play an important role in verbal motivation and affect. Habib et al. [16] have proposed that bilateral damage to the insula would disrupt the motivational mechanisms that lead to the motoric production of human communication, by depriving them of connections with various limbic structures. Hence, insular damage would result in an impairment in linguistic motivation. Anatomical studies suggest that the insula could act as a cortical representation of the limbic nervous system and may provide a direct input to the brain emotional system that could in turn monitor the affective tone of language output. Within the same vein, Beaty et al. [46] have included the right insula in an extended brain network involved in metaphor production.

It can be considered that the insula is not only involved in language, but also participates in emotional networks [47].

Conclusion

Since the nineteenth century, it has been known that the insula has a crucial role in language processes. However, the interest in the insula somehow decreased during the twentieth century, because it has not directly included the perisylvian language area proposed by Dejerine. Since the last decade of the twentieth century, the interest in understanding the participation of the insula in language reemerged. A crucial role on speech praxis was initially postulated. It was also emphasized that the insula damage frequently results in aphasia. Among the various language disturbances associated with damage in the left insula are motor planning and organization of speech in Broca's aphasia, repetition defects associated with conduction aphasia, and the word-deafness component of Wernicke's aphasia [28, 29]. Mutism and oral apraxia have been reported to be associated with left insula pathology. Recent neuroimaging studies have illustrated that the insula has extensive brain connection, including not only the frontal motor language area but also posterior language areas, including the temporal and parietal lobes. Anatomical connections of the insula point also to an important viscero-limbic role, and it may accordingly be suggested that the insula is involved in verbal motivation.

It can be concluded that the brain language area should be reconsidered to include not only the perisylvian area of the left hemisphere, but also the insula.

References

1. Freud S. On aphasia : a critical study. London: Imago Publishing Co. Ltd.; 1891.
2. Dejerine J. *Semiologie des Aàections du Systeme Nerveux*. Paris: Masson; 1914.
3. Head H. Aphasia and kindred disorders. Cambridge, UK: Cambridge University Press; 1920.
4. Nielsen JM. Agnosia, apraxia and aphasia: their value in cerebral localization. New York, NY: Hafner; 1936.
5. Luria AR. Higher cortical functions in man. New York, NY: Basic Books; 1966.
6. Benson DF, Geschwind N. The aphasias and related disturbances. In: Baker AB, Bake LH, editors. Clinical neurology, vol. 1. New York, NY: Harper and Row; 1971. p. 112–40.
7. Goodglass H, Kaplan E. The assessment of aphasia and related disorders. Philadelphia: Lippincott Williams & Wilkins; 1972.
8. Gasquoine PG. Contributions of the insula to cognition and emotion. Neuropsychol Rev. 2014;24(2):77–87.
9. Wernicke C. Der Aphasische Symtomenkomplex. Breslau: Cohn and Weigert; 1874.
10. Bernheim F. De l'Aphasie Motrice. Paris: These de Paris; 1900.
11. Liepmann H, Storck E. Ein Fall von reiner Sprachtaubheit. Manuschfrift Psychiatrie und Neurologie. 1902;17:289–311.
12. Alexander MP, Naeser MA, Palumbo CL. Correlations of subcortical CT lesions sites and aphasia profiles. Brain. 1987;110:961–91.
13. Nielsen JM, Friedman AP. The quadrilateral space of Marie. Bull Los Angeles Neurol Soc. 1942;8:131–6.
14. Starkstein SE, Berthier M, Leiguarda R. Bilateral opercular syndrome and crossed aphemia due to a right insular damage: a clinicopathological study. Brain Lang. 1988;34:253–61.
15. Fifer RC. Insular stroke causing unilateral auditory processing disorder: case report. J Am Acad Audiol. 1993;4:364–9.
16. Habib M, Daquin G, Milandre L, Royere ML, Rey M, Lanteri A, Slamanon G, Khalil R. Mutism and auditory agnosia due to bilateral insular damage—role of the insula in human communication. Neuropsychologia. 1995;33:327–39.
17. Baratelli E, Laiacona M, Capitani E. Language disturbances associated to insular and entorhinal damage: study of a patient affected by herpetic encephalitis. Neurocase. 2015;21(3):299–308.
18. Lemieux F, Lanthier S, Chevrier MC, Gioia L, Rouleau I, Cereda C, Nguyen DK. Insular ischemic stroke: clinical presentation and outcome. Cerebrovasc Dis Extra. 2012;2(1):80–7.
19. Schiff HB, Alexander MP, Naeser MA, Galaburda AM. Aphemia: clinical-anatomical correlations. Arch Neurol. 1983;40:720–7.
20. Alexander MP, Benson DF, Stuss DT. Frontal lobes and language. Brain Lang. 1989;37:656–91.

21. Sussman NM, Gur RC, Gur RF, O'Connor MJ. Mutism as a consequence of callosotomy. J Neurosurg. 1983;59:514–9.
22. Cappa SF, Guidotti M, Papagno C, Vignolo LA. Speechlessness with occasional vocalization after bilateral opercular lesions: a case study. Aphasiology. 1987;1:35–9.
23. Groswaser Z, Korn C, Groswaser-Reider I, Solzi P. Mutism associated with buccofacial apraxia and bihemispheric lesions. Brain Lang. 1988;34:157–68.
24. Pineda D, Ardila A. Lasting mutism associated with buccofacial apraxia. Aphasiology. 1992;6:285–92.
25. Shuren J. Insula and aphasia. J Neurol. 1993;240:216–8.
26. Duffau H, Bauchet L, Lehéricy S, Capelle L. Functional compensation of the left dominant insula for language. Neuroreport. 2001;12(10):2159–63.
27. Dronkers NN. A new brain region for coordinating speech articulation. Nature. 1996;384:159–61.
28. Ardila A. The role of insula in language: an unsettled question. Aphasiology. 1999;13(1):79–87.
29. Ardila A, Benson DF, Flynn FG. Participation of the insula in language. Aphasiology. 1997;11:1159–70.
30. Nagao M, Takeda K, Komori T, Isozaki E, Hirai S. Apraxia of speech associated with an infarct in the precentral gyrus of the insula. Neuroradiology. 1999;41(5):356–7.
31. Baldo JV, Wilkins DP, Ogar J, Willock S, Dronkers NF. Role of the precentral gyrus of the insula in complex articulation. Cortex. 2011;47(7):800–7.
32. Baker SC, Frith CD, Dolan RJ. The interaction between mood and cognitive function studied with PET. Psychol Med. 1997;27:565–78.
33. McCarthy G, Blamire AM, Rothman DL, Gruetter R, Shulman RG. Echo-plantar magnetic resonance imaging studies of frontal cortext activating during word generation in humans. Proc Natl Acad Sci. 1993;90:4952–6.
34. Price CJ, Moore CJ, Humphreys GW, Frackowiak RS, Friston KJ. The neural regions sustaining object recognition and naming. Proc Royal Soc Biol Sci. 1996;263:1501–7.
35. Rumsey JM, Horwitz B, Donohue BC, Nace K, Maisog JM, Andreason P. Phonological and orthographic components of word recognition. A PET-rCBF study. Brain. 1997;120:739–59.
36. Ackermann H, Riecker A. The contribution of the insula to motor aspects of speech production: a review and a hypothesis. Brain Lang. 2004;89(2):320–8.
37. Owen WJ, Borowsky R, Sarty GE. fMRI of two measures of phonological processing in visual word recognition: ecological validity matters. Brain Lang. 2004;90(1):40–6.
38. Borowsky R, Cummine J, Owen WJ, Friesen CK, Shih F, Sarty GE. FMRI of ventral and dorsal processing streams in basic reading processes: insular sensitivity to phonology. Brain Topogr. 2006;18(4):233–9.
39. Friederici AD, Rueschemeyer SA, Hahne A, Fiebach CJ. The role of left inferior frontal and superior temporal cortex in sentence comprehension: localizing syntactic and semantic processes. Cereb Cortex. 2003;13(2):170–7.
40. Rousseaux M, Steinling M, Griffie G, Quint S, Cabaret M, Lesoin F, Mazingue M, Destee A. Correlation of thalamic aphasia and cerebral blood flow. Rev Neurol. 1990;146:345–53.
41. Bamiou DE, Musiek FE, Luxon LM. The insula (Island of Reil) and its role in auditory processing: literature review. Brain Res Rev. 2003;42(2):143–54.
42. Zaccarella E, Friederici AD. Reflections of word processing in the insular cortex: a sub-regional parcellation based functional assessment. Brain Lang. 2015;142:1–7.
43. Fuertinger S, Horwitz B, Simonyan K. The functional connectome of speech control. PLoS Biol. 2015;13(7):e1002209.
44. Oh A, Duerden EG, Pang EW. The role of the insula in speech and language processing. Brain Lang. 2014;135:96–103.
45. Ardila A, Bernal B, Rosselli M. Participation of the insula in language revisited: a meta-analytic connectivity study. J Neurolinguistics. 2014;29:31–41.
46. Beaty RE, Silvia PJ, Benedek M. Brain networks underlying novel metaphor production. Brain Cogn. 2017;111:163–70.
47. Clos M, Rottschy C, Laird AR, Fox PT, Eickhoff SB. Comparison of structural covariance with functional connectivity approaches exemplified by an investigation of the left anterior insula. Neuroimage. 2014;99:269–80.

Lateralization of the Insular Cortex

Michael J. Montalbano and R. Shane Tubbs

The insula is a cortical brain region found deep to the lateral fissure, surrounded by the frontoparietal and superior temporal cortex [1]. It functions primarily to integrate interoceptive and viscerosensory input from the periphery. Ascending tracts from basal parasympathetics and posterior sympathetics of the ventromedial nucleus of the thalamus lead to lateralization of function. The lateralization of function can be broadly separated into positive or negative emotional valence, approach or avoidance behaviors, perception or experience of a certain phenomenon, as well as conscious perception of one's own affective state [2, 3].

A variety of autonomic functions are associated with insular activation. In general terms, findings suggest that the right insula is involved in sympathetic nervous system responses such as avoidance behaviors, subjective pain, and stimuli promoting negative affect. The right insula is hypothesized to process such visceral information in accordance with subjective relevance of bodily states. This includes sympathetic homeostatic afferents from pain to cognitive awareness of bladder sensation and even cardiopulmonary stimulation [4–7]. Evidence for these processes is further supported by Wada test lesions that produce arrhythmia and cardiac arrest. On the other hand, the left insula is associated with functions such as nourishment, safety, positive affect, and approach behaviors. In general terms, this suggests that the left insula is responsible for parasympathetic homeostatic afferents. This is supported by findings that demonstrate Wada test lesions performed on the left insula produces an increased sympathetic response, while Valsalva maneuvers and vagus nerve stimulation treatment of epilepsy and major depression work via activation of the left insula [2, 3].

However, there are also functions that involve the insula bilaterally, such as empathic pain and feedback of heart rate. Even attempted acts of micturition involve bilateral insula activation, with the activation showing increased response proportionally with increased volume [2, 6, 8]. This role for the insula is further demonstrated by the finding that pelvic floor training occurs in association with several structures including the insula [5]. This suggests overlap in autonomic function that may not clearly be demarcated into lateralized functions.

In conjunction with autonomic monitoring, the insula's integration of interoceptive signals may also play a role in emotional processing insofar as there is an embodied appraisal model of emotions whereby internal bodily states can be associated with emotional states [8, 9]. Specifically, visual stimuli of a negative valence lateralize to the left anterior, as do emotions

M. J. Montalbano, M.D. (✉)
Department of Anatomical Sciences, St. George's University, True Blue, Grenada
e-mail: mmontalb@sgu.edu

R. Shane Tubbs, Ph.D., P.A.-C.
Seattle Science Foundation, Seattle, WA, USA

leading to withdrawal behaviors. Experience of emotions is also left associated during active emoting [3, 8]. However, negative valence emotions that are perceived in others, not experienced by the individual, activate the left insula. It may be that negative valence stimuli activate bilaterally during experience but lateralize left during perception of others undergoing similar negative valence stimuli [10].

Other studies, however, have found right insula activation, and even bilateral responses, associated with subjective emotional response, as well as negative valence emotions such as nausea. Thus, it may be that the right insular cortex activity is more pronounced with awareness of emotional stimuli and representation of interoception and bodily responses [3, 4]. Likewise, unfair offers increased bilateral anterior insula activation for processing, perhaps as a way of balancing and processing the relevant emotions, even as anterior insula activation increased the chance of rejection of unfair offers in ultimatum games [11]. Additionally, the right insula has been found to be highly active in autism patients when viewing neutral images of the self, but not when viewing others, and an increase in right insula surface area is associated with increased impairment on multiple autism scales [3, 12]. This would support findings that show right insula activity corresponding to increased interoceptive attention, generated during subjective affective experience or otherwise [4].

The overlap between emotional states and cognitive control is highlighted by the role of the insula during information processing for the salience network. It is thought that once an incentive is encoded in the salience network, the insular cortex is responsible for retrieving those values. For this, the right fronto-insular cortex engages attentional working memory while disengaging task irrelevant systems [13]. The right anterior insula also activates with moving stimulus detection and novel changes [14]. Furthermore, the right anterior insula has a role in error processing for the salience network, modulating the default network during cognitive tasks, as well as predicting the need for avoidance and any compensatory negative feedback after a response [10, 15]. On the other hand, the left insula is involved with top-down modulation and shows increased activation during self-attribution in reward task pre-response conflicts and is also associated with behaviors of approach over withdrawal, self-referencing, proactive control, error monitoring, as well as attention orientation, response inhibition, emotions of others, perspective-taking, and empathic affiliative behaviors [10].

In contrast, bilateral activation of the insula is associated with cessation signaling, emotional interference, and risk decisions during gambling [10]. Similarly, bilateral insular cortex lesions abolish specific satiety-induced outcome devaluations. This implies that the insular cortex recalls incentive value of outcome of choice. This balancing of attention, self-reference, and incentive salience not surprisingly corresponds to a role in monitoring social-emotional functions [3, 16].

In addition to cognitive tasks already mentioned, the insula may also play a role in facilitating language. Word articulation and pronunciation has been correlated with the left anterior insula [2, 17]. Additionally, deficits such as dysarthria and apraxia have been associated with lesions in the left superior precentral gyrus of the insula [18, 19]. Thus, there is evidence that the left ventral and median insula are involved in language production, specifically syntax and semantics, while the right anterior is correlated more with control of rhythm and prosody, which means distorted speech comprehension involves, among other regions, the bilateral anterior insula [2, 19]. This is further evidenced by lesions bilaterally to the insula producing total auditory agnosia [19].

These bilateral findings are not uncontroversial however. Much of the extant literature accredits language functions primarily to the left hemisphere of the brain. Also, other studies have not found these corresponding correlations with deficits except in cases where lesions include other areas. Additionally, some researchers find the insula only tangentially linked to language production through a mediating third factor such as motor control of respiration [19, 20].

Despite the progress made in the above findings, there are still limitations that need to be overcome in future research. Certain confounds, such as patients' expectancies, personality traits, and handedness, may need to be separated in order to account for divergent results. The sex of the subjects investigated is one such important possible confound. For example, with negative and positive valence stimuli, females activate the left insula more than males [3]. However, there are also contrary findings to show that females predominantly perform affective processing bilaterally, while males perform the same activities in the left insula [10]. Additionally, certain autonomic functions, such as bladder testing, show increased insula activation overall in females [5]. Thus, further research is needed to clarify such findings and determine in more general terms the extent of lateralization of such functions and how much differentiation occurs across sex, handedness, and other factors.

This is not to say that there are not important clinical applications that can make use of these findings regarding the insula. Possible modulation of insular activity could be valuable in studying emotion regulation in biofeedback treatments [4]. This could lead to possible conditioning for mood disorders, as the right insula has been shown to be more amenable through emotional tasks [4, 8]. This is fortunate for psychiatric states such as mania, which is associated with higher right insula activation [21]. On the other hand, antisocial personality shows amygdala activation with hypoactive insula during classical fear conditioning. Additionally, increased psychopathy checklist scores are associated with decreased success at volitional insular control and emotional regulation [4, 22]. Further research therefore should also be aimed at efficacious clinical applications in addition to empirical clarification as outlined above.

References

1. Touroutoglou A, Hollenbeck M, Dickerson BC, et al. Dissociable large-scale networks anchored in the right anterior insula subserve affective experience and attention. Neuroimage. 2012;60(4):1947–58.
2. Craig AD. Forebrain emotional asymmetry: a neuroanatomical basis? Trends Cogn Sci. 2005; 9(12):566–71.
3. Duerden E, Arsalidou M, Lee M, et al. Lateralization of affective processing in the insula. Neuroimage. 2013;78:159–75.
4. Caria A, Veit R, Sitaram R, et al. Regulation of anterior insular cortex activity using real-time fMRI. Neuroimage. 2007;35:1238–46.
5. Kuhtz-Buschbeck JP, van der Horst C, Wolff S, et al. Activation of the supplementary motor area (SMA) during voluntary pelvic floor muscle contractions: an fMRI study. Neuroimage. 2007; 35:449–57.
6. Kuhtz-Buschbeck JP, van der Horst C, Pott C, Wolff S, et al. Cortical representation of the urge to void: a functional magnetic resonance imaging study. J Urol. 2009;174:1477–81.
7. Zaki J, Davis JI, Ochsner KN. Overlapping activity in anterior insula during interoception and emotional experience. Neuroimage. 2012;62:493–9.
8. Caria A, Sitaram R, Veit R, et al. Volitional control of anterior insula activity modulates the response to aversive stimuli. a real-time functional magnetic resonance imaging study. Biol Psychiatry. 2010;68:425–32.
9. James W. What is an emotion? Mind. 1884;9(34): 188–205.
10. Kann S, Zhang S, Manza P, et al. Hemispheric lateralization of resting-state functional connectivity of the anterior insula: association with age, gender, and a novelty-seeking trait. Brain Connect. 2016;6(9):724–34.
11. Harle KM, Chang LJ, van't Wout M, et al. The neural mechanisms of affect infusion in social economic decision-making: a mediating role of the anterior insula. Neuroimage. 2012;61:32–40.
12. Doyle-Thomas KAR, Kushki A, Duerden EG, et al. The effect of diagnosis, age, and symptom severity on cortical surface area in the cingulate cortex and insula in autism spectrum disorders. J Child Neurol. 2012;28(6):732–9.
13. Sridharan D, Levitin DJ, Menon V. A critical role for the right fronto-insular cortex in switching between central-executive and default-mode networks. PNAS. 2008;105(34):12,569–74.
14. Howard LR, Kumaran D, Olafsdottir HF, et al. Double dissociation between hippocampal and parahippocampal responses to object-background context and scene novelty. J Neurosci. 2011; 31(14):5253–61.
15. Ullsperger M, Harsay HA, Wessel JR, et al. Conscious perception of errors and its relation to the anterior insula. Brain Struct Funct. 2010;214:629–43.
16. Parkes SL, Balleine BW. Incentive memory: evidence the basolateral amygdala encodes and the insular cortex retrieves outcome values to guide choice between goal-directed actions. J Neurosci. 2013;33(20):8753–63.
17. Kuriki S, Isahai N, Takeuchi F, et al. Where do perception and recognition of words take place in the brain? A neuromagnetic approach. Electroencephalogr Clin Neurophysiol Suppl. 1999;49:179–83.

18. Borovsky A, Saygin AP, Bates E, et al. Lesion correlates of conversational speech production deficits. Neuropsychologia. 2007;45(11):2525–33.
19. Oh A, Duerden EG, Pang EW. The role of the insula in speech and language processing. Brain Lang. 2014;135:96–103.
20. Fedorenko E, Fillmore P, Smith K, et al. The superior precentral gyrus of the insula does not appear to be functionally specialized for articulation. J Neurophysiol. 2015;113(7):2376–82.
21. Hagele C, Friedel E, Schlagenhauf F, et al. Affective responses across psychiatric disorders: a dimensional approach. Neurosci Lett. 2016;623:71–8.
22. Sitaram R, Caria A, Veit R, et al. Volitional control of the anterior insula in criminal psychopaths using real-time fMRI neurofeedback: a pilot study. Front Behav Neurosci. 2014;8:344. https://doi.org/10.3389/fnbeh.2014.00344.

Gustatory Areas Within the Insular Cortex

14

Richard J. Stevenson, Heather M. Francis, and Cameron J. Ragg

14.1 Introduction

This chapter focuses on human gustatory processing in the insular cortex. We start by presenting a brief overview of human gustation (used interchangeably here with taste), covering perception, sensory physiology and the flow of information from receptor to the brain. This is followed by an examination of the insula's role in each of the three major domains of gustatory experience—perceptual quality, sensory intensity and hedonics. We then consider the degree to which gustatory processing is lateralised in the insula, before turning to a more pressing issue—the multimodal nature of both routine taste perception and more generally of the insular cortex. Taste forms part of the broader flavour system, which directly involves somatosensation and olfaction and indirectly vision and audition [1]. It is well established that the insula is a multimodal processor and not principally restricted to taste in any region [2, 3]. While the insula is the first point of cortical processing—so being primary in that meaning of the word—whether it is *primary* in the manner we speak of for other sensory systems (e.g. primary visual cortex) is a significant point of debate. Another is what gustatory processing may reveal about the broader function of the insular cortex.

14.1.1 Human Gustation

Human gustatory experience involves the perception of taste quality and intensity, as well as an affective (hedonic) dimension. Taste quality is generally agreed to involve five perceptual dimensions—sweet, salty, sour, bitter and umami (meaty) taste [4, 5]—although there may be more (e.g. metallic). Each of these discrete taste qualities is allied with a particular function: (1) sweet with carbohydrate energy detection, (2) bitter with poison detection, (3) sour with bacterial fermentation and ripeness detection, (4) salty with maintaining electrolyte balance, and (5) umami with detection of protein-based foods.

While the focus of this chapter is on taste, it is inevitably broader. This is because when humans encounter taste stimuli, it is always accompanied by somatosensation (from oral mechanical stimulation resulting from eating and drinking) and nearly always accompanied by retronasal olfaction (i.e. perception of food-based odorants in the mouth via the nasopharynx [1]). Thus, most of our experience with taste occurs in a multimodal context, because taste serves as just one part of the larger flavour system, which is focussed on identifying nutritious food and avoiding ingesting things that may make one sick.

R. J. Stevenson (✉) · H. M. Francis · C. J. Ragg
Department of Psychology, Macquarie University, Sydney, NSW, Australia
e-mail: dick.stevenson@mq.edu.au; richard.stevenson@psy.mq.edu.au

The five taste qualities are detected by two basic classes of receptor. The first class is G-protein-coupled receptors, with one specific for umami and another for sweet and with multiple types for bitter representing the potentially large array of plant-based poisons. The second class is the ion-based receptor channels, specific to sodium ions and protons, respectively [6]. These receptors are located on taste receptor cells, which in turn are grouped into structures called taste buds. Taste buds are located on the surface of the tongue, most densely on the anterior tip and less densely on the sides, back and center of the tongue, as well as at the back of the throat. G-protein-coupled taste receptors are also located in the gut [7], and while ultimately projecting to the same primary cortical region (insula via vagal afferents) as oral-based gustation, they do not appear to be consciously accessible as 'tastes'. Functionally, these gut-based receptors *may* be important for monitoring nutrient quality and toxins in the gut.

A major and as yet incompletely resolved problem in taste perception is whether the gustatory information used to generate a taste percept is represented discretely at all levels of the nervous system (i.e. a labelled line; e.g. [6]) or by some form of population coding across many neurons (i.e. pattern coding; e.g. [8, 9]) or variant thereof (e.g. [10]). Even after nearly half a century of work, there is evidence favourable to both theories, including at all levels of processing from receptor output to the cortex. It is perfectly possible that both accounts may be correct, with labelled lines connoting the basic quality (e.g. salty) and pattern coding the specific characteristics within a particular taste quality (e.g. metallic vs. mineral salty taste).

Once taste information passes out of the taste receptor cells on the tongue and throat, it flows along the three cranial nerves XII (chorda tympani), IX (glossopharyngeal) and X (lingual) to the nucleus of the solitary tract (NST). In rodents, information then flows to the parabrachial nucleus before splitting into an affective (hypothalamic, central grey, ventral striatum) and a sensory (thalamic) stream [11]. In contrast, in primates, as well as humans, information flows from the NST to the ventral posteromedial nucleus of the thalamus (VPMNT) as well as directly to the insula [12]. From the VPMNT the major projection is to the insular cortex and overlying opercula and from there to the orbitofrontal cortex.

It has been suggested that gustatory processing areas within the insula/opercula may differ somewhat between humans and primates, with processing occurring more broadly across these structures in humans [11]. While it is the norm to regard the insula (and relatedly contiguous tissue of the frontal and parietal opercula) as being the primary taste cortex, it is important to flag that this view may be ill-considered for several reasons—physiologically, psychologically and ecologically—as we discuss later. Nonetheless, the insula is the location where the first cortical processing of gustatory information takes place, and there is no doubt that it plays a key role in this regard (e.g. [11, 13, 14]).

Anatomically, it is the mid-insula that appears to receive the main projections from the VPMNT and the mid-/anterior insula that has afferent projections from the NST [12]. This mid-region of the insula is a transition zone (dysgranular) between the granular posterior insula and the agranular anterior. This difference in cellular types often reflects functional differences, as the granular cortex is normally associated with sensory processing and the agranular cortex with motor processing [15]. A further difference is that the anterior portion of the insula receives and projects to a quite different set of cortical and subcortical structures to the posterior insula [12]. Indeed, the breadth of inputs and outputs from all of the thalamic nuclei points, along with other evidence discussed later, to the multimodal nature of this structure [16]. This is also reflected in the remarkable statistic that only 6% of neurons in the primate insular cortex (i.e. primary taste cortex) are actually responsive to gustatory stimuli [17].

14.2 Gustatory Functions of the Insula

Until the 1940s there was considerable uncertainty about the locus of cortical taste processing in humans. Most of what was known came from neuropsychological cases, which generated three different processing theories. In order of popularity, these placed the emphasis on the parietal operculum, temporal lobe (uncus and hippocampus) and the insular cortex—the last and least popular option [18]. Over the years since this paper, evidence has accumulated from multiple sources revealing that the insula is the first point of cortical processing for taste. Apart from tracer studies, which have consistently pointed to the insula as being the first point of cortical processing in humans and monkeys [19, 20], the most important animal-based work has come from electrophysiology. While these studies clearly indicate that insula neurons are responsive to tastants, these represent only a small fraction (around 6%) of the large number tested (around 13,000 neurons; [17]).

In humans, evidence has accumulated from three sources. The first is from stimulation studies of patients being prepared for epilepsy surgery. These studies have found that stimulating neurons in the insula generate taste sensations, especially in the mid-region, alongside oral-based somatosensory sensations and, the most common of all, unpleasant tastes and nausea [21, 22].

A second stream of evidence comes from two types of patients: (1) those who have experienced occlusion or haemorrhage of the middle cerebral artery, which often results in indirect damage to the insula, and (2) those with more direct damage from tumours or mechanical injury (e.g. bullet wounds). Over many single and multiple case studies, a clear pattern emerges of impairment in taste perception associated with damage to this structure (e.g. [23–27]). At least one recent study has built upon this neuropsychological approach by studying people with insula-based impairments using functional magnetic resonance imaging (fMRI). Here, people with insula damage and abnormal taste perception were compared to healthy controls when exposed to tastants in the scanner. Imaging revealed abnormal patterns of activity characterised by significantly greater activation of the insula in patients with partial or total loss of taste perception (ageusia; [28]).

The third and most extensive body of data in humans comes from neuroimaging using positron emission tomography (PET) and fMRI in healthy participants. Most PET studies, of which there are relatively few, indicate that there are significant bilateral changes in cerebral blood flow in the insular cortex and opercula in response to tastants and to water (water as a control may make it more difficult to detect insula activation due to concurrent oral somatosensory stimulation; [29]). Typically, even when using water as a control, greater activity is seen with gustatory stimulation than for water alone [30, 31]. For fMRI the set of available studies is much larger, including now three meta-analyses of these data. The earlier meta-analysis by Small et al. [22] found that tastants (including water) bilaterally activated the insular cortex and parietal operculum. One more recent meta-analysis by Veldhuizen et al. [33] reported a more extensive set of activations in response to tastants in the anterior, mid-, and posterior insula and in the frontal and parietal opercula. The other by Kurth et al. [34] found bilateral activations to tastants mainly in the anterior and mid-insula. Many of these studies—fMRI and PET—also find activity in the orbitofrontal cortex (OFC), which is typically regarded as the secondary gustatory cortex.

All of this body of work—animal and human—indicates that the insular cortex is involved in taste processing. However, as we noted in "Introduction", taste processing involves at least three discrete components—the perception of different taste qualities, of differing levels of intensity and affective response. In the following three sections, we examine what is known about the insula's role in each of these three aspects of gustatory processing.

14.2.1 Taste Quality

A key question is whether there is some form of organisation of neurons that are responsive to particular tastants in the insula. Perhaps the easiest to detect would be a clustering of neurons responsive to one taste quality, with these being physically discrete from those responding to other taste qualities. Monkey electrophysiology has provided some initial evidence for this type of organisation. It finds that if say a sucrose-sensitive neuron is detected, it is more likely that neighbouring neurons will also be sucrose sensitive, than sensitive to other tastants—although this is not a large effect [17].

More detailed studies have been conducted in rats and mice. Using in vivo two-photon calcium imaging, Chen et al. [35] found that in anaesthetised mice each taste quality has its own discrete cortical field in the insula. More recent findings have confirmed the existence of an overlapping topography [36] but one that is still discrete enough to allow for the control of behaviour in awake animals. Accolla et al. [37] found that in rats, stimulation of the region in the insula associated with bitter tastants could suppress drinking in thirsty rats, while stimulation of the sweet-sensitive region could promote drinking and make rats work harder to obtain water. In sum, these animal findings suggest partially overlapping fields of sensitivity to particular taste qualities, neither confirming a strict labelled-line nor pattern-based approach to taste quality coding.

That some form of topographical organisation of taste quality in the human insula exists is initially suggested by neuropsychological data. If gustatory processing is solely dependent on temporal coding [38, 39] or some other distributed pattern-based form, then lesions to the insula would be most likely to produce a general impairment to taste perception. In contrast, if this is some form of topographical arrangement, a lesion could potentially disable part of this field resulting in an impairment of just one or two taste qualities. Although the literature is not large, there are certainly documented cases of patients with insula lesions who have a selective taste-quality impairment ([40]; and see [27] for related evidence). This is consistent with the findings emerging from the animal literature of some form of fuzzy topographic taste quality coding.

Human neuroimaging data using fMRI also suggests some form of fuzzy topographical taste quality coding in the insula. Schoenfled et al. [41] using examples of all five basic tastants generated activation maps spanning the insula and opercula, finding overlapping but clearly visible tastant-specific regions. A further and important observation here was that discrete topography was much clearer and reproducible in individual participants, than it was when averaging took place across the sample. Evidence favouring discrete topography was also observed by de Araujo et al. [42] between MSG/IMP (examples of umami taste) and glucose, but there was a considerable degree of overlap in their averaged data, suggesting only weak organisation by taste quality. Some even weaker evidence for topographical distribution comes from a study by Spetter et al. [43] with some small differences noted between two taste qualities (salty and sweet). Finally, O'Doherty et al. [44] found limited differences in topography between sweet and salty qualities, even when testing at the individual or group level, with much overlap and only a small sweet-only area.

Taken together these findings suggest that the insula is involved in taste quality coding and that it probably has some loose form of topographic organisation that relates to taste quality. The fMRI data indicate that evidence for topographic organisation of taste quality coding may be idiosyncratic, varying from individual to individual, but stable within a person, noting however, that this is based on limited study. Importantly, there is no compelling data as yet to indicate how the insula generates discrete taste qualia from different classes of tastant. That is, there is no empirical resolution to the question of labelled-line versus pattern-based coding. Both are probably used by the insula.

14.2.2 Taste Intensity

Changes in concentration of a tastant, assuming they are of sufficient magnitude, can be readily detected by participants and are perceived as an increase in sensory intensity. One immediate problem here is that increases in stimulus intensity are also likely to be accompanied by hedonic changes—intense sweet tastes may be unpleasant and intense sour, salty and bitter tastes especially so. To determine whether the insula is involved in taste intensity coding independent of any affective changes, Small et al. [45] conducted an fMRI study using sweet and bitter tastes. In this design increasing concentration was accompanied by divergent changes in affect, allowing for contrasts to explore brain areas involved in mediating intensity change independent of affective change. This revealed activations in the insula and operculum, as well as in several other brain areas (notably the cerebellum, pons, amygdala, anterior cingulate cortex). Indeed, neuropsychological data indicate that the amygdala may be involved in taste intensity processing in addition to the insula, but this has not been well explored (see [46]).

A more recent fMRI study examined changes in intensity perception for sweet and salty tastes. Changes in perceived intensity and in tastant concentration were both associated with activity in the mid-insula regions, but changes in the amygdala were only associated with changes in tastant concentration [43]. While this was the case for both sucrose and salt for the insula, only changes in salt concentration were associated with amygdala activation. So, while both of the studies that have examined the issue of intensity coding find evidence of insula involvement, the amygdala also seems implicated in processing this aspect of taste perception.

14.2.3 Taste Hedonics

While it was expected that the insula would be involved in sensory perception, it was not expected to be involved in affective processing. This is because animal electrophysiological studies had been used to argue that the OFC was responsible for the affective dimension of gustatory experience, and possibly the amygdala as well, at least for aversive tastes [47]. Human PET data also seemed to support this view, with subtractions between quinine and sucrose being used to locate affective processing differences between these stimuli and finding OFC activation, but not insula activity [48]. Similarly, affective judgments of sweet, salty and sour tastants seem to involve the OFC, but not the insula [49]. However, there is evidence that the insula's involvement in gustatory processing extends beyond the sensory into the affective realm.

As we noted earlier, human electrostimulation studies, which have involved stimulating particular regions of the insula in awake patients, have frequently found activation sites that yield either aversive taste experiences or aversive taste responses [21, 22]. Similarly, selective damage to the insula (e.g. by stroke) has frequently been associated with dysgeusia—unpleasant tastes [25, 50]. Of course, it is possible that electrostimulation and lesion damage have downstream effects in the OFC or other structures associated with affect (e.g. amygdala), so while this evidence is suggestive, it is not definitive.

Stronger evidence comes from fMRI. Returning to the Small et al. [45] study discussed above, recall that this study presented participants with sweet and bitter tastants at two concentration levels. The study is also informative about brain regions active in mediating pleasant and unpleasant taste sensations, by subtracting responses between the two tastants (sucrose and quinine independent of the effects of concentration) and of oral mouth movements in general. For the pleasant minus unpleasant contrast, activation was restricted to the OFC, with no insula involvement. For the unpleasant minus pleasant contrast, activation was restricted to the insula—although this effect was more evident with the intense (high concentration) tastants. This suggests some role for the insula in mediating taste hedonics.

In a further study, this time using PET, Zald et al. [51] examined participants' responses to the aversive taste of intense saline relative to the pleasant taste of chocolate. Noting that chocolate is both a taste and an olfactory stimulus, contrasting saline versus chocolate, again revealed insula activity, in addition to OFC and amygdala responding. A final fMRI study found that *expecting* an aversive taste allowed the participant to dampen activity within the insula and that activity within the insula and frontal operculum was associated with participants evaluations of taste unpleasantness [52]. These findings would seem to suggest that the insula is particularly involved in aversive taste perception, which is consistent with the human electrophysiology and neuropsychological studies.

While the focus so far has been on affective reactions to tastants independent of physiological state, much of our contact with taste stimuli starts when we are in one state (hungry) and ends when we are in another (full). Changes in state are already known to alter affective responses to tastants [53]. Interestingly in this regard, the human insula is known to possess in its mid-region neurons that are jointly sensitive to the bodies' interoceptive state of hunger/fullness and gustation [54]. The link between taste perception and state was explored by Haase, Cerf-Ducastel and Murphy [55]. They examined participants' response to a range of pleasant and unpleasant tastes when fasted and when sated after consuming food. The most interesting finding was that multiple brain regions changed their responsiveness to tastants, with greater activation typifying the fasted state. These areas were the insula, opercula, OFC, amygdala, hypothalamus, hippocampus and entorhinal cortices. Fasted activations tended to be greatest for pleasant tastes in all of these structures. So while the insula is state dependent in its response to taste stimuli, so too are most other brain regions involved in food-related processing.

14.2.4 Conclusion

Much remains to be known about the role of the insula in taste quality, intensity and affective processing and especially about the role of more specific regions within this structure. For taste quality, there appears to be a loose topographic structure reflecting taste quality, a finding based upon animal and human data. For taste intensity, insula responding is related to both perception of taste intensity and physical concentration, with greater activation correlated with intensity/concentration increases. For taste hedonics, as with many food-related brain processing areas, the insula appears to be state dependent. The insula also seems to be especially involved in modulating responses to aversive tastes.

14.3 Gustatory-Related Lateralisation

Two aspects of lateralisation are relevant here. The first concerns the general suggestion of greater right-sided cortical processing of gustatory information relative to left-sided processing. This notion first emerged in an early meta-analysis of human gustatory PET studies [32]. Based upon nine imaging studies, they found more right-sided than left-sided activation peaks in the insula and opercula, with this right-sided bias attributed to the greater processing of language in the left hemisphere. A further meta-analysis [33], drawing upon 15 imaging studies (PET and fMRI), found no right-sided bias. However, a more inclusive meta-analysis using data from 31 imaging studies again found right-sided bias [34]. Some of the ambiguity here may revolve around the type of task being used (e.g. passive tasting vs. some active task), some of which may favour processing on one side or the other [56]. Another factor is handedness, which may influence gustatory lateralisation [57].

Lateralisation has also been studied by presenting tastants to one side of the tongue and examining whether task performance for discrimination, naming, perception of taste quality, intensity and hedonics differs from when the tastant is presented to the other side of the tongue. Using a large sample of healthy participants, Stevenson, Miller and McGrillen [56] found significantly better discriminative performance on the right side of the tongue, significantly better perception/judgment of taste quality on the left side of the tongue, some varied effects for intensity judgments (sour stronger on the left, sucrose and quinine on the right) and no effects for hedonics or naming. While it is not possible to say at what level in the brain these effects manifest, the cortical level is likely to be important (especially with judgments being made), suggesting that there may be differences in lateralisation by task in healthy participants, something that needs to be further explored using neuroimaging.

A second and important and related issue, when considering lateralisation of gustatory function, is to consider how information reaches the insula from the tongue. Four different models have been proposed, utilising neuropsychological and/or neuroimaging data. First, Lee et al. [58], based upon a study of three stroke patients, suggested that gustatory information flows to each insula from the contralateral side of the tongue, being fully decussated at or below the midbrain. This model seems unlikely, as Aglioti et al.'s [59] study of a man born without a corpus callosum (i.e. the main presumed means of decussation) indicated he had fully intact gustation. A second model was suggested by Pritchard et al. [27] based upon a study of several patients with unilateral insula lesions. They proposed that information flow from each side of the tongue to the insula was ipsilateral, with an additional flow of information from the right to the left insula, making the right insula effectively dominant in processing gustatory information. Iannilli et al. [60] presented a third model, in which information flow from each side of the tongue to the insula was again ipsilateral, but this time with right-sided information flow being dominant at all levels of processing. Finally, Onoda et al. [61] suggested that each insula received both ipsilateral and contralateral inputs from each side of the tongue, with Aglioti et al. [62] offering a variant of this model, suggesting that the ipsilateral flow was the more dominant.

Stevenson, Miller and McGrillen [56] tested these models by examining taste processing on each half of the tongue (as described above) in seven patients with left-sided insula damage and in seven patients with right-sided insula damage. The overarching finding was that unilateral insula damage impaired taste naming, quality and intensity judgments irrespective of the side of the tongue that the tastant was applied. This suggests that information flow is likely to have an ipsilateral and contralateral component, as suggested by Onoda et al. [61], but with no obvious dominance of the ipsilateral or contralateral pathway. There was some limited evidence of lateralisation in patients with left-sided insula lesions. They reported salty stimuli as being less intense on their lesioned side, were especially poor at naming salt and had some hedonic abnormalities for sweet and bitter stimuli.

In conclusion, there is some neuroimaging evidence that right-sided processing may be more dominant at the cortical level—notably within the insula—when processing taste stimuli, although this may be affected by task and handedness. In healthy participants, applying tastants unilaterally on the tongue reveals clear lateralised processing benefits on certain tasks, but not on others, although this finding is based upon just one study. Finally, neuropsychological tests of models of information flow to the insula suggest that there are both ipsilateral and contralateral pathways arising from the tongue, with neither being obviously dominant.

14.4 Multimodal Processing in the Insula Involving Gustation

It was originally thought that multimodal processing relating to gustation, namely, the creation of flavour percepts from taste, retronasal smell and somatosensation, took place in the OFC [63, 64]. Very convincing evidence has now accumulated that the insula is involved in multimodal processing related to food flavour, both including anticipatory cues to food (i.e. 'tasting with the eyes') and the construction of multimodal flavour percepts [2, 3]. In this part of the chapter, we consider two issues. First, the evidence suggesting that the insula is a multimodal flavour processor. Second, its involvement in cross-modal anticipatory gustatory responses—odour-induced tastes.

14.4.1 Multimodal Flavour Processing

Electrophysiological recording from single neurons in the insular cortex of monkeys indicates that they have cells that are responsive to all combinations of gustatory, thermal, irritant, viscous and fatty stimuli [65, 66]. That the insula is involved in multimodal processing is also suggested by the finding that lesions/stimulation to this structure disrupts olfactory perception in rats [67, 68]. Human data also points to the multimodal nature of the insula. First, a meta-analysis of unimodal olfactory and unimodal gustatory studies revealed several specific areas within the insula that are activated by both types of stimulation [2]. Second, being asked to focus attention on unimodal tastes activates the anterior insula and frontal operculum, while focussing on unimodal smells activates primary olfactory cortex—as one would expect [69]. In addition, these investigators also found that attending to unimodal tastes or smells also activated a unique (relative to unimodal taste or smell) but common area of the anterior insula, further suggesting multimodal processing in this structure.

A third line of evidence comes neuroimaging studies. Pure olfactory stimuli activate the insula [70], as do a variety of somatosensory stimuli, including astringent chemicals, oral somatosensation [71] and tasteless viscous substances [72]. Indeed, in this last-mentioned study, they also demonstrated that activation in the anterior insula increased in lock step with increases in viscosity of tasteless carboxy-methyl cellulose (a food-grade thickening agent). Moreover, increases in activation were also observed in the anterior insula to increasing viscosity of sucrose concentrations and responses to fat.

A fourth and final piece of evidence comes from neuropsychological studies in humans. Insula damage has been shown to be associated with both impairment to gustatory and olfactory perception [26], and selective insula lesions can produce parosmia (abnormal smell perception) and disrupt flavour [73]. Together, all of these findings indicate that the insula is a point of convergence for all of the types of perceptual information necessary to construct a flavour percept. It also seems to be the first cortical point of convergence for all of these different sensory streams.

Two studies have directly explored the brain regions that are active during flavour perception. An important part of eating and drinking involves becoming familiarised with particular combinations of ingredients—food flavours, that is [1]. Chocolate, chilli and meat might have been a dish fit for Mayan royalty, but the flavour of this concoction would probably clash with the modern palate. Crucially, we learn those flavour combinations that we are routinely exposed to (e.g. strawberry and sugar) and regard with suspicion and neophobia those we have not (e.g. strawberry and salt or meat, chilli and chocolate). Thus, we appear to learn and retain flavour combinations, these forming a template to access when we want to anticipate what something will taste like and for comparison when we are actively consuming a food.

Small et al. [74] presented participants with familiar unimodal tastes and smells as well as congruent (i.e. familiar) and incongruent (i.e. unfamiliar) combinations of these stimuli. The brain area that was selectively active during processing the multimodal congruent flavour (i.e., [congruent flavour, components]—[incongruent flavour, components]) was the insula, frontal operculum and OFC. This suggests that these structures process and retain representations of familiar multimodal flavour percepts. The study also explored which brain areas were activated in learning about a new (and at this point incongruent) flavour composed of familiar things (i.e., [incongruent flavour—congruent flavour]), and this too revealed significant anterior insula activity (but no activity in the OFC). A second study, which only examined a congruent flavour relative to its components, found no insula activity and only revealed OFC-related processing [75]. Finally, a third study by Seo et al. [76] compared the processing of a congruent odour and taste (bacon and salt) with an incongruent odour and taste (strawberry and salt). As with the Small et al. [74] study, this investigation also observed significant activation in the insula, frontal operculum and OFC when processing the congruent flavour relative to the incongruent flavour.

These various studies suggest two general conclusions. First, the insula is a multimodal processor, with inputs from all of the senses—gustation, retronasal olfaction and oral somatosensation—involved in flavour perception. Second, but based upon a limited data set, the insula is involved in learning and generating multimodal flavour percepts.

14.4.2 Cross-Modal Anticipatory Gustatory Responses

It was noticed nearly fifty years ago that taste-based terms such as 'sweet' are commonly used to describe certain odours (e.g. vanilla, strawberry, chocolate; [77]). Since that time several investigators have explored whether the use of taste-based terms actually reflects some gustatory characteristic that has become linked to an olfactory percept (see [78] for review). The upshot of this now large body of work is, as described above, that people encode flavours they have experienced. One consequence of this flavour learning is that when the odorous component of the flavour is smelled orthonasally (e.g. when you smell a cake baking or sniff some food), that odorous component smells of the *flavour* (i.e. it retains its taste characteristics, [79]). This phenomenon has now been demonstrated extensively in humans for sweet, sour, salty and bitter tastes [78] and also in animals [80]. Functionally, it would make sense to be able to smell a food and get an idea of what it is going to taste like.

There is now a small body of work examining the neural basis of odour-induced tastes. In animals, lesions to the insula disrupt flavour acquisition and appear to prevent this phenomenon occurring [81]. The human data suggests a similar conclusion. Stevenson, Miller and Thayer [82] examined a group of patients, including several with focal insula lesions, to determine whether they had gustatory, olfactory and crucially odour-induced taste deficits. Using multiple measures to assess odour-induced tastes, they established that insula lesions, and especially selective ones (i.e. with the lesion restricted to the insula), led to an impaired ability to smell 'tastes' normally associated with these smells (e.g. 'sweet' strawberry or 'sour' lemon). Further studies of insula patients have tended to confirm these earlier findings [83], but not when testing patients with damage to OFC and the antero-medial temporal lobes [84], suggesting an insula-based effect.

14.4.3 Conclusion

The insula is a multimodal hub for food-based processing, combining sensory inputs to form flavour percepts. It seems likely that this is the first cortical structure to receive convergent inputs from all of the senses that go to make up flavour. It also appears likely that the insula encodes flavour percepts and that this information is used when we encounter the olfactory component of a flavour (e.g. the smell of our dinner cooking recovers the flavour encoding of that food). Finally, it would also seem likely that the sight of food can activate insula-based flavour encodings, not necessarily enough to generate a flavour percept but sufficient to trigger a desire to eat.

14.5 General Discussion

There is strong support from the animal and human literature for the claim that the insular cortex is involved in multiple aspects of gustatory perception, although there is still no clear indication of which part of the insula (anterior, mid or posterior) or which of the surrounding opercula may constitute 'primary taste cortex'. There is also well-developed evidence for its role in taste quality coding. There is support for some loose form of topographic mapping of taste quality, albeit with no resolution as yet as to whether taste coding is primarily by labelled line or some form of pattern coding. The insula also plays a role in taste intensity coding along with other structures, most notably the amygdala. For taste hedonics, the insula seems to be especially involved in supporting aversive responses to taste, with other structures too implicated in affective processing, including the OFC. There is some evidence that the right insula may be more involved in gustatory processing than the left. Each insula receives both ipsilateral and contralateral information flow from each side of the tongue, with neither route apparently dominant. The insula is multimodal, receiving inputs from all of the senses, especially those involved in flavour perception. The insula also appears to be the first place of convergence for the information necessary to build flavour percepts and for encoding these as well, although the OFC is also involved in these processes. In this final part of the chapter, we consider three issues: (1) whether the insula supports conscious experience of taste, (2) whether the insula can be considered a primary sensory cortex as it is often labelled, and (3) how gustatory processing fits into the broader picture of insula function.

The insula may be responsible for generating our conscious experience of taste. As discussed earlier, insula lesions disrupt all aspects of taste experience. However, it is unusual in humans to come across a person with bilateral lesions of the insula. The closest to this reported in the literature is a case described by Adolphs et al. [85] who in addition also had extensive bilateral damage to his medial temporal lobe, including the amygdala and orbitofrontal cortex. This patient had no ability to discriminate or recognise tastants and responded favourably (i.e. showing enjoyment) to all taste stimuli. However, the patient exhibited a phenomenon that seems to parallel blindsight (i.e. where a person has no conscious experience of seeing but can still make decisions about visual stimuli that are above chance). In this case when the patient was asked to choose between two drinks, they consistently selected the sucrose-flavoured one over the sour one, although not being able to apparently discriminate between them. This pattern of outcome suggests two things. First, affective taste decisions may be made based upon gustatory processing that occurs *prior* to conscious cortical processing, and second, an absent insula is associated with a loss of conscious taste experience. Of course, it is possible that the damaged OFC or amygdala might also fill this role (i.e. of supporting conscious awareness of taste), but a recent electroencephalogram study suggests otherwise. In this study, the temporal pattern of electrical activity in the insula was the most reliable correlate of qualitative taste experience [8]. This suggests that the insula may indeed be the seat of gustatory awareness.

There has been some discussion over whether the insula can be considered a primary sensory cortex [86]. We suggest this may be a problematic label for two reasons. First, as we noted in "Introduction", the insula is composed of three regions: an agranular anterior zone, a dysgranular mid-region and a granular posterior zone [12]. There is no clear indication that any of these areas are disproportionately devoted to gustatory processing, yet other sensory cortex is granular in nature [15]. Second, as we have frequently pointed out in this chapter, (1) the insula is multimodal; (2) it has a low proportion of taste-sensitive neurons; and (3) gustatory information is never experienced in isolation, being accompanied by somatosensory stimulation and almost always by retronasal olfaction [79]. Primary sensory cortex is characterised by its unimodal nature. We have argued elsewhere that taste experiences are not 'objects' in the way that many visual, olfactory and auditory experiences are [87]. We suggest here that taste may not be a discrete sense but rather a modular part of a broader sensory-affective system designed to support eating and drinking.

This brings us finally to consider the broader issue of insula function and how gustation fits within this. There are several theories of insula function, which we briefly summarise here: (1) it is a key neural hub for interoception and self-monitoring [88, 89], (2) it integrates internal and external information to direct action and attention [90], (3) it integrates multiple sources of information to create bodily feeling states [14], and (4) it coordinates internal and external information to generate conscious emotional states [13]. From these various models, certain common themes emerge. For appetitive behaviour, of which gustation is just one part, these various themes can be readily brought together. The brain needs to collate information about internal energy state and, utilising multimodal cues, external availability of nutrients. This then allows it to generate a sensory-affective bodily state on which to base a decision—to eat or not to eat? We suggest that the insula generates this sensory-affective bodily state.

References

1. Stevenson RJ. The psychology of flavour. Oxford: Oxford University Press; 2009.
2. Verhagen JV, Engelen L. The neurocognitive bases of human multimodal food perception: sensory integration. Neurosci Biobehav Rev. 2006;30(5):613–50.
3. Maffei A, Haley M, Fontanini A. Neural processing of gustatory information in insular circuits. Curr Opin Neurobiol. 2012;22(4):709–16.
4. Mcburney DH, Gent JF. On the nature of taste qualities. Psychol Bull. 1979;86(1):151–67.
5. Simon SA, de Araujo IE, Gutierrez R, Nicolelis MA. The neural mechanisms of gustation: a distributed processing code. Nat Rev Neurosci. 2006;7(11):890–901.
6. Chandrashekar J, Hoon MA, Ryba NJ, Zuker CS. The receptors and cells for mammalian taste. Nature. 2006;444(7117):288–94.
7. Calvo SS, Egan JM. The endocrinology of taste receptors. Nat Rev Endocrinol. 2015;11(4):213–27.
8. Crouzet SM, Busch NA, Ohla K. Taste quality decoding parallels taste sensations. Curr Biol. 2015;25(7):890–6.
9. Erickson RP. A study of the science of taste: on the origins and influence of the core ideas. Behav Brain Sci. 2008;31(1):59–105.
10. Di Lorenzo PM. The neural code for taste in the brain stem: response profiles. Physiol Behav. 2000;69(1–2):87–96.
11. Small DM. Taste representation in the human insula. Brain Struct Funct. 2010;214(5–6):551–61.
12. Flynn FG, Benson G, Ardila K. Anatomy of the insula functional and clinical correlates. Aphasiology. 1999;13(1):55–78.
13. Ibañez A, Gleichgerrcht E, Manes F. Clinical effects of insular damage in humans. Brain Struct Funct. 2010;214(5–6):397–410.
14. Jones CL, Ward J, Critchley HD. The neuropsychological impact of insular cortex lesions. J Neurol Neurosurg Psychiatry. 2010;81(6):611–8.
15. Shepherd GM. The synaptic organization of the brain. Cambridge, UK: ñ; 1976.
16. De Araujo IE, Geha P, Small DM. Orosensory and homeostatic functions of the insular taste cortex. Chemosens Percept. 2012;5(1):64–79.
17. Scott TR, Plata-Salaman CR. Taste in the monkey cortex. Physiol Behav. 1999;67(4):489–511.
18. Börnstein WS. Cortical representation of taste in man and monkey: II. The localization of the cortical taste area in man and a method of measuring impairment of taste in man. Yale J Biol Med. 1940;13(1):133–56.
19. Mesulam MM, Mufson EJ. The insula of Reil in man and monkey. In: Peters A, Jones EG, editors. Cerebral Cortex. New York, NY: Plenum; 1985. p. 179–226.
20. Pritchard TC, Hamilton RB, Morse JR, Norgren R. Projections of thalamic gustatory and lingual areas in the monkey, Macaca fascicularis. J Comp Neurol. 1986;244(2):213–28.
21. Penfield W, Faulk MR. The insula; further observations on its functions. Brain. 1955;78(4):445–70.
22. Stephani C, Fernandez-Baca Vaca G, Maciunas R, Koubeissi M, Lüders HO. Functional neuroanatomy of the insular lobe. Brain Struct Funct. 2011;216(2):137–49.
23. Kim JS, Choi-Kwon S, Kwon SU, Kwon JH. Taste perception abnormalities after acute stroke in postmenopausal women. J Clin Neurosci. 2009;16(6):797–801.
24. Kim JS, Choi S. Altered food preference after cortical infarction: Korean style. Cerebrovasc Dis. 2002;13(3):187–91.
25. Kocaeli H, Korfalı E, Doğan Ş, Savran M. Sylvian cistern dermoid cyst presenting with dysgeusia. Acta Neurochir. 2009;151(5):561–3.
26. Mak YE, Simmons KB, Gitelman DR, Small DM. Taste and olfactory intensity perception changes following left insular stroke. Behav Neurosci. 2005;119(6):1693–700.
27. Pritchard TC, Macaluso DA, Eslinger PJ. Taste perception in patients with insular cortex lesions. Behav Neurosci. 1999;113(4):663–71.
28. Hummel C, Frasnelli J, Gerber J, Hummel T. Cerebral processing of gustatory stimuli in patients with taste loss. Behav Brain Res. 2007;185(1):59–64.
29. Small DM, Jones-Gotman M, Zatorre RJ, Petrides M, Evans AC. A role for the right anterior temporal lobe in taste quality recognition. J Neurosci. 1997;17(13):5136–42.
30. Frey S, Petrides M. Re-examination of the human taste region: a positron emission tomography study. Eur J Neurosci. 1999;11(8):2985–8.
31. Kinomura S, Kawashima R, Yamada K, Ono S, Itoh M, Yoshioka S, et al. Functional anatomy of taste perception in the human brain studied with positron emission tomography. Brain Res. 1994;659(1–2):263–6.
32. Small DM, Zald DH, Jones-Gotman M, Zatorre RJ, Pardo JV, Frey S, et al. Human cortical gustatory areas: a review of functional neuroimaging data. Neuroreport. 1999;10(1):7–14.
33. Veldhuizen MG, Albrecht J, Zelano C, Boesveldt S, Breslin P, Lundström JN. Identification of human gustatory cortex by activation likelihood estimation. Hum Brain Mapp. 2011;32(12):2256–66.
34. Kurth F, Zilles K, Fox PT, Laird AR, Eickhoff SB. A link between the systems: functional differentiation and integration within the human insula revealed by meta-analysis. Brain Struct Funct. 2010;214(5–6):519–34.
35. Chen X, Gabitto M, Peng Y, Ryba NJ, Zuker CS. A gustotopic map of taste qualities in the mammalian brain. Science. 2011;333(6047):1262–6.
36. Peng Y, Gillis-Smith S, Jin H, Tränkner D, Ryba NJ, Zuker CS. Sweet and bitter taste in the brain of awake behaving animals. Nature. 2015;527(7579):512–5.
37. Accolla R, Bathellier B, Petersen CC, Carleton A. Differential spatial representation of taste modalities in the rat gustatory cortex. J Neurosci. 2007;27(6):1396–404.

38. Jones LM, Fontanini A, Katz DB. Gustatory processing: a dynamic systems approach. Curr Opin Neurobiol. 2006;16(4):420–8.
39. Katz DB. The many flavors of temporal coding in gustatory cortex. Chem Senses. 2005;30(Suppl 1):i80–1.
40. Cereda C, Ghika J, Maeder P, Bogousslavsky J. Strokes restricted to the insular cortex. Neurology. 2002;59(12):1950–5.
41. Schoenfeld M, Neuer G, Tempelmann C, Schüßler K, Noesselt T, Hopf JM, et al. Functional magnetic resonance tomography correlates of taste perception in the human primary taste cortex. Neuroscience. 2004;127(2):347–53.
42. De Araujo IE, Krinkelbach ML, Rolls ET, Hobden P. Representation of umami taste in the human brain. J Neurophysiol. 2003;90(1):313–9.
43. Spetter MS, Smeets PA, de Graaf C, Viergever MA. Representation of sweet and salty taste intensity in the brain. Chem Senses. 2010;s35(9):831–40.
44. O'Doherty JP, Rolls ET, Francis S, Bowtell R, McGlone F. Representation of pleasant and aversive taste in the human brain. J Neurophysiol. 2001;85(3):1315–21.
45. Small DM, Gregory MD, Mak YE, Gitelman D, Mesulam MM, Parrish T. Dissociation of neural representation of intensity and affective valuation in human gustation. Neuron. 2003;39(4):701–11.
46. Small DM, Zatorre RJ, Jones-Gotman M. Changes in taste intensity perception following anterior temporal lobe removal in humans. Chem Senses. 2001;26(4):425–32.
47. Rolls ET. The cortical representation of taste and smell. In: Rouby C, Schaal B, Dubois D, Gervais R, Holley A, editors. Olfaction, taste, and cognition. Cambridge, UK: Cambridge University Press; 2002.
48. Zald DH, Hagen MC, Pardo JV. Neural correlates of tasting concentrated quinine and sugar solutions. J Neurophysiol. 2002;87(2):1068–75.
49. Bender G, Veldhuizen MG, Meltzer JA, Gitelman DR, Small DM. Neural correlates of evaluative compared with passive tasting. Eur J Neurosci. 2009 Jul;30(2):327–38.
50. Finsterer J, Stöllberger C, Kopsa W. Weight reduction due to stroke-induced dysgeusia. Eur Neurol. 2004;51(1):47–9.
51. Zald DH, Lee JT, Fluegel KW, Pardo JV. Aversive gustatory stimulation activates limbic circuits in humans. Brain. 1998;121(Pt 6):1143–54.
52. Nitschke JB, Dixon GE, Sarinopoulos I, Short SJ, Cohen JD, Smith EE, et al. Altering expectancy dampens neural response to aversive taste in primary taste cortex. Nat Neurosci. 2006;9(3):435–42.
53. Cabanac M. The physiological role of pleasure. Science. 1971;173:1103–7.
54. Avery JA, Kerr KL, Ingeholm JE, Burrows K, Bodurka J, Simmons WK. A common gustatory and interoceptive representation in the human mid-insula. Hum Brain Mapp. 2015;36(8):2996–3006.
55. Haase L, Cerf-Ducastel B, Murphy C. Cortical activation in response to pure taste stimuli during the physiological states of hunger and satiety. Neuroimage. 2009;44(3):1008–21.
56. Stevenson RJ, Miller LA, McGrillen K. The lateralization of gustatory function and the flow of information from tongue to cortex. Neuropsychologia. 2013;51(8):1408–16.
57. Faurion A, Cerf B, Van De Moortele PF, Lobel E, Mac Leod P, Le Bihan D. Human taste cortical areas studied with functional magnetic resonance imaging: evidence of functional lateralization related to handedness. Neurosci Lett. 1999;277(3):189–92.
58. Lee BC, Hwang SH, Rison R, Chang GY. Central pathway of taste: clinical and MRI study. Eur Neurol. 1998;39(4):200–3.
59. Aglioti S, Tassinari G, Corballis MC, Berlucchi G. Incomplete gustatory lateralization as shown by analysis of taste discrimination after callosotomy. J Cogn Neurosci. 2000;12(2):238–45.
60. Iannilli E, Singh PB, Schuster B, Gerber J, Hummel T. Taste laterality studied by means of umami and salt stimuli: an fMRI study. Neuroimage. 2012;60(1):426–35.
61. Onoda K, Ikeda M, Sekine H, Ogawa H. Clinical study of central taste disorders and discussion of the central gustatory pathway. J Neurol. 2012;259(2):261–6.
62. Aglioti SM, Tassinari G, Fabri M, Del Pesce M, Quattrini A, Manzoni T, et al. Taste laterality in the split brain. Eur J Neurosci. 2001;13(1):195–200.
63. Price JL. Multisensory convergence in the orbital and ventrolateral prefrontal cortex. Chemosens Percept. 2008;1(2):103–9.
64. Rolls ET, Baylis LL. Gustatory, olfactory, and visual convergence within the primate orbitofrontal cortex. J Neurosci. 1994;14(9):5437–52.
65. Hanamori T, Kunitake T, Kato K, Kannan H. Responses of neurons in the insular cortex to gustatory, visceral, and nociceptive stimuli in rats. J Neurophysiol. 1998;79(5):2535–45.
66. Kadohisa M, Rolls ET, Verhagen JV. Neuronal representations of stimuli in the mouth: the primate insular taste cortex orbitofrontal cortex and amygdala. Chem Senses. 2005;30(5):401–19.
67. Fortis-Santiago Y, Rodwin BA, Neseliler S, Piette CE, Katz DB. State dependence of olfactory perception as a function of taste cortical inactivation. Nat Neurosci. 2010;13(2):158–9.
68. Maier JX, Blankenship ML, Li JX, Katz DB. A multisensory network for olfactory processing. Curr Biol. 2015;25(20):2642–50.
69. Veldhuizen MG, Small DM. Modality-specific neural effects of selective attention to taste and odor. Chem Senses. 2011;36(8):747–60.
70. Plailly J, Radnovich AJ, Sabri M, Royet JP, Kareken DA. Involvement of the left anterior insula and frontopolar gyrus in odor discrimination. Hum Brain Mapp. 2007;28(5):363–72.
71. Cerf-Ducastel B, Van de Moortele PF, MacLeod P, Le Bihan D, Faurion A. Interaction of gustatory and lingual somatosensory perceptions at the cortical level in

the human: a functional magnetic resonance imaging study. Chem Senses. 2001;26(4):371–83.
72. De Araujo IE, Rolls ET. Representation in the human brain of food texture and oral fat. J Neurosci. 2004;24(12):3086–93.
73. Dutta TM, Josiah AF, Cronin CA, Wittenberg GF, Cole JW. Altered taste and stroke: a case report and literature review. Top Stroke Rehabil. 2013;20(1):78–86.
74. Small DM, Voss J, Mak YE, Simmons KB, Parrish T, Gitelman D. Experience-dependent neural integration of taste and smell in the human brain. J Neurophysiol. 2004;92(3):1892–903.
75. De Araujo IE, Rolls ET, Kringelbach ML, McGlone F, Phillips N. Taste-olfactory convergence, and the representation of the pleasantness of flavour, in the human brain. Eur J Neurosci. 2003;18(7):2059–68.
76. Seo HS, Iannilli E, Hummel C, Okazaki Y, Buschhüter D, Gerber J, et al. A salty-congruent odor enhances saltiness: functional magnetic resonance imaging study. Hum Brain Mapp. 2013;34(1):62–76.
77. Harper R, Bate-Smith EC, Lad DG. Odour descriptions and odour classification. London: Churchill; 1968.
78. Stevenson RJ. Multisensory interactions in flavor perception. In: Calvert G, Spence C, Stein B, editors. The new handbook of multisensory processes. Cambridge, MA: MIT Press; 2012. p. 283–300.
79. Stevenson RJ. Flavor binding: its nature and cause. Psychol Bull. 2014;140(2):487–510.
80. Gautam SH, Verhagen JV. Evidence that the sweetness of odors depends on experience in rats. Chem Senses. 2010;35(9):767–76.
81. Sakai N, Imada S. Bilateral lesions of the insular cortex or of the prefrontal cortex block the association between taste and odor in the rat. Neurobiol Learn Mem. 2003;80(1):24–31.
82. Stevenson RJ, Miller LA, Thayer ZC. Impairments in the perception of odor-induced tastes and their relationship to impairments in taste perception. J Exp Psychol Hum Percept Perform. 2008; 34(5):1183–97.
83. Stevenson RJ, Miller LA, Mcgrillen K. Perception of odor-induced tastes following insular cortex lesion. Neurocase. 2015;21(1):33–43.
84. Stevenson RJ, Miller LA. Taste and odour-induced taste perception following unilateral lesions to the anteromedial temporal lobe and the orbitofrontal cortex. Cogn Neuropsychol. 2013;30(1):41–57.
85. Adolphs R, Tranel D, Koenigs M, Damasio AR. Preferring one taste over another without recognizing either. Nat Neurosci. 2005;8(7):860–1.
86. Vincis R, Fontanini A. A gustocentric perspective to understanding primary sensory cortices. Curr Opin Neurobiol. 2016;40:118–24.
87. Stevenson RJ. Object concepts in the chemical senses. Cognit Sci. 2014;38(7):1360–83.
88. Craig AD. How do you feel? Interoception: the sense of the physiological condition of the body. Nat RevNeurosci. 2002;3:655–66.
89. Spinazzola L, Pia L, Folegatti A, Marchetti C, Berti A. Modular structure of awareness for sensorimotor disorders: evidence from anosognosia for hemiplegia and anosognosia for hemianaesthesia. Neuropsychologia. 2008;46(3):915–26.
90. Menon V, Uddin LQ. Saliency, switching, attention and control: a network model of insula function. Brain Struct Funct. 2010;214(5–6):655–67.

Role of the Insula in Human Cognition and Motivation

Oreste de Divitiis, Teresa Somma,
D'Urso Giordano, Mehmet Turgut,
and Paolo Cappabianca

15.1 Introduction

The insula, known as the fifth lobe of the brain and hidden by the operculum in the Sylvian fissure, is the most commonly activated region in functional MRI with mapping and tractography. According to the results of cytoarchitectonic and functional connectivity studies and meta-analyses, the insula has a mechanistic instrumental role in the integration of the complex brain networks regarding cognition, emotion, sensory, and motor systems [1, 2]. Its posterior part receives and interprets the sensory-motor information, while its anterior part receives and interprets the emotional and cognitive information [3]. In particular, it has been reported that the dorsal insula is activated in tasks related with goal-directed cognition [4]. The insula, also called the limbic or paralimbic cortex, is divided into anterior insular cortex and middle and posterior the regions [5]. Recently, Chang et al. [6] suggested a new parcellation for insula as ventro-anterior insula (a zone involved in the processing of olfaction and gustation), posterior insula (a zone for sensory, exteroceptive, interoceptive, auditory, and vestibular information), and dorso-anterior insula.

The most studied function of the insula is interoception, i.e., the perception and integration of visceral, hormonal, autonomic, and immunological ascending information that describes the state of the body [7]. These sensorial inputs reach the posterior section of the insula, where they are first processed at a lower level and are redirected to the anterior insular section where many connections with other cortical regions take place [8].

In recent years, cognitive and clinical neuroscience has given more and more value to the function of the insula and now considers this region central in human behavior [9]. In fact, the conscious perception of the body states are elicited by interoceptive stimuli (feeling states), which is now considered the basis of our perception of ourselves as individuals. In this view, the self is first and foremost a "physical self"; i.e., without the subjective feeling states deriving from the insular function, we could not be aware of ourselves [10]. As a matter of fact, the anterior insula is one of the most differentially expanded regions when we compare the neocortex of

O. de Divitiis, M.D. (✉) · T. Somma
P. Cappabianca
Division of Neurosurgery, Department of Neurosciences, Reproductive and Odontostomatological Sciences, Università degli Studi di Napoli Federico II, Naples, Italy
e-mail: oreste.dedivitiis@unina.it

D. Giordano
Division of Psychiatry, Department of Clinical Neurosciences, Anesthesiology and Pharmacoutilization, Università degliStudi di Napoli Federico II, Naples, Italy

M. Turgut
Department of Neurosurgery, Adnan Menderes University School of Medicine, Aydın, Turkey

humans with that of nonhuman primates. This differential anatomy could be the neural basis of the well-known notion that humans are the only self-conscious living organisms [11].

Moreover, owing to the direct connections with the dorsolateral prefrontal cortex, the insula strongly influences cognition and guides the allocation of such cognitive resources as attention, working memory, and conscious reasoning.

In fact, the subjective feeling states arising from the insula determine the salience of a stimulus and are responsible for the prioritization of a stimulus over the competing ones, based on the intensity of the associated feeling. In other words, the brain tends to give more attention, memory, and reasoning to a particular stimulus depending on the interoceptive marker provided by the insula [12].

Furthermore, the insula is considered a crucial region for the explicit motivation, i.e., the conscious willingness to undertake a specific behavior. In fact, the insular cortex evaluates the interoceptive pleasantness of a stimulus and, as a consequence, encodes its incentive value. Hence, a pleasant stimulus induces an action aimed at the persistence of that feeling, while unpleasant interoceptive information leads to avoidant behaviors. This influence on motivation is mediated by the strong connections between the anterior insula and the ventromedial prefrontal cortex, which is the brain area in charge of planning goal-directed behaviors [13].

In addition, many studies on psychiatric and neurological patients have found structural and/or functional abnormalities of the insula, confirming the involvement of this region in human emotion, cognition, and behavior [9].

15.2 Functional Specificity

As a rule, different insula regions have functional specificity [6]. The dorsal part of the anterior insula has a cognitive role, while the ventral part of the anterior insula has an emotional role [9, 14]. The anterior part of the insula adjacent to the frontal operculum has bipolar spindle cells named as "von Economo neurons." They produce peptides which are responsible for pain and immunity [15]. The activation of the dorsal part of the posterior insula produces pain, temperature, and autonomic changes [8]. The activation of the middle insula produces subjective experience, while the activation of the anterior insula produces emotional evaluation [9]. The right insula coordinates autonomic functions such as heart rate, while the left insula is related with emotional experiences [9, 16].

Insula has an integrative role at the interface of cognitive and affective domains: frontotemporal dementia (the number of "von Economo neurons" is reduced), bipolar disorder, major depressive disorder (increased activation of the left insula and decreased activation of the right insula), attention deficit hyperactivity disorder, and autism (hypoactivity in the right anterior insula, abnormal development of von Economo neurons) [2].

Based on findings from animal and human studies, the insula has a role in the detection of salience (cognition and empathy): the insula has a role in the networks related with cognitive processes [2]. The insula has a hub role in integration of internal states regarding cognition, emotion, sensory, and motor functions and external (environmental) states [2].

15.3 Connectivities of the Insula

Insula has efferent and afferent projections to the cingulate cortex, orbitofrontal cortex, olfactory cortex, and the amygdala [17]. The connectivities of the insula are (1) the posterior insula and the sensory-motor cortex, (2) the anterior insula and the amygdala and the thalamus, and (3) the middle insula and the thalamus [2]. The insula integrates the feeling of pain, temperature, touch, hunger, and thirst [9].

15.4 Functional MRI and PET Studies

Functional MRI studies revealed that the subregions of the insula (anterior, middle, and posterior) have different functions in pain perception

[2]. The anterior part of the insula is related with cognitive and affective aspects of the pain, while middle and posterior subregions are related with sensory and discriminative aspects of the pain [18].

The insula may be used as a biomarker for the selection of any treatment or pharmacotherapy option in major depressive disorders [19]. In 2013, using a PET study, McGrath et al. [19] reported that insular hypermetabolism is an imaging marker of pharmacological treatment, while insular hypometabolism is an imaging marker of cognitive behavioral therapy. In another study, increased insular activity was found in patients with schizophrenia [20].

Conclusion

In conclusion, beyond being the terminal of interoceptive information, the insula should be conceived as a "hub" where the subjective feelings of humans serve to regulate cognition and motivation.

References

1. Kurth F, Zilles K, Fox PT, Laird AR, Eickhoff SB. A link between the systems: functional differentiation and integration within the human insula revealed by meta-analysis. Brain Struct Funct. 2010;214(5–6):519–34.
2. Pavuluri M, May A. I feel, therefore, I am: the insula and its role in human emotion, cognition and the sensory-motor system. AIMS Neurosci. 2015;2:18–27.
3. Eckert MA, Menon V, Walczak A, Ahlstrom J, Denslow S, Horwitz A, Dubno JR. At the heart of the ventral attention system: the right anterior insula. Hum Brain Mapp. 2009;30:2530–41.
4. Yarkoni T, Barch DM, Gray JR, Conturo TE, Braver TS. BOLD correlates of trial-by-trial reaction time variability in gray and white matter: a multi-study fMRI analysis. PLoS One. 2009;4:e4257.
5. Augustine JR. Circuitry and functional aspects of the insular lobe in primates including humans. Brain Res Brain Res Rev. 1996;22:229–44.
6. Chang LJ, Yarkoni T, Khaw MW, Sanfey AG. Decoding the role of the insula in human cognition: functional parcellation and large-scale reverse inference. Cereb Cortex. 2013;23:739–49.
7. Aiba T, Tanaka R, Koike T, Kameyama S, Takeda N, Komata T. Natural history of intracranial cavernous malformations. J Neurosurg. 1995;83:56–9.
8. Craig AD. How do you feel? Interoception: the sense of the physiological condition of the body. Nat Rev Neurosci. 2002;3:655–66.
9. Craig AD. How do you feel-now? The anterior insula and human awareness. Nat Rev Neurosci. 2009;10:59–70.
10. Namkung H, Kim SH, Sawa A. The insula: an underestimated brain area in clinical neuroscience, psychiatry, and neurology. Trends Neurosci. 2017;40:200–7.
11. Damasio A. Feelings of emotion and the self. Ann N Y Acad Sci. 2003;1001:253–61.
12. Bauernfeind AL, de Sousa AA, Avasthi T, Dobson SD, Raghanti MA, Lewandowski AH, Zilles K, Semendeferi K, Allman JM, Craig AD, Hof PR, Sherwood CC. A volumetric comparison of the insular cortex and its subregions in primates. J Hum Evol. 2013;64:263–79.
13. Uddin LQ. Salience processing and insular cortical function and dysfunction. Nat Rev Neurosci. 2015;16:55–61.
14. Stephani C, Fernandez-Baca Vaca G, Maciunas R, Koubeissi M, Lüders HO. Functional neuroanatomy of the insular lobe. Brain Struct Funct. 2011;216:137–49.
15. Cauda F, Torta DM, Sacco K, D'Agata F, Geda E, Duca S, Geminiani G, Vercelli A. Functional anatomy of cortical areas characterized by Von Economo neurons. Brain Struct Funct. 2013;218:1–20.
16. Santos M, Uppal N, Butti C, Wicinski B, Schmeidler J, Giannakopoulos P, Heinsen H, Schmitz C, Hof PR. Von Economo neurons in autism: a stereologic study of the frontoinsular cortex in children. Brain Res. 2011;1380:206–17.
17. Uddin LQ, Menon V. The anterior insula in autism: under-connected and under-examined. Neurosci Biobehav Rev. 2009;33:1198–203.
18. Wiech K, Jbabdi S, Lin CS, Andersson J, Tracey I. Differential structural and resting state connectivity between insular subdivisions and other pain-related brain regions. Pain. 2014;155:2047–55.
19. McGrath CL, Kelley ME, Holtzheimer PE, Dunlop BW, Craighead WE, Franco AR, Craddock RC, Mayberg HS. Toward a neuroimaging treatment selection biomarker for major depressive disorder. JAMA Psychiat. 2013;70:821–9.
20. Rolland B, Amad A, Poulet E, Bordet R, Vignaud A, Bation R, Delmaire C, Thomas P, Cottencin O, Jardri R. Resting-state functional connectivity of the nucleus accumbens in auditory and visual hallucinations in schizophrenia. Schizophr Bull. 2015;41:291–9.

Role of the Insula in Visual and Auditory Perception

Matthew Protas

16.1 Multimodal Function of the Insula

Multimodal responsiveness to changes in how we perceive stimulus auditory and visual stimulus are reliant on a right-lateralized network of cortices including the insula, temporoparietal junction, inferior frontal gyrus, left cingulate, and supplementary motor areas. It is thought that injury to this network is important in the underlying pathology of hemineglect syndromes [1]. In a study of six volunteers who passively viewed pairs of blue and red stereoscopic drifting grating images, it was determined by fMRI that when humans were exposed to bistable viewing conditions, subjective visual perception was related to multiple cortices including the insula [1]. Lumer and Rees [1] also determined that the coordination of the activity was not linked to the sensory or motor events but instead internal changes in perception that varied in strength with the frequency of perceptual events.

Multimodal areas of the brain such as the insula have also been implied in the binding of audio and visual stimuli even when they are not related [2]. The right claustrum/insula region is differentially activated through multisensory integration of conceptually related common objects. Grapheme-color synesthesia (GCS), a phenomenon of multisensory integration, has been associated with an increased activity of the insula as well as surrounding areas such as the precentral motor cortex, supplementary motor cortex, and intraparietal cortex. Compared to the controls ($n = 20$), patients who experienced GCS ($n = 20$) were shown to have increased responsiveness of the left hemisphere to phenomenological localization as well as automaticity/attention, while the right hemisphere was only activated for localization [3].

16.1.1 Speech Perception

Of all the multimodal sensory integrative functions of the insula, one of the most imperative functions is the role it plays in the ability to perceive speech. In a study by Kim et al. [4], linkage dissonance was used to demonstrate that auditory speech perception, by congruent audiovisual stimulation in 15 patients, had tighter coupling of the left anterior temporal gyrus-anterior insula and right premotor-visual than when using auditory or visual speech cues, respectively. They also showed that in white noise conditions, visual speech is perceived through tight negative coupling in areas such as the left inferior frontal region, right anterior cingulate, and left anterior insula. These findings help support that the insula is useful in the efficient and effortful processes of natural audiovisual integration or lip reading in speech perception [4].

M. Protas
Department of Anatomical Sciences, Saint George's University School of Medicine, True Blue, Grenada
e-mail: mprotas@sgu.edu

16.1.2 Multimodal Illusion Perception

Audiovisual illusions have been implied to be derived from the physiological limitations of the insula. Sekuler et al. [5] demonstrated through fMRI that an increase in insular cortex activity plays a role in interpreting whether two objects traveling toward each other collided based on if a sound is made at the same time of collision. This further supports the role of the insula in making decisions based on stimulus to shape how we perceive the different audio and visual stimulus. Perception of both auditory and visual stimulus can fuse into a merged perception best known as the McGurk effect. This phenomenon has been attributed to the insula as well as the superior temporal sulcus, intraparietal sulcus, and precentral cortex [6].

16.2 Auditory Perception

16.2.1 Perception of Sound Localization

As far back as 1959, it has been demonstrated and known that stimulation of the right posterior superior temporal gyrus and insula has elicited auditory motion perception [7]. The perception of auditory motion is an imperative survival function which involves many cortices and neural networks. The right insula and associated neuronal networks have been more associated with the perception of auditory movement than the left hemisphere [7]. These results are supported by a study conducted by Griffiths et al. [8] who determined that when patients received moving auditory stimulus compared to stationary stimulus, there was a differentially increased blood flow to the area as measured by PET ($p < 0.001$). These findings support the role that the insula acts as an auditory association area analogous to V5 of the visual cortex which is directly responsible for auditory motion perception. Ducommun et al. [9] study supported these findings by stating that there is a separation of the where and motion pathways of auditory perception. Contrary to Griffiths et al. [8], Ducommun et al. (2004) described a patient with an intact insula who has auditory motion deftness. They hypothesized from these findings that the auditory association area is in the superior temporal gyrus not the insula. Due to the complex nature of the insula and its intricate network of neuronal pathways, it is impossible to say who is right due to the probable role they both play in the perception of auditory motion.

16.2.2 Auditory Pattern Recognition

Perception of auditory pattern is an important process that is dependent on the insular cortex. Lesions of the insula and temporal cortex caused retrograde degeneration of the medial geniculate body demonstrating distinct connectivity between areas associated with secondary auditory processing [10]. Ablation of the insular temporal region in cats produced a deficit in perceiving tones as being part of a pattern as well as the inability to relearn them [11]. A similar finding is also seen in perceiving whether an auditory frequency is discriminative from another. Kelly and Whitefield [12] trained cats to recognize different ranges of auditory frequencies. After ablation of the insula as well as areas in the temporal and auditory cortex, the cats were unable to recognize and distinguish them. After significant retraining and through much difficulty, they were however able to relearn it unlike the patterns. Experiments on humans have not been able to demonstrate this yet. Colavita [13] further demonstrated that auditory pattern recognition is impaired with insular lesion. When different amplitudes of a set frequency were presented, such as "loud, soft, loud" vs. "soft, loud, soft," before and after bilateral lesions of the insula temporal cortices, relearning was not possible. The study also described that relearning was possible when presented "loud, soft" vs. "soft, loud," but even after learning this, they failed to relearn "loud, soft, loud" compared to the original experiment.

16.2.3 Comprehension and Perception of Speech

The insula is thought to play a major role in the interaction of auditory search and comprehension. When perceiving normal stimulus, the left insula's activation peaks initially then decrease for intelligible and non-intelligible noises. Activation of the insula however remains active for broadband speech envelope noises which have a complex temporal envelop and could be understood after listening to it [14]. This role can be demonstrated functionally in its input from the auditory cortex and output to the frontal lobe [15]. Functional significance can also be derived from the fact that as the complexity of the perceived acoustic stimulus increased, so would activity of the anterior insular cortex to help perceive and determine meaning [14]. These studies help to show that the anterior insular cortex helps to play an imperative role in perceiving auditory stimulus derived from its' functional decision-making capacity which is proportional to the amount of "cognitive effort" required to derive meaning [16]. Patients who are blind, compared to the normal sited, perceive human produced sounds differently [17]. It was demonstrated through fMRI that the blind individuals had to use different mechanisms (left anterior insula, bilateral anterior calcarine, and medial occipital regions) to retrieve and perceive memories of sound compared to sighted who relied more heavily on bilateral parietal as well as medial and frontal networks [17].

16.2.4 Perception of Music

The left insula is particularly imperative for perceiving whether audio stimulus is pleasing or not. In a patient with a lesion of the insula due to a right hemisphere infraction of the middle and posterior cerebral artery, there was a lack of musical appreciation. This was determined to be due to a dissociative receptive musical deficit in the presence of normal appreciation of sound and speech [18]. This patient also suffered from a perception of sound-source movement; however, he did not lack the ability to detect sound frequency in the form of sinusoidal frequency modulation [18]. This study helped to demonstrate that insular infarctions can lead to amusia separate from auditory agnosia. fMRI imaging has also demonstrated that the insula is imperative in the emotional perception to auditory stimulus. Nguyen et al. [19] described that the anterior insula was "specifically tuned" to salient moments of an audio stream which serves as a relay for interoceptive processing in the posterior insula. This ability is not unique to audio processing as it serves for multiple stimulus experiences.

16.2.5 Pathological Auditory Perception

Tinnitus distress has been associated with an increased activation of neural networks such as the insula, parahippocampal area, and the amygdala. This pathway is described to be involved in the perception of auditory pain. Increasing perceived tinnitus intensity and distress has been shown to increase the activation of this network [20, 21].

16.3 Visual Perception

Visual motion is typically associated with the V5–V6 areas of the brain; however, a study by Banca et al. [22] described that the anterior insula and other associated cortices are imperative to perceiving visual motion imagery neurofeedback. The role of the insula is thought to be initiating the process of visual motion imagery neurofeedback as insular activity was present significantly in failed and successful attempts [22].

The frontoparietal network consisting of the anterior insula cortex, dorsal anterior cingulate cortex, intraparietal sulcus, and lateral prefrontal cortex has been implied to have a functional role in visual perception. Activation of the insular cortex is imperative for determining whether an object is perceived as falling or moving horizontally [23]. Visual-motor perception deficits have also been seen in patients with low preterm birth weight. These patients have been shown to have decreased cortical thickness of the lateral occipito-temporal-parietal junction, temporal gyrus, insula, and superior parietal regions which correlated with a significantly decreased visual-motor integration score [24]. Visual processing of words that are flashed "unmasked visible words" compared to those masked by presenting them in close spatial and temporal proximity with each other is reliant on the frontoparietal network's activation. Activation of the insula in the processing of visible words was bilateral compared to specific activation of the left parietal and fusiform gyrus for masked words [25]. At the threshold between visible and masking, perception as measured by an fMRI was lateralized to the right anterior insula cortex, middle frontal gyrus, and intraparietal sulcus [26]. Reasoning for this difference in processing between the interval of presentation in the study conducted by Dehaene et al., (2010) and Carmel et al. [26] is unknown. From these findings, they could conclude that the anterior insular cortex plays a role in the perception of temporal visual stimulus [16, 25, 26].

Activation of the anterior insular cortex has also been implicated in perceiving speech. Insular activation was not observed when the words were audibly spoken as well as when random lip gestures were shown to subjects, but when the speech was barely audible with correct lip movement, the insula was activated suggesting that the insular cortex plays an implicit role in understood speech processing [27]. Christensen et al. [28] described that there was an increase in insular activity when clear visual stimuli was perceived as well as graded stimuli based on when subjects were asked to rate stimuli based on clarity. Among the increase in insular activity in Christensen et al. [28], there was activation of the parietal cortex, prefrontal cortex, premotor cortex, supplementary motor cortex, insula, and thalamus. Vague perception on the other hand is concentrated in a wider spread activation pattern [28]. Unlike the previous experiment, when 20 patients were asked to discriminate whether 1 of 2 parallel vertical lines were longer, Deary et al. [29] showed through fMRI that there was a decreased activation of the insula proportional to an increased difficulty of the task. They did show however that there was an increase in the activity of the insula with brief presentation of the stimuli [29]. It is suggested that the insular cortex activity plays a greater role with increasing perceptual demand of the task instead of just with tonic awareness of the target stimulus [16].

16.3.1 Decision-Making in Visual Perception

Perceptual recognition and decision-making of visual stimuli is a complex process that is believed to involve the insula. Ploran et al. [30] provided pivotal insight into the role of the insula and visual stimulus perception. They showed that prior to the time of recognition of a stimulus, the anterior insular cortex was not activated, but once it was recognized, there was a drastic increased in activity as measured by fMRI. The specific role of the insula in this processing is unknown; however they speculated that it could possibly be due to the capture of focal attention or it could possibly be feedback processes of whether we should perceive this object [31, 32].

16.3.2 Role in Facial Perception

The fusiform face area (FFA) has always been associated with interpreting whether a perceived face is recognized or thought to be a stranger. Recently, however, the anterior insular cortex has been thought to play a role in the perception of recognized faces. Pessoa et al. [33] explained that when patients viewed fearful faces vs. a bar-oriented task, there was an increased activity of the amygdala as well as the insula. The intensity of this response was found to be further modulated by attentional resources and cognitive modulation [33]. These findings help to further imply a direct connectivity between the FFA, insula, and amygdala. Further study of the insula's role in the perception of the face revealed that the right insula reacted not only to fearful faces but was specifically involved in the perception of disgusted faces [34]. Thielscher and Pessoa [34] findings further supported the notion that an increased activity, as measured by an fMRI, of the insula plays a role in the perceptual interpretation or decision-making of whether the stimulus is threatening or not. Patients with lesion of the right anterior insula have also shown deficits in perceiving disgust, while visual fear perception of faces remained intact [35]. Further proof of the insula's role in facial perception is demonstrated when one selectively stimulated the anterior insula in an epileptic subject he perceived disgusted and fearful faces [36]. Interesting enough a similar finding was found in children (age 7–11) with decreased sleep. Decreased sleep was significantly associated with a proportional increase in left insular, bilateral amygdala, and left temporal pole activation [37]. Cheng et al. [38] further supported this and noted that the insula will activate by the mere expression of someone else's pain and send this information to the amygdala.

Activation of the mid-insula bilaterally is thought to be responsible for deciding whether the visual image will be perceived in our consciousness [39]. Adolphs et al. [40] confirmed this notion by studying a patient "B" who had extensive bilateral brain injuries including the insula. They discovered that B lacked the ability to visually perceive a person's emotions other than happiness due to the extensive damage, though it is impossible to determine whether this functional defect is due solely to the insular lesion.

The bilateral insula plays an important role in distinguishing whether a face is visually perceived as babyish or as an adult which is a process highly reliant on oxytocin [41]. Lastly, the insula has been associated with an abnormal visual perception of one's own face. Increased functional connectivity of the insula, fusiform face area, and precuneus/posterior cingulate gyrus has been implicated in psychopathologies such as how one visually perceives themselves in anorexia nervosa [42].

References

1. Lumer ED, Rees G. Covariation of activity in visual and prefrontal cortex associated with subjective visual perception. Proc Natl Acad Sci U S A. 1999;96(4):1669–73.
2. Bushara KO, Hanakawa T, Immisch I, Toma K, Kansaku K, Hallett M. Neural correlates of cross-modal binding. Nat Neurosci. 2003;6(2):190–5.
3. Gould van Praag CD, Garfinkel S, Ward J, Bor D, Seth AK. Automaticity and localisation of concurrents predicts colour area activity in grapheme-colour synaesthesia. Neuropsychologia. 2016;88:5–14.
4. Kim H, Hahm J, Lee H, Kang E, Kang H, Lee DS. Brain networks engaged in audiovisual integration during speech perception revealed by persistent homology-based network filtration. Brain Connect. 2015;5(4):245–58.
5. Sekuler R, Sekuler AB, Lau R. Sound alters visual motion perception. Nature. 1997;385(6614):308.
6. Benoit MM, Raij T, Lin FH, Jaaskelainen IP, Stufflebeam S. Primary and multisensory cortical activity is correlated with audiovisual percepts. Hum Brain Mapp. 2010;31(4):526–38.
7. Mullan S, Penfield W. Illusions of comparative interpretation and emotion; production by epileptic discharge and by electrical stimulation in the temporal cortex. AMA Arch Neurol Psychiatry. 1959;81(3):269–84.
8. Griffiths TD, Bench CJ, Frackowiak RS. Human cortical areas selectively activated by apparent sound movement. Curr Biol. 1994;4(10):892–5.
9. Ducommun CY, Murray MM, Thut G, Bellmann A, Viaud-Delmon I, Clarke S, et al. Segregated processing of auditory motion and auditory location: an ERP mapping study. NeuroImage. 2002;16(1):76–88.
10. Goldberg JM, Neff WD. Frequency discrimination after bilateral ablation of cortical auditory areas. J Neurophysiol. 1961;24:119–28.
11. Neff WD. Behavioral studies of auditory discrimination. Ann Otol Rhinol Laryngol. 1957;66(2):506–13.
12. Kelly JB, Whitfield IC. Effects of auditory cortical lesions on discriminations of rising and falling frequency-modulated tones. J Neurophysiol. 1971;34(5):802–16.
13. Colavita FB. Insular-temporal lesions and vibrotactile temporal pattern discrimination in cats. Physiol Behav. 1974;12(2):215–8.
14. Giraud AL, Kell C, Thierfelder C, Sterzer P, Russ MO, Preibisch C, et al. Contributions of sensory input, auditory search and verbal comprehension to cortical activity during speech processing. Cereb Cortex. 2004;14(3):247–55.
15. Augustine JR. Circuitry and functional aspects of the insular lobe in primates including humans. Brain Res Brain Res Rev. 1996;22(3):229–44.
16. Sterzer P, Kleinschmidt A. Anterior insula activations in perceptual paradigms: often observed but barely understood. Brain Struct Funct. 2010;214(5-6):611–22.
17. Lewis JW, Frum C, Brefczynski-Lewis JA, Talkington WJ, Walker NA, Rapuano KM, et al. Cortical network differences in the sighted versus early blind for recognition of human-produced action sounds. Hum Brain Mapp. 2011;32(12):2241–55.

18. Griffiths TD, Rees A, Witton C, Cross PM, Shakir RA, Green GG. Spatial and temporal auditory processing deficits following right hemisphere infarction. A psychophysical study. Brain. 1997;120(Pt 5):785–94.
19. Nguyen VT, Breakspear M, Hu X, Guo CC. The integration of the internal and external milieu in the insula during dynamic emotional experiences. NeuroImage. 2016;124(Pt A):455–63.
20. Vanneste S, Plazier M, der Loo E, de Heyning PV, Congedo M, De Ridder D. The neural correlates of tinnitus-related distress. NeuroImage. 2010;52(2):470–80.
21. Brooks JC, Zambreanu L, Godinez A, Craig AD, Tracey I. Somatotopic organisation of the human insula to painful heat studied with high resolution functional imaging. NeuroImage. 2005;27(1):201–9.
22. Banca P, Sousa T, Duarte IC, Castelo-Branco M. Visual motion imagery neurofeedback based on the hMT+/V5 complex: evidence for a feedback-specific neural circuit involving neocortical and cerebellar regions. J Neural Eng. 2015;12(6):066003.
23. Rousseau C, Fautrelle L, Papaxanthis C, Fadiga L, Pozzo T, White O. Direction-dependent activation of the insular cortex during vertical and horizontal hand movements. Neuroscience. 2016;325:10–9.
24. Sripada K, Lohaugen GC, Eikenes L, Bjorlykke KM, Haberg AK, Skranes J, et al. Visual-motor deficits relate to altered gray and white matter in young adults born preterm with very low birth weight. NeuroImage. 2015;109:493–504.
25. Dehaene S, Naccache L, Cohen L, Bihan DL, Mangin JF, Poline JB, et al. Cerebral mechanisms of word masking and unconscious repetition priming. Nat Neurosci. 2001;4(7):752–8.
26. Carmel D, Lavie N, Rees G. Conscious awareness of flicker in humans involves frontal and parietal cortex. Curr Biol. 2006;16(9):907–11.
27. Campbell R, MacSweeney M, Surguladze S, Calvert G, McGuire P, Suckling J, et al. Cortical substrates for the perception of face actions: an fMRI study of the specificity of activation for seen speech and for meaningless lower-face acts (gurning). Brain Res Cogn Brain Res. 2001;12(2):233–43.
28. Christensen MS, Ramsoy TZ, Lund TE, Madsen KH, Rowe JB. An fMRI study of the neural correlates of graded visual perception. NeuroImage. 2006;31(4):1711–25.
29. Deary IJ, Simonotto E, Meyer M, Marshall A, Marshall I, Goddard N, et al. The functional anatomy of inspection time: an event-related fMRI study. NeuroImage. 2004;22(4):1466–79.
30. Ploran EJ, Nelson SM, Velanova K, Donaldson DI, Petersen SE, Wheeler ME. Evidence accumulation and the moment of recognition: dissociating perceptual recognition processes using fMRI. J Neurosci. 2007;27(44):11912–24.
31. Wheeler ME, Petersen SE, Nelson SM, Ploran EJ, Velanova K. Dissociating early and late error signals in perceptual recognition. J Cogn Neurosci. 2008;20(12):2211–25.
32. Nelson SM, Dosenbach NU, Cohen AL, Wheeler ME, Schlaggar BL, Petersen SE. Role of the anterior insula in task-level control and focal attention. Brain Struct Funct. 2010;214(5-6):669–80.
33. Pessoa L, Padmala S, Morland T. Fate of unattended fearful faces in the amygdala is determined by both attentional resources and cognitive modulation. NeuroImage. 2005;28(1):249–55.
34. Thielscher A, Pessoa L. Neural correlates of perceptual choice and decision making during fear-disgust discrimination. J Neurosci. 2007;27(11):2908–17.
35. Calder AJ, Lawrence AD, Young AW. Neuropsychology of fear and loathing. Nat Rev Neurosci. 2001;2(5):352–63.
36. Krolak-Salmon P, Henaff MA, Isnard J, Tallon-Baudry C, Guenot M, Vighetto A, et al. An attention modulated response to disgust in human ventral anterior insula. Ann Neurol. 2003;53(4):446–53.
37. Reidy BL, Hamann S, Inman C, Johnson KC, Brennan PA. Decreased sleep duration is associated with increased fMRI responses to emotional faces in children. Neuropsychologia. 2016;84:54–62.
38. Cheng Y, Yang CY, Lin CP, Lee PL, Decety J. The perception of pain in others suppresses somatosensory oscillations: a magnetoencephalography study. NeuroImage. 2008;40(4):1833–40.
39. Critchley HD, Mathias CJ, Dolan RJ. Fear conditioning in humans: the influence of awareness and autonomic arousal on functional neuroanatomy. Neuron. 2002;33(4):653–63.
40. Adolphs R, Tranel D, Damasio AR. Dissociable neural systems for recognizing emotions. Brain Cogn. 2003;52(1):61–9.
41. Doi H, Morikawa M, Inadomi N, Aikawa K, Uetani M, Shinohara K. Neural correlates of babyish adult face processing in men. Neuropsychologia. 2017;97:9–17.
42. Moody TD, Sasaki MA, Bohon C, Strober MA, Bookheimer SY, Sheen CL, et al. Functional connectivity for face processing in individuals with body dysmorphic disorder and anorexia nervosa. Psychol Med. 2015;45(16):3491–503.

The Anterior Insula and Its Relationship to Autism

Seong-Jin Moon, Lara Tkachenko, Erick Garcia-Gorbea, R. Shane Tubbs, and Marc D. Moisi

17.1 Autism Spectrum Disorders

Autism spectrum disorders (ASD) are a disease process that affect the development of children, primarily typified by impairment in social settings with either interaction or communication, as well as repetitive movements or behaviors. It is described as occurring in approximately 1 of every 68 children currently within the United States, emphasizing its vast prevalence [1]. ASD generally can arise during early childhood with solidification of symptoms by ages 2 to 3. While the direction of neurobiological research has provided different theories as to the cause behind ASD onset, the advent of new diagnostic modalities such as functional MRI has greatly advanced the neuroimaging basis of ASD in an anatomical context.

17.2 The Notion of Connectivity

Theoretical treatment of the connection between anterior insula and ASD has lately focused on the concept of "salience networks." This notion relies on the premise that signals within the brain are modulated between a focus and "hub" of networks. Each network can have both externally directed and self-directed internal processes. In ASD, these "salience networks" have been shown to be either hyperactive or hypoactive – abnormal activity within these networks have been associated with ASD [2–5]. One such "salience network" has looked at the anterior insula and its abnormal connectivity in either hyper- or hypofunctioning state in ASD patients, further emphasizing the notion of a contributory aspect of anterior insula relationship to ASD [6–8]. Other authors have pointed to a neuroimaging basis of frontostriatal dysfunction in ASD [9, 10]. Still others have stated that it is hyper-connectivity within the frontal lobes with consequent hypoconnectivity to other salient networks that may predispose to ASD [11, 12].

17.3 The Anterior Insula

The insula is nestled deep within the lateral sulcus and bordered by the frontal, parietal, and temporal opercula. It is divided into two primary parts by the central sulcus, that of the anterior

S.-J. Moon · E. Garcia-Gorbea
Department of Neurosurgery, Wayne State University, Detroit, MI, USA

L. Tkachenko · R. Shane Tubbs
Seattle Science Foundation, Seattle, WA, USA

M. D. Moisi, M.D. (✉)
Department of Neurosurgery, Wayne State University, Detroit, MI, USA

Seattle Science Foundation, Seattle, WA, USA

insula and the posterior insula. The anterior insula itself is composed of an anterior, middle, and posterior short gyrus [13]. It serves as a central hub receiving both direct and indirect projections and concurrently sending out signals, notably from the thalamus as well as amygdala (Fig. 17.1).

The anterior insula may exhibit both overall hypoactivity and abnormal connections with other functional entities which may predispose to ASD. Craig suggested in his 2002 paper that the right anterior insula was behind one's notion of sense of physiological condition, that of perceived interoception [14]. Different authors have also stressed the notion that the anterior insula cortex provides one's sense of bodily state. Critchley stated that the anterior insula cortex "critically contributes to emotional and social processing by supporting the neural representation of the own physiological state" [15]. Interoception was not the only role the insula was associated with, however. As a portion of the paralimbic circuit, the anterior insula serves as a "hub" and both receives and sends information to various critical structures in the neighboring region, including the amygdala, anterior cingulate cortex, and orbital and olfactory cortices.

The insula was also known to be a center for management of natural impulses [16, 17].

17.4 The Advent of fMRI and Experimental Studies

Recent research has highlighted the importance of the anterior insula in ASD, with different studies pointing out the relative reduction in functional connectivity of the anterior insula with other brain networks, while other studies have highlighted the relative hypoactivity of the anterior insula in ASD patients. Many of these studies and experiments have incorporated fMRI into their armamentarium, further bolstering their discussion and results. Silani and colleagues demonstrated that in comparing ASD individuals with matched controls, there was noticeable reduced activation within self-reflection/mentalizing regions [18]. Di Martino and colleagues demonstrate in their landmark 2009 paper that a consistent area of hypoactivity that was demonstrated across 24 studies was within the right anterior insula [19]. This was not only exclusive to the Di Martino group but was also reproduced in other groups as

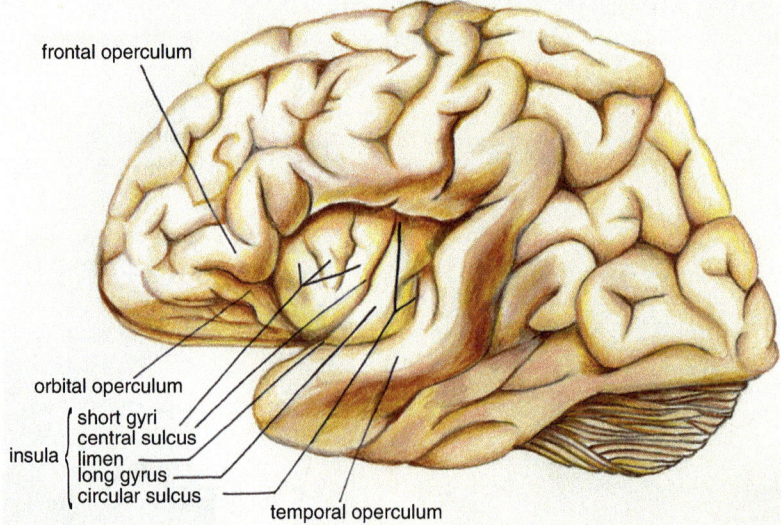

Fig. 17.1 Representation of the left cerebral hemisphere and the surface anatomy of the insula

well, including empathy, visualizing emotional facial expressions, as well as unequal eye gazing [20, 21].

17.5 Future Directions

Future directions of research on anterior insula and its relationship to ASD should focus on functional connectivity, anatomical basis, and hypoactivity versus hyperactivity. The specific nature of how ASD manifests may have an anatomical basis, but this should be further explored in the context of whether hypoactivity versus ineffective input and output processes govern how ASD appears clinically. If this is explored further, there is the potential for therapeutic advances to govern the activity and function of the AI to benefit patients with ASD. Further research could potentially focus within neuromodulation/feedback, which could help explain which specific signals and how they are received at the AI are demonstrated as ASD. Still other research can focus on how different diagnostic imaging modalities can be utilized and implemented for the purposes of characterizing AI activity.

References

1. Christensen D, Baido J, Van Naarden Braun K, Bilder D, Charles J, Constantino J, et al. Prevalence and characteristics of autism spectrum disorder among children aged 8 years—autism and developmental disabilities monitoring network, 11 sites, United States, 2012. MMWR Surveill Summ. 2016;65:1–23.
2. Just M, Cherkassky V, Keller T, Kana R, Minshew N. Functional and anatomical cortical underconnectivity in autism: evidence from an fMRI study of an executive function task and corpus callosum morphometry. Cereb Cortex. 2007;4:951–61.
3. Uddin L. Salience processing and insular cortical function and dysfunction. Nat Rev Neurosci. 2015;16:55–61.
4. Yamada T, Itahashi T, Nakamura M, Watanabe H, Kuroda M, Ohta H, et al. Altered functional organization within the insular cortex in adult males with high-functioning autism spectrum disorder: evidence from connectivity-based parcellation. Mol Autism. 2016;7:1–15.
5. Caria A, de Falco S. Anterior insular cortex regulation in autism spectrum disorders. Front Behav Neurosci. 2015;9:1–9.
6. Uddin L, Menon V. The anterior insula in autism: under-connected and under-examined. Neurosci Biobehav Rev. 2009;33:1198–203.
7. Gogolla N, Takesian A, Feng G, Fagiolini M, Hensch T. Sensory integration in mouse insular cortex reflects GABA circuit maturation. Neuron. 2014;83:894–905.
8. Zhou Y, Shi L, Cui X, Wang S, Luo X. Functional connectivity of the caudal anterior cingulate cortex is decreased in autism. PLoS One. 2016;11:1–14.
9. Bradshaw J. Developmental disorders of the frontostriatal system: neuropsychological, neuropsychiatric, and evolutionary perspectives. 1st ed. UK: Taylor & Francis; 2002.
10. Russel J. Autism as an executive disorder. 1st ed. Oxford: Oxford University Press; 1998.
11. Frith C. Is autism a disconnection disorder? Lancet Neurol. 2004;3:577.
12. Geschwind D, Levitt P. Autism spectrum disorders: developmental disconnection syndromes. Curr Opin Neurobiol. 2007;17:103–11.
13. Stephani C, Fernandez-Baca Vaca G, Maciunas R, Koubeissi M, Lüders H. Functional neuroanatomy of the insular lobe. Brain Struct Funct. 2011;216:137–49.
14. Craig A. How do you feel? Interoception: the sense of the physiological condition of the body. Nat Rev Neurosci. 2002;3:655–66.
15. Critchley H, Wiens S, Rotshtein P, Ohman A, Dolan R. Neural systems supporting interoceptive awareness. Nat Neurosci. 2004;7:189–95.
16. Naqvi N, Rudrauf D, Damasio H, Bechara A. Damage to the insula disrupts addiction to cigarette smoking. Science. 2007;315:531–4.
17. Lerner A, Bagic A, Hanakawa T, Boudreau E, Pagan F, Mari Z, et al. Involvement of insula and cingulate cortices in control and suppression of natural urges. Cereb Cortex. 2009;1:218–23.
18. Silani G, Bird G, Brindley R, Singer T, Frith C, Frith U. Levels of emotional awareness and autism: an fMRI study. Soc Neurosci. 2008;3:97–112.
19. Di Martino A, Ross K, Uddin L, Sklar A, Castellanos F, Milham M. Functional brain correlates of social and non-social processes in autism spectrum disorders: an ALE meta-analysis. Biol Psychiatry. 2009;65:63–74.
20. Hubl D, Bölte S, Fieneis-Matthews S, Lanfermann H, Federspiel A, Strik W, et al. Functional imbalance of visual pathways indicates alternative face processing strategies in autism. Neurology. 2003;61:1232–7.
21. Dichter G, Belger A. Social stimuli interfere with cognitive control in autism. NeuroImage. 2007;35:1219–30.

Role of the Insular Cortex in Emotional Awareness

18

Fareed Jumah

18.1 Emotional Awareness: Existing Findings and Theories

18.1.1 Conscious vs. Unconscious Emotional Processes

Emotion is a very complex concept. Just like a cube has six faces, emotions can have six different aspects: physiological, biological, experiential, psychological, expressive, and social [1, 2]. On the other hand, emotional awareness has five different levels, as defined by Lane and Schwartz: awareness of bodily sensations, the body in action, individual feelings, and blends of blends of feelings [2]. Gu et al. [3] provide a more simplified definition of emotional awareness: the conscious experience of emotions which occurs during the supraliminal processing of affective stimuli [4].

But why is emotional awareness important? As eloquently stated by A.D. Bud Craig: *"An organism must be able to experience its own existence as a sentient being before it can experience the existence and salience of anything else in the environment"* [5]. Myriad evidence suggests that emotional perception, evaluation, and behavior can be processed with or without conscious awareness [6] and that emotional awareness is a necessary, but not sufficient, condition for successful emotional processing. Nonetheless, studies have shown that only coarse affective properties can be registered without conscious awareness [4] and that the more significantly one experiences emotions, the more likely they are to make an appropriate decision or action [2].

18.1.2 Interoception and Emotional Awareness

Interoception and emotional awareness are tightly intertwined concepts. Interoception is the perception of afferent information arising anywhere and everywhere within the body [7, 8] or, simply, the sense of the physiological condition of the body [9, 10]. The current discussion on the relationship between interoception and emotional awareness dates back to the 1880s, where the James-Lange peripheral theory of emotion originated [11, 12]. Carl Lange [12] considered cardiovascular responses as the basis of emotional awareness, whereas William James [11] added to this view by including autonomic functions other than cardiovascular responses. This theory was challenged by an opposing one, the Cannon-Bard theory [13, 14], which suggests that a central nervous system is needed to generate emotions; therefore, bodily responses are a result, not the cause, of emotions. On the other hand, radical behaviorism gave birth to the self-perception

F. Jumah
School of Medicine, An-Najah National University, Nablus, Palestine

theory which states that emotional feelings follow behavior [15, 16].

Recent studies [17–22] have proposed that emotional awareness can be attributed to the reactivation of bodily and neural responses that are involved in lower-level sensorimotor processes. Such embodiment of high-level emotions gives a physical dimension to subjective awareness, often termed the "somatic marker" [23]. Furthermore, Seth et al. [24] suggested that predictive coding of interoceptive information is essential in awareness, where subjective feelings depend on the active interpretation of changes in the body's physiology.

Craig et al. [5, 25, 26], who has studied the insula extensively, proposed that the insula works in a posterior-to-anterior fashion, where the physical features of interoception are processed in the posterior insula, whereas the integration of interoception with cognition and motivation occurs in the AIC, with the right AIC being more dominant than left AIC. This gives the insular cortex its complex processing capabilities, where the AIC holds the end of the axis. This anatomical posterior-to-anterior progression of information ultimately ends in the AIC, where the representation of one's feelings (i.e., the sentient self) resides [9, 71].

18.1.3 Brain Mechanisms of Emotional Awareness

The insular cortex is considered the powerhouse of interoceptive and emotional awareness [5, 24–28], working harmoniously with other brain regions like the anterior cingulate gyrus (ACC) [29–34], amygdala [35], and somatosensory cortex [36]. This is especially evident in patients with bilateral insular damage who preserve certain aspects of emotional awareness, suggesting that emotional feelings might first arise from the brainstem and hypothalamus, which are later enhanced and refined by the insula [37].

Each part of the insula serves a distinct function. The posterior insular cortex has been associated with somatotropic bodily representations like pain, temperature, itch, and touch [5, 9, 17, 38]. The anterior insular cortex (AIC), in contrast, contributes to a myriad of functions including and beyond bodily sensations like touch [39, 40] and pain [41, 42]. It is involved in regulation of vital cardiac [43, 44] and respiratory [43, 45] centers, thermosensory awareness [46], disgust [47–49], interoceptive awareness [30, 50], general emotional processing [50, 51], cognitive control [52, 53], empathy [3, 54–59], intuition [60], unfairness [61, 62], risk and uncertainty [63–66], trust and cooperation [67], and norm violations [68, 69]. Interestingly, patients with epileptic activity emerging from the AIC report enhance sense of well-being and emotional awareness [70].

18.2 The Insular Cortex and Empathy

Empathy is the awareness and understanding of the sensory and emotional states of others [55]. Investigations of emotion rely mainly on visual stimuli depicting another person's emotions (e.g., images of facial expressions) [51, 72–74]. In a recent meta-analysis of 47 fMRI studies of brain activity during empathy in healthy adults, Gu et al. [3] proposed that the AIC is necessary for empathetic emotions using empathy as a test case for emotional awareness. In their study, multiple brain regions are involved in empathetic processing: AIC, ACC, middle and superior temporal gyri, somatosensory cortices S1/S2, dorsal frontoparietal regions, medial prefrontal cortex (MPFC), amygdala, thalamus, and midbrain structure—substantia nigra and red nucleus [3]. Furthermore, it has been noted that different brain regions activated during empathy for positive versus negative emotions and empathy for pain summarized in Table 18.1. Moreover, it has been shown that AIC, not the ACC, is consistently involved in all categories of empathetic processing, suggesting a unique role of AIC in processing affective visual stimuli. Furthermore, the left AIC specifically encodes "energy-nourishing" positive emotions, whereas the right AIC is responsible for "energy-consuming" negative emotions [3, 26], especially in males [75].

Table 18.1 Gu et al. [3]

Empathy for positive emotions	Empathy for negative emotions	Empathy for pain
AIC	AIC	Bilateral insula
S1	ACC	ACC
MPFC, LPFC	LPFC, MPFC	LPFC, MPFC
Superior temporal gyrus	Red nucleus, substantia nigra	S1
Inferior parietal lobule	Putamen, caudate nucleus	Middle occipital gyrus
		Fusiform gyrus
		Inferior parietal lobule
		Amygdala
		Globus pallidus
		Claustrum
		Thalamus
		Cerebellum

18.3 Functional Dissociation Between AIC and ACC

Research has commonly demonstrated that AIC and ACC are coactivated during resting states [76–79] and in many other behaviors [29, 32, 34, 80]. Interestingly, though, researchers have observed that the two structures are actually functionally dissociable [36, 52, 57, 81], a phenomenon fundamental to understanding the true relationship between AIC and ACC. Just like the AIC, studies have shown that ACC is also an important structure in emotional awareness [82, 83], but it is not critical in *generating* awareness. Instead, the ACC seems to receive input from the AIC and relay the "feeling" information to brain regions that serve voluntary control functions [5, 34, 84, 85]. This is further supported by the fact that ACC has connections with lateral prefrontal, primary, and supplementary motor areas [86]. Therefore, it sounds reasonable to imagine the ACC as a limbic motor structure and the ACC as a limbic sensory region [5, 34].

18.4 Top-Down and Bottom-Up Integration

Adding on previous models of AIC [5, 24, 28], Gu et al. proposed a new dual-process model [3] in which the AIC serves two major functions (Fig. 18.1). First, it integrates bottom-up interoceptive input with top-down signals from higher cortical regions (i.e., prefrontal cortex and ACC). This process is similar to the function of sensory cortices where visual and auditory areas integrate bottom-up exteroceptive input with top-down output signals. This integration is what translates into the being's current state of awareness [87–89] and consciousness [89].

Second, the AIC provides output of interoceptive predictions to visceral systems (i.e., via smooth muscles) which serve as a point of reference for autonomic reflexes and sympathetic/parasympathetic outflow and for generating future awareness states. This is similar to how the motor cortex controls proprioceptive reflexes via striated muscles. Furthermore, AIC can be considered as an extension of the sensorimotor cortex that is responsible not for proprioception (and exteroception) but for interoception [5, 9, 26]. To put it simply, the AIC both responds to and regulates the internal milieu or literally the "gut feelings," where it is capable of perceiving and causing changes in the physiological state of the body.

18.5 Clinical Considerations

Emotional awareness is crucial to mental health and overall social well-being. Since the AIC acts as a hub between spatially distinct regions to achieve a state of emotional awareness, AIC injuries understandably lead to diminished ability to make quick and intuitive judgments in continuously changing social situations [90]. Many fMRI studies have shown that perception of another person's feelings activates AIC and ACC; therefore, injuries to these structures cause *alexithymia* [91], deficits in emotional awareness.

In the absence of any neurological or psychiatric disorder, alexithymia can be prevalent in the young and elderly populations (10% and 34%, respectively) [92, 93]. However, these percentages are much higher in conditions with degeneration of the von Economo neurons (VENs) and in conditions causing functional impairment of the AIC, such as frontotemporal dementia (FTD) [94–96], callosal agenesis [97], autism [98, 99], and depersonalization syndrome [100].

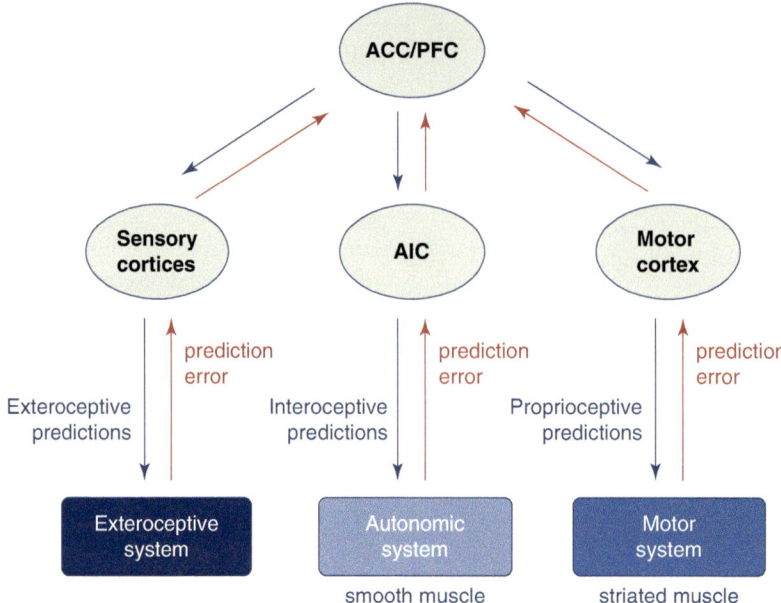

Fig. 18.1 Hypothetical model of insula and awareness. In this hierarchical scheme, each lower-level structure receives descending predictions from and sends ascending prediction errors to higher-level regions. Anterior insular cortex (AIC) serves two major processes in this model (center): (1) integrating bottom-up interoceptive prediction errors with top-down predictions from high-order brain regions such as the anterior cingulate cortex (ACC) and prefrontal cortex (PFC), analogous to the role of sensory cortices (e.g., visual and auditory areas) in exteroceptive processing (left) and (2) sending descending predictions to the autonomic system via smooth muscles to provide a point of reference for autonomic reflexes, similar to the role of motor cortex in generating proprioceptive output via striated muscles (right)

A useful tool implemented by researchers is the 20-item Toronto Alexithymia Scale [101] (TAS-20) to measure three aspects of emotional deficits: difficulty in describing emotions, difficulty in identifying emotions, and externally oriented thinking. Hill et al. have shown alexithymia to be present in around 85% of patients with autism spectrum disorder [102], where lower AIC activations are correlated with higher TAS-20 scores [103]. Similarly, patients with FTD were found to be more alexithymic than their matched controls [95, 104].

Gu et al. [55] compared empathetic pain perception in patients with AIC and ACC lesions and found that patients with AIC lesions, but not ACC lesions, exhibited impairments in implicit and explicit empathy for pain. Moreover, patients with insular lesions were shown to have diminished emotional sensitivity across all emotions rather than to specific emotions like disgust [105]. A recent study on 15 cases with insular cortex lesions showed affected ability to recognize and experience fear, happiness, and surprise but unaffected sense of disgust [106]. However, one case of selective left middle insula lesion due to stroke exhibited impairments in recognizing and experiencing disgust, although memory and intelligence were intact [107].

On the contrary, Damasio et al. [37] challenged these findings when they reported intact emotion and self-awareness in a patient with completely destroyed insular cortices due to herpes simplex encephalitis. Similarly, others have shown that insular cortical lesions had minimal impact when compared to subcortical lesion cases [108]. These findings again suggest that the insular cortex is not solely responsible for emotional awareness but, instead, that subcortical connections between the insula and other regions (e.g., thalamus) are involved.

References

1. Dolan RJ. Emotion, cognition, and behavior. Science. 2002;298(5596):1191–4.
2. Lane RD, Schwartz GE. Levels of emotional awareness: a cognitive-developmental theory and its application to psychopathology. Am J Psychiatry. 1987;144(2):133–43.
3. Gu X, Hof PR, Friston KJ, Fan J. Anterior insular cortex and emotional awareness. J Comp Neurol. 2013;521(15):3371–88.
4. Pessoa L. To what extent are emotional visual stimuli processed without attention and awareness? Curr Opin Neurobiol. 2005;15(2):188–96.
5. Craig AD. How do you feel—now? The anterior insula and human awareness. Nat Rev Neurosci. 2009;10(1):59–70.
6. Ohman A, Soares JJ. "Unconscious anxiety": phobic responses to masked stimuli. J Abnorm Psychol. 1994;103(2):231–40.
7. Sherrington CS. The integrative action of the nervous system. New Haven: Yale University Press; 1906.
8. Cameron OG. Interoception: the inside story—a model for psychosomatic processes. Psychosom Med. 2001;63(5):697–710.
9. Craig AD. How do you feel? Interoception: the sense of the physiological condition of the body. Nat Rev Neurosci. 2002;3(8):655–66.
10. Craig AD. Interoception: the sense of the physiological condition of the body. Curr Opin Neurobiol. 2003;13(4):500–5.
11. James W. What is an emotion? Mind. 1884;19:188–205.
12. Carl Lange G. The emotions: a psychophysiological study. Baltimore: William and Wilkins; 1885/1992.
13. Bard P. A diencephalic mechanism for the expression of rage with special reference to the sympathetic nervous system. Am J Physiol. 1928;84:490–515.
14. Cannon WB. The wisdom of the body. New York: W.W. Norton and Co.; 1932.
15. Laird JD. Self-attribution of emotion: the effects of expressive behavior on the quality of emotional experience. J Pers Soc Psychol. 1974;29(4):475–86.
16. Bem DJ. Self-perception: an alternative interpretation of cognitive dissonance phenomena. Psychol Rev. 1967;74(3):183–200.
17. Harrison NA, Gray MA, Gianaros PJ, Critchley HD. The embodiment of emotional feelings in the brain. J Neurosci. 2010;30(38):12878–84.
18. Niedenthal PM. Embodying emotion. Science. 2007;316(5827):1002–5.
19. Oosterwijk S, Lindquist KA, Anderson E, Dautoff R, Moriguchi Y, Barrett LF. States of mind: emotions, body feelings, and thoughts share distributed neural networks. NeuroImage. 2012;62(3):2110–28.
20. Pollatos O, Fustos J, Critchley HD. On the generalised embodiment of pain: how interoceptive sensitivity modulates cutaneous pain perception. Pain. 2012;153(8):1680–6.
21. Gray MA, Beacher FD, Minati L, Nagai Y, Kemp AH, Harrison NA, et al. Emotional appraisal is influenced by cardiac afferent information. Emotion. 2012;12(1):180–91.
22. Thompson E, Varela FJ. Radical embodiment: neural dynamics and consciousness. Trends Cogn Sci. 2001;5(10):418–25.
23. Damasio AR. The somatic marker hypothesis and the possible functions of the prefrontal cortex. Philos Trans R Soc Lond Ser B Biol Sci. 1996;351(1346):1413–20.
24. Seth AK, Suzuki K, Critchley HD. An interoceptive predictive coding model of conscious presence. Front Psychol. 2011;2:395.
25. Craig AD. The sentient self. Brain Struct Funct. 2010;214(5–6):563–77.
26. Craig AD. Significance of the insula for the evolution of human awareness of feelings from the body. Ann N Y Acad Sci. 2011;1225:72–82.
27. Jones CL, Ward J, Critchley HD. The neuropsychological impact of insular cortex lesions. J Neurol Neurosurg Psychiatry. 2010;81(6):611–8.
28. Singer T, Critchley HD, Preuschoff K. A common role of insula in feelings, empathy and uncertainty. Trends Cogn Sci. 2009;13(8):334–40.
29. Critchley HD. The human cortex responds to an interoceptive challenge. Proc Natl Acad Sci U S A. 2004;101(17):6333–4.
30. Critchley HD, Wiens S, Rotshtein P, Ohman A, Dolan RJ. Neural systems supporting interoceptive awareness. Nat Neurosci. 2004;7(2):189–95.
31. Denny BT, Kober H, Wager TD, Ochsner KN. A meta-analysis of functional neuroimaging studies of self- and other judgments reveals a spatial gradient for mentalizing in medial prefrontal cortex. J Cogn Neurosci. 2012;24(8):1742–52.
32. Fan J, Gu X, Liu X, Guise KG, Park Y, Martin L, et al. Involvement of the anterior cingulate and frontoinsular cortices in rapid processing of salient facial emotional information. NeuroImage. 2011;54(3):2539–46.
33. Lindquist KA, Wager TD, Kober H, Bliss-Moreau E, Barrett LF. The brain basis of emotion: a meta-analytic review. Behav Brain Sci. 2012;35(3):121–43.
34. Medford N, Critchley HD. Conjoint activity of anterior insular and anterior cingulate cortex: awareness and response. Brain Struct Funct. 2010;214(5–6):535–49.
35. Etkin A, Wager TD. Functional neuroimaging of anxiety: a meta-analysis of emotional processing in PTSD, social anxiety disorder, and specific phobia. Am J Psychiatry. 2007;164(10):1476–88.
36. Gu X, Liu X, Van Dam NT, Hof PR, Fan J. Cognition-emotion integration in the anterior insular cortex. Cereb Cortex. 2013;23(1):20–7.
37. Damasio A, Damasio H, Tranel D. Persistence of feelings and sentience after bilateral damage of the insula. Cereb Cortex. 2013;23(4):833–46.
38. Damasio AR, Grabowski TJ, Bechara A, Damasio H, Ponto LL, Parvizi J, et al. Subcortical and cortical

brain activity during the feeling of self-generated emotions. Nat Neurosci. 2000;3(10):1049–56.
39. Keysers C, Wicker B, Gazzola V, Anton JL, Fogassi L, Gallese V. A touching sight: SII/PV activation during the observation and experience of touch. Neuron. 2004;42(2):335–46.
40. Lindgren L, Westling G, Brulin C, Lehtipalo S, Andersson M, Nyberg L. Pleasant human touch is represented in pregenual anterior cingulate cortex. NeuroImage. 2012;59(4):3427–32.
41. Treede RD, Kenshalo DR, Gracely RH, Jones AK. The cortical representation of pain. Pain. 1999;79(2–3):105–11.
42. Wager TD, Rilling JK, Smith EE, Sokolik A, Casey KL, Davidson RJ, et al. Placebo-induced changes in FMRI in the anticipation and experience of pain. Science. 2004;303(5661):1162–7.
43. Henderson LA, Macey PM, Macey KE, Frysinger RC, Woo MA, Harper RK, et al. Brain responses associated with the Valsalva maneuver revealed by functional magnetic resonance imaging. J Neurophysiol. 2002;88(6):3477–86.
44. King AB, Menon RS, Hachinski V, Cechetto DF. Human forebrain activation by visceral stimuli. J Comp Neurol. 1999;413(4):572–82.
45. Banzett RB, Mulnier HE, Murphy K, Rosen SD, Wise RJ, Adams L. Breathlessness in humans activates insular cortex. Neuroreport. 2000;11(10):2117–20.
46. Craig AD, Chen K, Bandy D, Reiman EM. Thermosensory activation of insular cortex. Nat Neurosci. 2000;3(2):184–90.
47. Calder AJ, Beaver JD, Davis MH, van Ditzhuijzen J, Keane J, Lawrence AD. Disgust sensitivity predicts the insula and pallidal response to pictures of disgusting foods. Eur J Neurosci. 2007;25(11):3422–8.
48. Phillips ML, Young AW, Senior C, Brammer M, Andrew C, Calder AJ, et al. A specific neural substrate for perceiving facial expressions of disgust. Nature. 1997;389(6650):495–8.
49. Wicker B, Keysers C, Plailly J, Royet JP, Gallese V, Rizzolatti G. Both of us disgusted in my insula: the common neural basis of seeing and feeling disgust. Neuron. 2003;40(3):655–64.
50. Zaki J, Davis JI, Ochsner KN. Overlapping activity in anterior insula during interoception and emotional experience. NeuroImage. 2012;62(1):493–9.
51. Davidson RJ, Irwin W. The functional neuroanatomy of emotion and affective style. Trends Cogn Sci. 1999;3(1):11–21.
52. Eckert MA, Menon V, Walczak A, Ahlstrom J, Denslow S, Horwitz A, et al. At the heart of the ventral attention system: the right anterior insula. Hum Brain Mapp. 2009;30(8):2530–41.
53. Menon V, Uddin LQ. Saliency, switching, attention and control: a network model of insula function. Brain Struct Funct. 2010;214(5–6):655–67.
54. Ebisch SJ, Ferri F, Salone A, Perrucci MG, D'Amico L, Ferro FM, et al. Differential involvement of somatosensory and interoceptive cortices during the observation of affective touch. J Cogn Neurosci. 2011;23(7):1808–22.
55. Gu X, Gao Z, Wang X, Liu X, Knight RT, Hof PR, et al. Anterior insular cortex is necessary for empathetic pain perception. Brain. 2012;135(Pt 9):2726–35.
56. Gu X, Han S. Attention and reality constraints on the neural processes of empathy for pain. NeuroImage. 2007;36(1):256–67.
57. Gu X, Liu X, Guise KG, Naidich TP, Hof PR, Fan J. Functional dissociation of the frontoinsular and anterior cingulate cortices in empathy for pain. J Neurosci. 2010;30(10):3739–44.
58. Lamm C, Meltzoff AN, Decety J. How do we empathize with someone who is not like us? A functional magnetic resonance imaging study. J Cogn Neurosci. 2010;22(2):362–76.
59. Singer T, Seymour B, O'Doherty J, Kaube H, Dolan RJ, Frith CD. Empathy for pain involves the affective but not sensory components of pain. Science. 2004;303(5661):1157–62.
60. Kuo WJ, Sjostrom T, Chen YP, Wang YH, Huang CY. Intuition and deliberation: two systems for strategizing in the brain. Science. 2009;324(5926):519–22.
61. Kirk U, Downar J, Montague PR. Interoception drives increased rational decision-making in meditators playing the ultimatum game. Front Neurosci. 2011;5:49.
62. Sanfey AG, Rilling JK, Aronson JA, Nystrom LE, Cohen JD. The neural basis of economic decision-making in the Ultimatum Game. Science. 2003;300(5626):1755–8.
63. Bach DR, Dolan RJ. Knowing how much you don't know: a neural organization of uncertainty estimates. Nat Rev Neurosci. 2012;13(8):572–86.
64. Bossaerts P. Risk and risk prediction error signals in anterior insula. Brain Struct Funct. 2010;214(5–6):645–53.
65. Preuschoff K, Quartz SR, Bossaerts P. Human insula activation reflects risk prediction errors as well as risk. J Neurosci. 2008;28(11):2745–52.
66. Ullsperger M, Harsay HA, Wessel JR, Ridderinkhof KR. Conscious perception of errors and its relation to the anterior insula. Brain Struct Funct. 2010;214(5–6):629–43.
67. King-Casas B, Sharp C, Lomax-Bream L, Lohrenz T, Fonagy P, Montague PR. The rupture and repair of cooperation in borderline personality disorder. Science. 2008;321(5890):806–10.
68. Montague PR, Lohrenz T. To detect and correct: norm violations and their enforcement. Neuron. 2007;56(1):14–8.
69. Xiang T, Lohrenz T, Montague PR. Computational substrates of norms and their violations during social exchange. J Neurosci. 2013;33(3):1099–8a.
70. Picard F. State of belief, subjective certainty and bliss as a product of cortical dysfunction. Cortex. 2013;49(9):2494–500.
71. Barrett LF, Quigley KS, Bliss-Moreau E, Aronson KR. Interoceptive sensitivity and self-reports

72. Adolphs R. Neural systems for recognizing emotion. Curr Opin Neurobiol. 2002;12(2):169–77.
73. Phelps EA. Emotion and cognition: insights from studies of the human amygdala. Annu Rev Psychol. 2006;57:27–53.
74. Pessoa L, Adolphs R. Emotion processing and the amygdala: from a 'low road' to 'many roads' of evaluating biological significance. Nat Rev Neurosci. 2010;11(11):773–83.
75. Stevens JS, Hamann S. Sex differences in brain activation to emotional stimuli: a meta-analysis of neuroimaging studies. Neuropsychologia. 2012;50(7):1578–93.
76. Britz J, Van De Ville D, Michel CM. BOLD correlates of EEG topography reveal rapid resting-state network dynamics. NeuroImage. 2010;52(4):1162–70.
77. Cauda F, D'Agata F, Sacco K, Duca S, Geminiani G, Vercelli A. Functional connectivity of the insula in the resting brain. NeuroImage. 2011;55(1):8–23.
78. Fan J, Xu P, Van Dam NT, Eilam-Stock T, Gu X, Luo YJ, et al. Spontaneous brain activity relates to autonomic arousal. J Neurosci. 2012;32(33):11176–86.
79. Seeley WW, Menon V, Schatzberg AF, Keller J, Glover GH, Kenna H, et al. Dissociable intrinsic connectivity networks for salience processing and executive control. J Neurosci. 2007;27(9):2349–56.
80. Dosenbach NU, Visscher KM, Palmer ED, Miezin FM, Wenger KK, Kang HC, et al. A core system for the implementation of task sets. Neuron. 2006;50(5):799–812.
81. Sridharan D, Levitin DJ, Menon V. A critical role for the right fronto-insular cortex in switching between central-executive and default-mode networks. Proc Natl Acad Sci U S A. 2008;105(34):12569–74.
82. Lane RD, Reiman EM, Axelrod B, Yun LS, Holmes A, Schwartz GE. Neural correlates of levels of emotional awareness. Evidence of an interaction between emotion and attention in the anterior cingulate cortex. J Cogn Neurosci. 1998;10(4):525–35.
83. Lieberman MD. Social cognitive neuroscience: a review of core processes. Annu Rev Psychol. 2007;58:259–89.
84. Posner MI, Rothbart MK. Attention, self-regulation and consciousness. Philos Trans R Soc Lond Ser B Biol Sci. 1998;353(1377):1915–27.
85. Valentini E. The role of anterior insula and anterior cingulate in empathy for pain. J Neurophysiol. 2010;104(2):584–6.
86. Fan J, Hof PR, Guise KG, Fossella JA, Posner MI. The functional integration of the anterior cingulate cortex during conflict processing. Cereb Cortex. 2008;18(4):796–805.
87. Friston K. Beyond phrenology: what can neuroimaging tell us about distributed circuitry? Annu Rev Neurosci. 2002;25:221–50.
88. Friston K. The free-energy principle: a unified brain theory? Nat Rev Neurosci. 2010;11(2):127–38.
89. Tononi G, Koch C. The neural correlates of consciousness: an update. Ann N Y Acad Sci. 2008;1124:239–61.
90. Allman JM, Watson KK, Tetreault NA, Hakeem AY. Intuition and autism: a possible role for Von Economo neurons. Trends Cogn Sci. 2005;9(8):367–73.
91. Taylor GJ. Recent developments in alexithymia theory and research. Can J Psychiatr. 2000;45(2):134–42.
92. Joukamaa M, Saarijarvi S, Muuriaisniemi ML, Salokangas RK. Alexithymia in a normal elderly population. Compr Psychiatry. 1996;37(2):144–7.
93. Kokkonen P, Karvonen JT, Veijola J, Laksy K, Jokelainen J, Jarvelin MR, et al. Prevalence and sociodemographic correlates of alexithymia in a population sample of young adults. Compr Psychiatry. 2001;42(6):471–6.
94. Kim EJ, Sidhu M, Gaus SE, Huang EJ, Hof PR, Miller BL, et al. Selective frontoinsular von Economo neuron and fork cell loss in early behavioral variant frontotemporal dementia. Cereb Cortex. 2012;22(2):251–9.
95. Seeley WW. Anterior insula degeneration in frontotemporal dementia. Brain Struct Funct. 2010;214(5–6):465–75.
96. Seeley WW, Carlin DA, Allman JM, Macedo MN, Bush C, Miller BL, et al. Early frontotemporal dementia targets neurons unique to apes and humans. Ann Neurol. 2006;60(6):660–7.
97. Kaufman JA, Paul LK, Manaye KF, Granstedt AE, Hof PR, Hakeem AY, et al. Selective reduction of Von Economo neuron number in agenesis of the corpus callosum. Acta Neuropathol. 2008;116(5):479–89.
98. Butti C, Santos M, Uppal N, Hof PR. Von Economo neurons: clinical and evolutionary perspectives. Cortex. 2013;49(1):312–26.
99. Santos M, Uppal N, Butti C, Wicinski B, Schmeidler J, Giannakopoulos P, et al. Von Economo neurons in autism: a stereologic study of the frontoinsular cortex in children. Brain Res. 2011;1380:206–17.
100. Simeon D, Giesbrecht T, Knutelska M, Smith RJ, Smith LM. Alexithymia, absorption, and cognitive failures in depersonalization disorder: a comparison to posttraumatic stress disorder and healthy volunteers. J Nerv Ment Dis. 2009;197(7):492–8.
101. Taylor GJ, Bagby RM, Parker JD. The 20-Item Toronto Alexithymia Scale. IV. Reliability and factorial validity in different languages and cultures. J Psychosom Res. 2003;55(3):277–83.
102. Hill E, Berthoz S, Frith U. Brief report: cognitive processing of own emotions in individuals with autistic spectrum disorder and in their relatives. J Autism Dev Disord. 2004;34(2):229–35.
103. Bird G, Silani G, Brindley R, White S, Frith U, Singer T. Empathic brain responses in insula are modulated by levels of alexithymia but not autism. Brain. 2010;133(Pt 5):1515–25.
104. Sturm VE, Levenson RW. Alexithymia in neurodegenerative disease. Neurocase. 2011;17(3):242–50.

105. Terasawa Y, Kurosaki Y, Ibata Y, Moriguchi Y, Umeda S. Attenuated sensitivity to the emotions of others by insular lesion. Front Psychol. 2015;6:1314.
106. Boucher O, Rouleau I, Lassonde M, Lepore F, Bouthillier A, Nguyen DK. Social information processing following resection of the insular cortex. Neuropsychologia. 2015;71:1–10.
107. Calder AJ, Keane J, Manes F, Antoun N, Young AW. Impaired recognition and experience of disgust following brain injury. Nat Neurosci. 2000;3(11):1077–8.
108. Couto B, Sedeno L, Sposato LA, Sigman M, Riccio PM, Salles A, et al. Insular networks for emotional processing and social cognition: comparison of two case reports with either cortical or subcortical involvement. Cortex. 2013;49(5):1420–34.

Alterations of Reil's Insula in Alzheimer's Disease

Paul Choi, Emily Simonds, Marc Vetter, Charlotte Wilson, and R. Shane Tubbs

Abbreviations

AD	Alzheimer's disease
FTD	Frontotemporal dementia
IC	Insular cortex
MDD	Major depressive disorder
NFPA	Nonfluent progressive aphasia
NFTs	Neurofibrillary tangles

19.1 Background

Alzheimer's disease (AD) is a progressive neurodegenerative disease that impacts about one-third of the people over 85 years of age. AD affects many individuals and is considered a serious medical challenge given today's aging population [1]. AD involves numerous cellular and biomechanical processes which result in synaptic alterations. AD initially presents as a short-term memory loss problem; as the disease progresses, it begins to interfere with safety and competency in daily life [1]. Patients often experience mood disturbance, lack of sleep regulation, confusion, social withdrawal, loss of language, inability to perform cognitive tasks, issues with gait, and eventually bodily function [1]. On average a patient will live 7 years after diagnosis. The insula is thought to be a key target of AD, though the role of the insula in AD is not well understood.

The most commonly impacted areas of the brain in patients with AD are the frontal lobe which impacts intelligence, judgment, governance, and socially acceptable behavior [1]. Other areas impacted include the temporal region and parietal brain which are active in memory and language. It is apparent that the insular cortex (IC) plays an important role in the clinical manifestation of AD.

The IC plays a role in memory, emotion, sense of taste and smell, and autonomic control [2]. Therefore, an atrophic IC would result in not only cognitive deficits, which is a typical presentation of an early AD, but it may display additional three main symptomatologies: neuropsychiatric symptoms, special sensory dysfunction, and autonomic instability [2, 3].

19.2 First Historical Mention of Insula and Alzheimer's Disease

19.2.1 Case Reports

Several previous studies have suggested a neuropathological relationship between the insular cortex and AD [2, 4]. Arnold et al. [5] found that

P. Choi · E. Simonds · M. Vetter · C. Wilson
R. Shane Tubbs (✉)
Seattle Science Foundation, Seattle, WA, USA
e-mail: shanet@seattlesciencefoundation.org

limbic periallocortex and allocortex, including the insula, contained more neurofibrillary tangles than in any other cortical region. Sandor et al. [3] reported a wider Sylvian fissure in patients with AD on axial MRI scans; however, they did not conclude that the widening of the fissure was due to insular cortex atrophy.

It is well understood that the nucleus basalis of Meynert (a region of the basal forebrain) is involved in AD [6]. Mesulam and Mufson's [7] research involving rhesus monkeys found that neurons of the basal forebrain receive cortical input from the anterior insula. This may suggest a pathological involvement of the basal forebrain, and therefore the nucleus basalis indirectly affects the insular cortex in patients with AD. These findings suggest that the insular cortex is critically involved in integrating higher cognitive behaviors. The insular cortex also has connections with the hippocampus, basal forebrain, and parietal polymodal association cortex, suggesting atrophy of the insular cortex may give way to partial functional degeneration of these regions [7–10]. It should be noted, however, that behavioral implications of insular atrophy are inconclusive and require further study [11].

19.3 Clinical Diagnosis

Patients suspected of AD are usually given a series of cognitive tests and memory assessments over a certain amount of time [1]. Currently, clinicians must rule out a variety of other diseases before diagnosing AD. MRI and CT scans may be ordered and a spinal tap typically performed [11]. Amyloid protein builds up in CSF; this is a benchmark for diagnosis of AD [12].

19.3.1 Imaging

Imaging done over the course of AD will show progressive nerve cell death and dramatic shrinking of the brain as a whole which ultimately affects all of its functions. Shrinkage is most prevalent in the hippocampus and the ventricles dilate [1].

Neuroimaging of AD patients has shown focal neurodegenerative processes in the structures around the hippocampal region such as the insular cortex (IC) [12–14]. Moreover, neurofibrillary tangle (NFT) deposition, with its amount directly proportional to the duration of the degenerative process, in an altered blood flow to the IC, is seen in AD [12–16].

According to Petrides et al. [17] the loss of dendritic spines in AD is related to synaptic dysregulation and dysfunction, memory issues, cognition, and executive functioning [17]. It is difficult to distinguish all the facets that play a role in the pathophysiology of AD and their role in dysfunction. According to Petrides et al. [17], a decrease in the synaptic contacts may lead to impairment of function of the pyramidal–intraneuronal connectivity as well as partial loss of contact with other cortical and subcortical areas [17]. These may be implicated in control of the insular cortex including autonomic control over bodily functions, mental decision-making, and relation to self and general well-being [11, 12, 17].

19.3.2 Histology

Alzheimer's nervous tissue has fewer neuronal cells and synapses. The classic neuropathological signs of AD are nervous tissues which have buildup of abnormal plaques (senile plaques) and proteins. These tissues may also be the dying neurofibrillary tangles [12]. The plaques that are present often consist of fragment beta-amyloids [11, 12]. Tangles consist of tau which is a protein involved in maintaining the internal architecture of a neuron [12]. In AD excessive phosphorylation occurs forming a tangle which arrests the nerve cell from carrying out cellular functions [17]. Oxidative stress can also be seen in histology due to reactive oxidative stress at oxygen-binding sites. Brain tissue is generally also inflamed. Many Alzheimer's patients also show signs of cerebrovascular disease [17].

19.3.3 Neuropsychiatric Symptoms

The neuropsychiatric symptoms of AD such as aggression, agitation, anxiety, and phobia are associated with structural changes of and reduced metabolism within the IC [16]. Patients may present with fidgeting, wandering, and restlessness [16]. Moreover, loss of insular function has been described as the underlying mechanism of major depressive disorder (MDD) [18]. Lee et al. performed a series of voxel-based morphometry and measured a decreased amount of gray matter in the insular gyrus of a patient with MDD [19]. This finding is crucial in the context of AD since depression is one of its most common comorbidities [16]. This depressive component of IC atrophy may also explain apathy displayed by AD patients, which is the disease's most common emotional feature [20]. Lastly, loss of self-awareness may result since the IC receives all somatosensory inputs and is closely coupled with both the limbic system and the autonomic system [12].

19.3.4 Special Sensory Dysfunction

Impaired special sensory functions, i.e., olfaction and gustation, combined with the lack of self-awareness and the neuropsychiatric problem may result in an inability to perceive the feeling of disgust [12, 21]. For instance, AD patients often eat objects or substances that are deemed inedible [12]. Further, patients with frontotemporal dementia (FTD), which is also characterized by IC atrophy, present with hyperorality, and certain obese populations have been reported to have diminished basal insular responses [21–23]. Similarly, a greater insular response to food has been documented to cause hyperphagia in Prader-Willi syndrome [21]. Hyperorality associated with special sensory defects due to IC atrophy may have a role in the increased body mass index (BMI) and waist-to-hip ratio of AD patients [13]. Luchsinger et al. claim AD patients are at a high risk of insulin resistance, elevated blood pressure, dyslipidemia, and acute coronary syndromes due to central obesity [24–27].

19.3.5 Autonomic Instability

The autonomic instability resulting from IC atrophy often causes cardiovascular problems, irregular respiration, and abnormal gastrointestinal peristalses [12]. AD patients may suffer from chronic hypertension, heart failure, myocardial infarction, and pneumonia, which can be fatal [12].

19.4 Links Between Insular Pathology and Alzheimer's

Reil's insula is an important structure in the regulation of autonomic physiological processes such as visceral control and the ability to form complex social emotions—both of which are deficient in patients suffering from Alzheimer's disease (AD) [28]. De Morree et al. have shown that damage to the insula can severely affect parasympathetic autonomous regulation leading to cardiac arrhythmia in patients [29]. Myocardial infarction, cardiac failure, and bronchopneumonia are all leading causes of death in AD patients. As a result, there is a growing research interest in analyzing the specific relationship between insular pathology and AD progression [12]. A growing list of clinical symptoms, physiological symptoms, and histological characteristics present in AD patients is helping medical professionals to better understand how Reil's insula is impacted by this disease.

19.4.1 Clinical Symptoms Linking AD and Insula Pathology

Multiple clinical symptoms associated with AD have been specifically linked to insular cortex pathology. One example is the frequent presentation of dysosmia in AD patients. AD is known to affect olfactory pathways, and clinical studies show that the insula plays an important part in olfaction and gustation [30]. Furthermore, dementia and AD are often linked to an inability on the part of patients to react conventionally to images and scenarios that would generally produce a feeling of disgust [31]. The insula has been identified as a cortical structure which helps generate such responses. In a study by O'Doherty et al., the insula was shown to be strongly active in participants asked to discriminate between aversive and pleasant gustatory stimuli [32]. Another clinical symptom relating AD and insular pathology is nonfluent progressive aphasia (NFPA), a condition associated with dementia and which causes slow, hesitant speech. In a series of imaging studies, Gorno-Tempini et al. found that NFPA was correlated to the atrophy of the anterior insula and left precentral gyrus of the insula [33].

Emotional contagion, a primitive form of empathy which causes an individual's emotions to be influenced by those expressed around them, is often noticeably more pronounced in patients with AD. According to Choi et al., a higher amyloid plaque burden in the anterior left insula, one of the hallmark histological characteristics of AD, likely increases the degree to which emotional contagion influences the emotional state of an individual [34].

19.4.2 Physiological Symptoms Linking AD and Insula Pathology

One of the major physiological changes to the insula noted in several imaging studies of AD patients was the presence of high levels of cerebrospinal fluid, indicating a significant amount of atrophy [11, 12]. In one of the first such imaging studies, carried out by Foundas et al., atrophy in the insular and parietal cortex of AD patients was remarkably more advanced than in other parts of the cortical brain, such as the striate cortex [11]. Such findings suggest that the insula is among a variety of cortical structures that are specifically more vulnerable to AD, bearing a higher degree of pathological burden. Further studies are needed in order to understand why structures such as the insula are more vulnerable to changes in the brain brought about by AD.

19.4.3 Histological Characteristics of Insular Pathology in AD

In patients with AD, the insular cortex has a high burden of neurofibrillary tangles (NFTs) and neuritic plaques [12]. According to Bonthius et al., the insular density of NFTs is strongly correlated with an earlier onset of clinical AD symptoms. Not all regions of the insula are equally impacted by the NFTs and neuritic plaques. The insula is composed of three distinct cytoarchitectonic regions: granular, agranular, and dysgranular. The agranular region has the strongest connections to the olfactory cortex and is also the most heavily burdened in terms of NFT density [12]. The dysgranular portion of the insula, which has a stronger impact on the emotion of disgust, is also heavily burdened by NFT formation, though not as much as the agranular portion. Relative to the other two cytoarchitectonic regions, the granular portion of the insula is not heavily burdened by NFTs [12]. Besides NFT formation, researchers have noted that in AD patients, dendritic polarity in the pyramidal neurons of the insula is significantly lower, inhibiting the ability of these cells to send coordinated signals efficiently [17]. Furthermore, these cells exhibited significantly less terminal branching in the insula, decreasing the interconnectivity of the insular neural network [17].

19.5 Need for Further Studies

The combined effect of depression, apathy, loss of self-awareness, insulin resistance, hypertension, and hyperlipidemia may predispose AD patients to or accelerate their preexisting cardiovascular morbidity and mortality. To our knowledge, the correlation of insular neurodegeneration and cardiovascular disease has never been studied. In order to improve the quality of life of both AD patients and their caregivers, who are already under a tremendous disease burden, further research is needed to prevent addition of potential comorbidities.

References

1. Ballard C, Gauthier S, Corbett A, Brayne C, Aarsland D, Jones E. Alzheimer's disease. Lancet. 2011;377(9770):1019–31.
2. Braak H, Braak E. Neuropathological stageing of Alzheimer-related changes. Acta Neuropathol. 1991;82(4):239–59.
3. Sandor T, Jolesz F, Tieman J, Kikinis R, Jones K, Albert M. Comparative analysis of computed tomographic and magnetic resonance imaging scans in Alzheimer patients and controls. Arch Neurol. 1992;49(4):381–4.
4. Braak H, Braak E. Staging of Alzheimer's disease-related neurofibrillary changes. Neurobiol Aging. 1995;16(3):271–8; discussion 8–84.
5. Arnold SE, Hyman BT, Flory J, Damasio AR, Van Hoesen GW. The topographical and neuroanatomical distribution of neurofibrillary tangles and neuritic plaques in the cerebral cortex of patients with Alzheimer's disease. Cereb Cortex. 1991;1(1):103–16.
6. Whitehouse PJ, Price DL, Clark AW, Coyle JT, DeLong MR. Alzheimer disease: evidence for selective loss of cholinergic neurons in the nucleus basalis. Ann Neurol. 1981;10(2):122–6.
7. Mesulam MM, Mufson EJ. Neural inputs into the nucleus basalis of the substantia innominata (Ch4) in the rhesus monkey. Brain. 1984;107(Pt 1):253–74.
8. Mesulam MM, Mufson EJ. Insula of the old world monkey. I. Architectonics in the insulo-orbito-temporal component of the paralimbic brain. J Comp Neurol. 1982;212(1):1–22.
9. Mufson EJ, Mesulam MM. Insula of the old world monkey. II: afferent cortical input and comments on the claustrum. J Comp Neurol. 1982;212(1):23–37.
10. Seltzer B, Pandya DN. Post-rolandic cortical projections of the superior temporal sulcus in the rhesus monkey. J Comp Neurol. 1991;312(4):625–40.
11. Foundas AL, Leonard CM, Mahoney SM, Agee OF, Heilman KM. Atrophy of the hippocampus, parietal cortex, and insula in Alzheimer's disease: a volumetric magnetic resonance imaging study. Neuropsychiatry Neuropsychol Behav Neurol. 1997;10(2):81–9.
12. Bonthius DJ, Solodkin A, Van Hoesen GW. Pathology of the insular cortex in Alzheimer disease depends on cortical architecture. J Neuropathol Exp Neurol. 2005;64(10):910–22.
13. Weiner MW, Veitch DP, Aisen PS, Beckett LA, Cairns NJ, Green RC, et al. The Alzheimer's disease neuroimaging initiative: a review of papers published since its inception. Alzheimers Dement. 2013;9(5):e111–94.
14. Bakkour A, Morris JC, Wolk DA, Dickerson BC. The effects of aging and Alzheimer's disease on cerebral cortical anatomy: specificity and differential relationships with cognition. NeuroImage. 2013;76:332–44.
15. Serrano-Pozo A, Frosch MP, Masliah E, Hyman BT. Neuropathological alterations in Alzheimer disease. Cold Spring Harb Perspect Med. 2011;1(1):a006189.
16. Van Dam D, Vermeiren Y. Neuropsychiatric disturbances in Alzheimer's disease: what have we learned from neuropathological studies? Curr Alzheimer Res. 2016;13:1145–64.
17. Petrides FE, Mavroudis IA, Spilioti M, Chatzinikolaou FG, Costa VG, Baloyannis SJ. Spinal alterations of Reil insula in Alzheimer's disease. Am J Alzheimers Dis Other Demen. 2017;32(4):222–9.
18. Sliz D, Hayley S. Major depressive disorder and alterations in insular cortical activity: a review of current functional magnetic imaging research. Front Hum Neurosci. 2012;6:323.
19. Lee HY, Tae WS, Yoon HK, Lee BT, Paik JW, Son KR, et al. Demonstration of decreased gray matter concentration in the midbrain encompassing the dorsal raphe nucleus and the limbic subcortical regions in major depressive disorder: an optimized voxel-based morphometry study. J Affect Disord. 2011;133(1–2):128–36.
20. Ossenkoppele R, Pijnenburg YA, Perry DC, Cohn-Sheehy BI, Scheltens NM, Vogel JW, et al. The behavioural/dysexecutive variant of Alzheimer's disease: clinical, neuroimaging and pathological features. Brain. 2015;138(Pt 9):2732–49.
21. de Araujo IE, Geha P, Small DM. Orosensory and homeostatic functions of the insular taste cortex. Chemosens Percept. 2012;5(1):64–79.
22. Avery JA, Powell JN, Breslin FJ, Lepping RJ, Martin LE, Patrician TM, et al. Obesity is associated with altered mid-insula functional connectivity to limbic regions underlying appetitive responses to foods. J Psychopharmacol. 2017;31(11):1475–84.
23. Mandelli ML, Vitali P, Santos M, Henry M, Gola K, Rosenberg L, et al. Two insular regions are differentially involved in behavioral variant FTD and nonfluent/agrammatic variant PPA. Cortex. 2016;74:149–57.
24. Gu Y, Scarmeas N, Cosentino S, Brandt J, Albert M, Blacker D, et al. Change in body mass index before and after Alzheimer's disease onset. Curr Alzheimer Res. 2014;11(4):349–56.
25. Luchsinger JA, Cheng D, Tang MX, Schupf N, Mayeux R. Central obesity in the elderly is related to late-onset Alzheimer disease. Alzheimer Dis Assoc Disord. 2012;26(2):101–5.
26. Luchsinger JA, Gustafson DR. Adiposity, type 2 diabetes, and Alzheimer's disease. J Alzheimers Dis. 2009;16(4):693–704.
27. Dineley KT, Jahrling JB, Denner L. Insulin resistance in Alzheimer's disease. Neurobiol Dis. 2014;72(Pt A):92–103.
28. Stephani C, Fernandez-Baca Vaca G, Maciunas R, Koubeissi M, Luders HO. Functional neuroanatomy of the insular lobe. Brain Struct Funct. 2011;216(2):137–49.
29. de Morree HM, Rutten GJ, Szabo BM, Sitskoorn MM, Kop WJ. Effects of insula resection on autonomic nervous system activity. J Neurosurg Anesthesiol. 2016;28(2):153–8.

30. Colurso GJ, Kan RK, Anthony A. Microdensitometric measures of cytoplasmic RNA and total protein in pyramidal neurons of the insular cortex and midfrontal gyrus in patients with Alzheimer's disease. Cell Biochem Funct. 1995;13(4):287–92.
31. Philippi N, Kemp J, Constans-Erbs M, Hamdaoui M, Monjoin L, Ehrhard E, et al. Insular cognitive impairment at the early stage of dementia with Lewy bodies: a preliminary study. Geriatr Psychol Neuropsychiatr Vieil. 2017;15(3):329–38.
32. O'Doherty J, Rolls ET, Francis S, Bowtell R, McGlone F. Representation of pleasant and aversive taste in the human brain. J Neurophysiol. 2001;85(3):1315–21.
33. Gorno-Tempini ML, Dronkers NF, Rankin KP, Ogar JM, Phengrasamy L, Rosen HJ, et al. Cognition and anatomy in three variants of primary progressive aphasia. Ann Neurol. 2004;55(3):335–46.
34. Choi J, Jeong Y. Elevated emotional contagion in a mouse model of Alzheimer's disease is associated with increased synchronization in the insula and amygdala. Sci Rep. 2017;7:46262.

Contributions of the Insula to Speech Production

20

Christoph J. Griessenauer and Raghav Gupta

20.1 Speech Apraxia and the Insula

Speech production is a complex physiological process involving the rapid and accurate coordination of nearly 100 different muscles [1]. Several muscle groups have been implicated in the production of speech including the respiratory muscles which enable voluntary expiration and maintain constant pressure below the glottis; the vocal cords which adduct and abduct to produce voiced and unvoiced sounds, respectively; and the articulators which modify expired air [2]. Speech apraxia is a motor speech disorder in which the muscles involved in speech production fail to produce the correct sounds of words in the appropriate order and with the appropriate timing; this differs from dysarthria which refers to weakness of the muscles involved in speech production [1]. It is characterized by speech with abnormal intonation or rhythm, significant phonetic variability (i.e., articulation errors or distortions), syllable segmentation, impairments in initiation, and frequent false starts [3]. Given that muscle strength and range is unaffected, speech apraxia is believed to be a higher-level disorder of speech motor control distinct from Broca's aphasia, conduction aphasia, and dysarthria [4]. It has been reported as the first manifestation of neurodegenerative disorders such as corticobasal degeneration and nonfluent progressive aphasia [5]. The role of the insula in the production of speech has been investigated in several studies in the past, particularly to understand the pathophysiology of speech apraxia [1–4, 6]. The identification of a singular region in the brain resulting in speech apraxia has been controversial. Speech apraxia has been associated with lesions in Broca's area, the left frontal or temporoparietal cortices, subcortical structures such as the basal ganglia, and the left anterior insula [5]. Patients suffering from nonfluent progressive aphasia with speech apraxia show atrophy in all the aforementioned structures [7].

C. J. Griessenauer, M.D. (✉)
Department of Neurosurgery, Geisinger Medical Center, Danville, PA, USA

Department of Neurosurgery, Paracelsus Medical University, Salzburg, Austria

Research Institute of Neurointervention, Paracelsus Medical University, Salzburg, Austria

R. Gupta, B.S.
Department of Neurosurgery, Rutgers New Jersey Medical School, Newark, NJ, USA

20.2 Clinical Studies of Insular Speech and Language Function

Dronkers analyzed strokes in patients with and without speech apraxia. In the study published in *Nature* in 1996, he demonstrated that in 25 patients with speech apraxia, the location of the stroke overlapped with a region in the left precentral gyrus of the insula. In comparison to a group of 19 patients with left middle cerebral artery infarctions without speech apraxia, this area was completely spared, indicating an important role of the insular in the pathophysiology of speech apraxia.[1] The major limitation of the study is that it is not the likelihood of speech apraxia that is assessed in patients with the lesion but the likelihood of the lesion in patients with the deficit. Thus, it may be that all patients with apraxia of speech have insular infarction, but few patients with insular damage have speech apraxia [8]. Cerebral infarctions limited to the insular cortex may provide additional insight into the relationship with speech apraxia. These infarctions are incredibly rare, however. A registry of 4800 first-ever acute strokes only identified 4 such cases. In two of those cases, the lesion was located in the posterior intrasylvian cortex in the left hemisphere and resulted in articulatory deficits coupled with fluent and nonfluent aphasia [9]. Most strokes involving the insular cortex are embedded in larger middle cerebral artery territory infarctions. Thus, current evidence is limited to few case reports of isolated insula lesions reporting on speech and language deficits. Most do not fit the concept of speech apraxia [3]. In a study of patients with acute left-sided middle cerebral artery territory strokes with and without insular involvement, no association of speech apraxia and injury to the left insula was noted. Apraxia of speech was encountered in patients with damage or ischemia of the left posterior inferior frontal gyrus regardless of insula involvement [8]. In addition to cerebrovascular lesions, brain tumors limited to the insula represent another disease entity to elucidate the role of the insula in speech and language function. In a study of 30 patients with intrinsic tumors involving the insula, 3 had isolated involvement of the left insular cortex. No speech and language deficits were recorded after surgical resection [10]. Even removal of the entire language-dominant insula in a glioma patient did not result in speech apraxia [11].

Slowly evolving pathologies may trigger a compensatory mechanism in the contralateral insula, however [3]. In summary, reports of speech and language deficits associated with insular pathologies paint a rather inconclusive picture and may in part be related to injury of the adjacent frontal operculum and its connections.

20.3 Functional Imaging Studies of Insular Speech and Language Function

Initial positron emission tomography (PET) investigations of speech motor control showed hemodynamic activation in bilateral primary motor and sensory cortices, the supplemental motor area, the cerebellar hemispheres, and, surprisingly, in the depth of the Sylvian fissure [12]. This spot was later linked to the anterior insula in a more recent follow-up PET study [2]. Functional imaging from PET and MRI (fMRI) provides evidence for a "minimal brain network" critical for motor aspects of speech and language production. This network is lateralized to the left at the level of the insula and also includes the mouth region of the bilateral motor cortices, Broca's area and the left inferior precentral gyrus, the left supplemental motor cortex, the basal ganglia, and the cerebellum [3, 13]. Another fMRI study assessing the temporal aspects of blood oxygen level-dependent (BOLD) activation during syllable repetition identified two clusters of cerebral structures involved in speech motor control. The preparative loop, consisting of BOLD signal in Broca's area, the supplemental motor cortex, anterior insula, and superior cerebellum, showed significantly earlier hemodynamic activation than the executive loop made up by the primary sensory and motor cortices, thalamus, basal ganglia, and inferior cerebellum [14]. Another fMRI study localized the response to the junction of the insular cortex and the frontal operculum [13]. In summary, functional imaging studies found fairly consistent activation of the insula, predominantly in the dominant hemisphere, with tasks targeting higher-order articulation.

20.4 Autonomic Aspects of Speech Motor Control and the Insula

Functional imaging studies identified other distinct area in the insula, apart from the transition zone between the insula and the frontal operculum, during speech production. This raises the question as to whether the insula is involved in speech and language function not related to higher-order articulation. There is convincing evidence that the insula is an important regulator of autonomic functions such as the control of blood pressure or heart rate, for instance [15]. In speech motor control, the link to the autonomic system is its intimate relationship to the respiratory system [3]. Information on the state of the respiratory system is relayed to the insula as evidenced by functional imaging obtained in patients experiencing dyspnea [16]. In animal models, stimulation of the insular cortex has shown to result in respiratory slowing and respiratory arrest [17]. Still, respiratory depression is not a feature of speech apraxia [3]. Conclusively, insular cortical projections may have modulatory autonomic effects on respiratory activity while speaking and may act in parallel with existing corticobulbar and corticospinal systems that are under voluntary control. While motor and premotor cortical input may be responsible for phasic or brief linguistic adjustments of respiration, such as in between sentences, the tonic state of hyperventilation required while speaking may be under control by the insula [3, 18].

Conclusions

Speech apraxia, a higher-order articulation disorder, has been associated with injury to Broca's area, the left frontal or temporoparietal cortices, subcortical structures, and the left anterior insula. Isolated lesions of the insular cortex are very rare and, in most instances, do not resemble the symptomatology of speech apraxia. Functional imaging studies, however, demonstrate conclusive evidence for activation of the left anterior insula adjacent to the frontal operculum with articulatory linguistic tasks. In addition to its role in articulation, the insula may also function as a relay center for autonomic signals pertaining to the linguistic respiratory state and may serve as a corollary to voluntary control of respiration during speech mediated by corticobulbar and corticospinal projections.

References

1. Dronkers NF. A new brain region for coordinating speech articulation. Nature. 1996;384(6605): 159–61.
2. Wise RJ, Greene J, Büchel C, Scott SK. Brain regions involved in articulation. Lancet. 1999; 353(9158):1057–61.
3. Ackermann H, Riecker A. The contribution(s) of the insula to speech production: a review of the clinical and functional imaging literature. Brain Struct Funct. 2010;214(5–6):419–33.
4. Ackermann H, Riecker A. The contribution of the insula to motor aspects of speech production: a review and a hypothesis. Brain Lang. 2004;89(2):320–8.
5. Ogar J, Slama H, Dronkers N, Amici S, Gorno-Tempini ML. Apraxia of speech: an overview. Neurocase. 2005;11(6):427–32.
6. Oh A, Duerden EG, Pang EW. The role of the insula in speech and language processing. Brain Lang. 2014;135:96–103.
7. Gorno-Tempini ML, Rankin KP, Woolley JD, Rosen HJ, Phengrasamy L, Miller BL. Cognitive and behavioral profile in a case of right anterior temporal lobe neurodegeneration. Cortex. 2004;40(4–5): 631–44.
8. Hillis AE, Work M, Barker PB, Jacobs MA, Breese EL, Maurer K. Re-examining the brain regions crucial for orchestrating speech articulation. Brain J Neurol. 2004;127(Pt 7):1479–87.
9. Cereda C, Ghika J, Maeder P, Bogousslavsky J. Strokes restricted to the insular cortex. Neurology. 2002;59(12):1950–5.
10. Zentner J, Meyer B, Stangl A, Schramm J. Intrinsic tumors of the insula: a prospective surgical study of 30 patients. J Neurosurg. 1996;85(2):263–71.
11. Duffau H, Bauchet L, Lehéricy S, Capelle L. Functional compensation of the left dominant insula for language. Neuroreport. 2001;12(10):2159–63.
12. Petersen SE, Fox PT, Posner MI, Mintun M, Raichle ME. Positron emission tomographic studies of the processing of singe words. J Cogn Neurosci. 1989;1(2):153–70.
13. Bohland JW, Guenther FH. An fMRI investigation of syllable sequence production. NeuroImage. 2006;32(2):821–41.

14. Riecker A, Mathiak K, Wildgruber D, Erb M, Hertrich I, Grodd W, et al. fMRI reveals two distinct cerebral networks subserving speech motor control. Neurology. 2005;64(4):700–6.
15. Oppenheimer SM, Gelb A, Girvin JP, Hachinski VC. Cardiovascular effects of human insular cortex stimulation. Neurology. 1992;42(9):1727–32.
16. Banzett RB, Mulnier HE, Murphy K, Rosen SD, Wise RJ, Adams L. Breathlessness in humans activates insular cortex. Neuroreport. 2000;11(10): 2117–20.
17. Sugar O, Chusid JG, French JD. A second motor cortex in the monkey, *Macaca mulatta*. J Neuropathol Exp Neurol. 1948;7(2):182–9.
18. Shea SA. Behavioural and arousal-related influences on breathing in humans. Exp Physiol. 1996;81(1): 1–26.

21. Processing Internal and External Stimuli in the Insula: A Very Rough Simplification

Alfonso Barrós-Loscertales

A grand unified theory of the brain [...]. Like the von Neumann architecture, it would be a unified mechanistic framework for how the brain processes information, organizes perception, drives action, and supports cognition and consciousness. (Paul King, 2014; Quora)

Let's say that the brain works as a present prediction organ. Its on/off function depends on any unpredicted/predicted fact. While off, the brain works according to inner and outer homeostatic goals. While on, the brain works by seeking information to adjust prediction and to include the necessary adjustments in the present prediction. What are signaling predicted facts? Motivation. What are signaling unpredicted facts? Emotion. What's going on when brain is on? It processes or "digests" information. What's the role of the insula in this present prediction organ? Craig [1] proposed the human insula's role in the "progressive" integration of all brain networks, and inner and outer salience feelings, which lead to emotion signaling of unpredicted changes. The general metaphor of the brain working as a present prediction organ may help us to understand brain working at the cortex level. Thus, brain prediction may be a main force that drives the neural function linked to cognition, emotion, motivation, action processes, and consciousness [2–8].

An everyday situation to exemplify this fact is when we are in a waiting room at a hospital. We arrive at the hospital in time for our appointment. We may expect a slight delay in the doctors' appointment, but, beyond an outer ("watch") and inner ("patience") internal error prediction, we will start to look for information that solves the unpredicted delay in our appointment. In experimental research, unpredicted facts are relevant and immediately processed by our brain as they are important for our goals. For example, errors in prediction are important when identifying any sudden change in our body or environment but are irrelevant when the current task or attentional set does not involve such a prediction, despite being processed [9]. Our brain seems to work by being guided by a predictive code that works efficiently and allows individuals conflict resolution or change adaptation by deviance mediated by the insula and its connections [10]. The significance of unpredicted facts is reflected in how our brain activates to deviant or novel events [3, 11], and, therefore, any deviant—unexpected or new—event usually increases brain activation over any other expected event, depending on its relevance for individual's inner and outer goals.

A. Barrós-Loscertales
Departamento Psicología Básica,
Clínica y Psicobiología, Universitat Jaume I,
Castelló de la Plana, Spain
e-mail: barros@uji.es

Researchers focus on brain studies to define several cognitive, emotional, and motivational processes as being present in the brain from our analysis of human behavior. However, I wonder whether our behavioral analysis might merely be products of a present prediction organ rather than a correct way to describe its actual functioning. Accordingly, the analysis of behavior may both guide and blur our understanding of brain working insomuch as we may be biasing, lacking, or omitting an important part of information biased by our preconception on brain organization according to the way we study behavior. We should not forget that interoception—the sensing and perception of internal bodily changes—affects brain working as much as external stimuli or contexts. However, given the difficulties of operationalizing it, interoception is usually missed as a variable that leads to inconsistencies in how it is defined and quantified. Interoception is differentiated from exteroception, or the sensing of stimulation from outside our body, proprioception, or kinesthesia, as the sensing of the body's position in space. Interoception may be defined as the sense of stimulation from visceral organs and vascular system in conscious or unconscious terms from low-level homeostatic control processes in the peripheral, brainstem, and subcortical brain structures to conscious visceromotor sensations [12]. Nowadays, we can combine the analysis of central and peripheral nervous system measures with self-report or reaction time measures in an attempt to view a whole environment-body-brain picture in which the insula is suggested to be responsible for signaling and integrating any mismatch. However, measuring interoception is far from a precise variable and gives way to unexpected behavioral outputs, insofar as we cannot experimentally measure or control inner stimulation or states, that is, interoception. More often than not, we apply procedures using a wide range of stimuli, e.g., pain, temperature, emotional or motivational stimuli, or mood induction, from which we assume interoception. Otherwise, there are proposals for the way we can attempt to directly measure interoception in three different dimensions: *accuracy*, measuring performance in behavioral tests of heartbeat detection; *sensitivity*, by self-assessments of subjective interoception using interviews and questionnaires; and *awareness* as metacognitive awareness of interoceptive accuracy by analyzing confidence and accuracy correspondence. All three dimensions are suggested as being distinct and dissociable [13]. The insula seems eloquent at structural and functional levels for interoception accuracy, sensitivity, and awareness [14, 15] and serves as a hub to mediate from interoception to emotional processing, to decision-making, to individual differences, and to social behavior [12, 15]. Meanwhile at a grounding level, we make attempts to simplify how the insula and the brain work as a way to minimize prediction errors by perceptually sampling the environment and encoding new events, tested from perception to action, once new information is integrated as a guiding principle for brain work at the cortical level [16]. This description has been previously defined from physics models of free energy, which assume hierarchies and dynamics sensitive to environmental changes. This approach may serve to reduce functional segregation in the insula—and the brain—to prediction checking and adaptation processes by involving perception, emotion, motivation, cognition, action, and consciousness and, therefore, by simplifying the description of brain working to prediction coding. Likely, this simple approach is incomplete and needs further development, adaptation, and restructuration, but it is worth considering that a main principle guides brain-way-of-working. Otherwise, we usually attempt to map mental processes onto the brain, but, what "mental processes"? And, what task do we use to manipulate and measure mental processes at several levels and, in the end, to map those mental processes to brain functioning? Experimental psychology and cognitive neuroscience have developed a wide range of behavioral tasks over 100 years (e.g., the Sternberg paradigm, a visual attention task, the N-Back task) to measure the mental processes between the behavioral and brain levels and to make great contributions to human behavior and brain understanding. However, mental processes are defined in several different ways, and experimental paradigms vary in the stimuli and the instructions given to the participants and their expected response. Initiatives, like the Cognitive Atlas [17], accessible online at http://www.cognitiveatlas.org, and the Cognitive Paradigm Ontology [18], accessible online at http://www.cogpo.org, work and lead to a unified framework to develop experimental psychology and cognitive neuroscience. Ultimately, these and other initiatives will overcome the human limitation in "cognitive resources." For a researcher, the research results—and even data sets—are more available than ever, thanks to the World Wide Web. At the same time, it may be discouraging to attempt to view a picture of the state of the art in any brain function or structure, such as the insula. It is so easy to miss a paper that may be crucial for our experimental design and find it when we are running our experiment that it is important to integrate discipline and transdiscipline research results. I expect that neuroinformatics, among other disciplines, will help to change the way we run our experiments, explore the literature, and synthesize knowledge to reach new and informed conclusions (see [19]). By way of example, new web-based large-scale informatic approaches have already led to an unknown functional partition of the insula's structure and mapping functions [20]. Likewise unified theories, such as the free-energy theory of brain response, involve the insula as a main hub on a brain wide web [10], in which the same inferential mechanisms would apparently underlie apparently different processes, such as perception, emotion, motivation, cognition, and action [21].

21.1 The Free-Energy Model and Prediction: A General Overview

The brain may be an energy-efficient system in a random, fluctuating environment. To be efficient, it must to be selective. Thus it should not remember everything that happens, but it may balance memory against prediction. So, it uses similar memory systems for remembering and future prediction [22]. In this context, the insula is suggested to provide a metrics for the amodal computation of homeostatic efficiency. However, these energetic models, like other prediction models, suggest a hierarchy in brain function remembering behavioral approaches that suggest serial processes between sensory stimulation and overt behavior [2]. The rationale of a hierarchical conception stems from evidence at the neurophysiological level, which suggests that the feedback connection from "high-level areas" shapes "lower-level areas" properties by tuning the neuron's response specificity through processing stages [16, 23]. This brain processing conception involves a limitation by encompassing a categorical brain organization conception, even in terms of brain networks. In free-energy models, hierarchical processing involves top-down testing of predictions against sensory input across different levels. Once a prediction error is detected, lower levels in the information flow return such information at higher levels to update predictions [8, 16]. However, free-energy predictions are called in terms of templates [24], which have been considered too rigid and limited in human vision, while human vision models usually rely on another kind of hierarchical organization [7, 23].

An alternative perspective may be more generally agreed as being parallel recurrent and transparallel processing in brain function [8]. Therefore, it should be difficult to find a beginning and end from stimulation to action in cortical brain function. Someone may argue that, although interrupted by new information, processing is serial from inner or outer stimulation to action, although cognitive processing may work in parallel. However, the modulation of motor on sensory processes may serve as a cue to rule out the serial order of processing, or even hierarchies, inherited from a categorical perception in brain research. Otherwise, we may argue that modality-specific systems are hierarchically organized but work according to the same underlying imperative minimizing prediction errors [21] as a present prediction organ. Nonetheless, top-down influences may be a useful metaphor to understand the predictive working of brain processing, like the "hierarchical" influence, which suggest only the brain prediction code. Thus, we know that these top-down influences are changed by inner and outer stimulation and top-down and bottom-up processes. For example, repeated drug administration involves brain changes that dysregulate the brain prediction code, and we know that insula damage involves major changes in addiction phenomena [25]. Brain way-of-working may be considered a continuous, rather than a categorical, way of working in which top-down and bottom-up processes are complimentary [26]. It suggests that brain function at the insula is a continuous function that changes continuously in a global neuronal workspace.

21.2 Emotion and Motivation, Insula, and Prediction

Emotion has been described as a response system and a feeling state or process that motivates and organizes cognition and action [27]. Emotion works a reactive system to unpredicted facts as an error prediction system. Motivation may involve an inner goal that arises from emotion reactivity to unpredicted moments and serves to delineate a new prediction code that had better adjusted to the environment. Indeed, emotion has been considered both, input for motivated goal behavior and output of motivation when goal conflicts are encountered [28]. Otherwise, is the insula a good interface (between) for feelings, cognition and action? The insula is considered a relay structure whose role involves the joint processing of inner and outer information processing. How the insula detects unpredicted facts may involve the ascending sensory pathways that underlie feelings from body signaling environmental unpredicted changes by serving as homeostatic sensory integration [1], which coincides with James-Lange's theory of emotion and the somatic marker hypothesis. Therefore, it has been suggested that the insula underpins emotions from feelings and awareness, while the anterior cingulate underpins motivation to action. Both regions seems to co-work in the salience network, particularly the anterior insula [29], but other networks involve this and other portions of the insula with other brain regions [20] in human cognition and behavior. Therefore, the insula is likely to interface the connection among emotion, motivation, and action, by linking detection to reaction as a node in the continuum between emotion and motivation if they cannot be considered to differ from one another [30–32].

21.3 Cognition, Insula, and Prediction

Attention, memory, language, and cognitive control are among those processes that we refer to with cognition. Cognition may come about to solve any error prediction by "digesting information" in a way that solves dissonance, the prediction error (Fig. 21.1). Memory

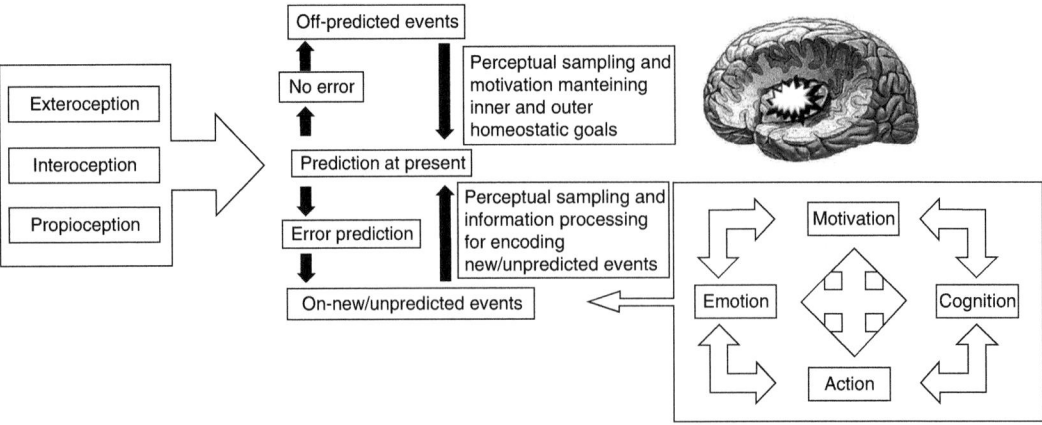

Fig. 21.1 Schematic representation of the insula's involvement in brain function as described in the chapter

may serve to remember the past and to image the future by creating a present prediction [22]. Likewise prediction may involve reward and pain and, therefore, basal ganglia and insula as the archetype for these functions, respectively. However, the whole brain may be involved in predictive processing, although it has been suggested that different aspects can be captured by the various roles played in distinct regions and that insula may play a role in these roles [20]. A different question is whether distinct cognitive and motivational roles are involved in error processing, independently of its consciousness. Another question is whether consciousness is necessary to run information processing after error processing. Error prediction-related brain activity involves a large set of regions. Among them, the rostral ACC, the supplementary motor area, and the insula (ventral) are considered hubs that process goal-directed cognition [20, 33, 34]. Regarding its necessity for error awareness, Klein et al. [35] have shown insula differential activation in error awareness, particularly in its anterior-inferior part. These authors have discussed that the role of the insula in error awareness may be mediated by visceral reactions to error and, therefore, autonomic responses and interoceptive processing that mediate external error consciousness. Thus, the insula may be considered a central hub at the brain cortex and body-brain interactions.

Another question is to what extent does the detection of bodily sensation guide our cognitive processing? Following the somatic marker hypothesis, the insula has been shown to be a central contribution of autonomic response to brain activity that underlies decision-making and other cognitive processes. In this context, the insula has been involved in individual differences in the risky decision-making that leads subjects to a bias in their certain perception and, therefore, to making safe decisions [36, 37]. Likewise its connectivity with the dorsolateral prefrontal cortex suggests that it plays a role in emotional regulation [38] as it has been shown to be affected in high-anxiety individuals [39].

Historically, cognition and emotion have been seen as separate entities [40, 41]. The behavioral analysis of mental processes has likely led to this differentiation based on construct validation, experimental analysis, and statistical analysis purposes. Chang et al. [20] observed anatomical insula organization. They observed the insula's parcellation in three parts: dorsoanterior, ventroanterior, and posterior. The dorsoanterior insula is more consistently involved in human cognition. This insula region has been functionally connected to the anterior cingulate cortex, the dorsolateral prefrontal cortex, the dorsal striatum, and the temporoparietal junction. The dorsoanterior insula has shown more consistent activations than the other two insula subdivisions in relation to 200 other psychological processes, including cognition, emotion, chemosensation, autonomic function, pain, language, and sensorimotor processes. All this agrees with previous results which have shown that the insula displays higher activation rates than other regions in neuroimaging studies [34]. However, when decoding mental states from brain activation, the dorsoanterior insula and its functionally connected regions have been associated with psychological processes related to cognitive tasks and executive control.

Experts generally agree that connections exist among emotion, cognition, and action [42]. They also agree that different factors activate emotion or a change in the ongoing emotion from innate, conditioned, cognitions, social interactions, and ongoing goals to spontaneous changes in neurobiological systems. Emotion, motivation, and cognition are commonly associated with goal-directed actions. Goals make behavior vary adaptively thanks to our cognitive control. But why does our behavior need to vary adaptively according to our goals? We may lead our behavior according to inner or outer goals [1]. These goals motivate our actions, but we likely encounter error prediction during the process to meet our goals. Inzlicht et al. [42] have directly suggested that goal conflicts lead to negative affect, which makes them salient and motivates goal-directed actions to resolve the conflict throughout cognition and action in order to minimize its recurrence and repetition.

In the last few years, some proposals have deleted the boundaries among cognition, motivation, and emotion. It has been suggested that cognitive and emotion/motivation segregation is lost at higher levels of specialization [32, 43]. In emotion, the insula is considered an extension of the core emotional circuit structurally defined as the limbic system, and the insula has been included as a paralimbic structure. Similarly, the insula is defined as a core region in cognitive processing, including higher-order functions under the umbrella of cognitive control or executive functions [20, 33]. Therefore, the emotional and cognitive dissociation of the insula depend on the variables involved in our experimental processes and in the conceptual context where we define our experiment. Maybe we should reflect that keeping behavioral construction in our understanding of brain function and structure forces the brain function interpretation to shift to a limited context [20]. The analysis of our behavior has been capital to understand brain function to date, although new approaches may overcome this approach and be welcomed and tested, even in the absence of a behavioral comparison [44, 45]. We may note that as we open the categories in which we attempt to define brain function, we go from functional segregation to functional integration. However, when we reach at the next integration level, for example, from the brain regions involved in interoception to the brain networks involved in interoception, once again, we categorize brain networks which we state at the same level. In the end, I suggest looking for a fully integrated approach to brain functioning or beyond (e.g., peripheral nervous system, body states). While we continue to define the insula from a separated emotion, motivation, cognitive, action, and conscious point of view, without integrating all these categories, we might be looking at what is the same, and we may not see what is actually going on. The brain works as interconnected brain regions, interconnected brain networks, and, finally, interconnected brains in social contexts, but where a definition of its works as a system restricts its own understanding.

21.4 Where Subjectivity Comes From

Subjective experience of this brain-way-of-working as a present prediction organ involves the consciousness of always being in the past (memory) or in the future (anticipation) by recovering information and testing its application to unpredicted facts. Therefore, our conscious experience may emerge for brain prediction work as a more efficient way to test future conditions according to current and previous experiences but always in the present and with continuous changes. Thus, subjective experience may involve only a more evolved way of pattern testing when previous ones failed. A question that comes to my mind is how to find the way that subjectivity works. In my opinion, psychology has been doing this for more than 100 years, and neuroscience suggests the insula as a "cinemascope model of human awareness and subjectivity" [1].

The insula is considered a critical area for the subjective experience of emotion [1, 46]. As previously noticed, the insula is related to the processing of interoceptive and environmental information. Likewise, the subjective experience of interoception or interoceptive awareness relates to the insula's function [14]. The association between the insula's activation and emotional traits is important to understand the role that the insula plays in subjective experience while it works during present prediction, be it based on present, but also future consequences.

Paulus and Stein [47] proposed that increased anxiety and depression affect are triggered by predictive signals of the aversive bodily state associated with negative consequences. Therefore, the anxiety that involves a body state with exaggerated interoceptive predictive sensitivity and the insula seems to play an important role in these processes. However, the studies that have applied emotional stimuli have not generated a clear condition to differentiate the insula's role in interoceptive sensitivity or as an accompanying response to the emotional autonomic response. Otherwise, Terasawa et al. [15] have shown that greater sensitivity to interoceptive information, as measured by anxiety trait scores, is related to the anterior insula during a task-involved self-report assessment of emotional and body awareness—as either current or usual—rather than emotional stimuli. These authors have also shown that insula activation mediates the association between thalamus activation during emotional awareness and social fear, as well as the insula's association with personality traits during interoception. Thus, anxiety and personality traits are thought to represent individual differences to emotional experience [47], and the insula may mediate this subjective experience by endorsing its interdependence with internal physiology.

21.5 Prediction Error, Insula, and Disease

Inefficient energy function may lead to brain "indigestion"; in other words, inefficient prediction or information processing may lead to brain dysfunction and disease. Both strict organization and disorganization may be manifested as an inefficient energy function that leads to brain-related disease or to different degrees of individual adaptation to the environment [48]. Hence the hyper- or hypoactivation observed in fMRI studies into different kinds of diseases may involve even hyperorganization or disorganization. Thus, anxiety disorders show hyperactivation of the insula, which suggests its responsibility for generating fear responses to symptom-provoking stimuli [49]. The right and left insula are suggested to play different roles in emotions like empathy, with the left insula being activated by all categories (pain, positive, and negative) but the right insula only by negative emotions [46]. This scenario suggests different energy consumptions by a "negative" right insula in contrast to a "positive" left insula [1] (but also see [39]).

The insula is suggested to generate a template for homeostatic sensory-motor conditions, which identify information about the unpredicted fact. Thus, emotion should be considered a signal to action, rather than a fact that needs identification and processing by itself, except in case of psychopathology or dysadapted behaviors in the therapeutic context. For example, posttraumatic stress disorder may be interpreted as an unsolved emotional reaction to unpredicted changes or facts. Thus unpredicted changes involve the traumatic fact, while the strong emotional reaction keeps attempting to integrate an unpredicted fact that is difficult to integrate. In such cases, individual accompaniment therapies, as well as group or individual meetings, are important to help find the information needed to reply to the emotional reaction which will, ultimately, guide reaction and action [50]. The process of emotion perception, valuation, and action may be conscious [51] or unconscious [52]. In this context, the insula and the amygdala are suggested to be key components of a common neurobiological pathway for several anxiety disorders [49, 53] through which, given its extended connections to other cortical and subcortical structures, its hub role in these disorders is suggested. In the affective disorders or psychopathology context, emotional unaware perception, valuation and consequent motivation, and action processes fail due to either rigidity or disorganization. Therefore, consciousness of emotion may not suffice for successful emotional processing [46] but may be necessary in some kinds of psychotherapies.

In the disease context, the concept of presence is used to refer to the subjective sense of the reality of the world and the self within the world [5]. Therefore, the manifestation of dissociative psychiatric disorders, such as derealization and depersonalization as loss of subjective sense of the world and the self in the world, respectively, are conditions that have been defined as clinical that may help us to understand the presence concept [5]. Likewise, the virtual reality condition may help to identify how presence can be generated even in situations in which it would normally be lacking; e.g., terror movies as a condition to generate another condition in which an individual reacts to an unpredicted reality, even when it is known as be false, but they react as if it were real. In this context, the anterior insula and associated brain regions are suggested as the virtual location of predictive coding for prediction and error prediction. It should be noticed that the anterior insula is particularly suggested for interoceptive coding, e.g., understanding interoception as the perception of the body's physiological condition, based on, for example, its limited activity in depersonalization disorders characterized by reduced activation in emotion awareness regions, but also increased activation in emotional regulation regions [54]. However, I would suggest considering exteroception and proprioception prediction coding given its role as a multimodal/transmodal sensory region [55]. The insula will act as prediction coding for "inside-in," "outside-in," and bottom-up processing streams for survival. Simultaneously, the role of expectations built and refined in the brain act as an "inside-out" and top-down processing stream with the same survival role in a complementary process. The insula and the brain constantly act by predicting incoming flows of information. However, but when a prediction error occurs, the body and the brain then detect and react through action that is motivated to look for information, which solves the prediction error or at least reduces "dissonance." In this context, schizophrenia is related to imprecise predictions about the sensory consequences of actions, and this condition may serve as another example of brain dysfunction as present prediction organ; then there is anxiety propensity to detect internal bodily states by forming the basis of a misinterpreted signal which may be associated with potential aversive or negative outcomes.

21.6 Insula Modularity and Connectivity

Given our difficulties to see a process as a continuum and to see instead extract inferences, we usually focus on discrete categories, modules, or spatial representations of connectivity patterns that, once again, draw a single picture from a process operationalized in, for example, time series. This approach has been fruitful for understanding the brain [56] and science development in general. Thus, we talk about parallel/serial processing, afferent/efferent, unimodal/multimodal, space/time, analytical/emotional, and external/internal top-down/bottom-up levels of processing. Some of these categorical distinctions serve to guide our analysis of insula organization—and the brain, by extension—through different kinds of analyses. Chang et al. [57] and Deen et al. [58] have proposed an optimum tripartite subdivision of the insula based on voxels that share similar patterns of connectivity with the rest of the brain by applying time/space independence-based analyses and semantic categorical approaches to analyze of the insula's organization. Therefore, the segregated subdivision of the insula involves different connectivity patterns by implicating a posterior medial subdivision, a dorsoanterior subdivision, and a ventroanterior subdivision.

The segregated functional connectivity of these three regions slightly differs in both studies. The ventroanterior insula has been functionally connected to the amygdala, the ventral tegmental area, the superior temporal sulcus, the medial prefrontal cortex, the posterolateral orbitofrontal cortex [20], and the pregenual anterior cingulate cortex [57]. The dorsoanterior insula has been functionally connected to the anterior cingulate cortex, the dorsolateral prefrontal cortex, the dorsal striatum, and the temporoparietal junction [20, 57]. Thirdly, the posterior insula has been connected to the supplementary motor area, the somatosensory cortex, the posterior temporal lobes, the right hippocampus, and the pregenual anterior cingulate cortex ([20, 57]. In terms of processes, Chang et al. [20] have shown by decoding analysis that the ventroanterior insula is related to emotion, chemosensation, and autonomic function; the dorsoanterior insula network is related to cognitive task and executive control; and the posterior insula network is related to pain, sensorimotor, and language. All these sets of results have been added to previous evidence to support a modular subdivision of the insula based on its connectivity. In fact the subdivision of the anterior insula had not been suggested before these studies. At the same time, this new result—just like any other—allowed us to notice that we may be unaware of a thousand ways of functional subdivision at the structural and functional levels, and we may wonder whether our way to analyze the brain structure and function may involve twofold limitation in the brain's understanding: first, statistical and methodological data approaches, although data driven, are applied to divide structure and function and second, the way we categorize behavior affects our understanding of the brain function. These two reasons are strong and weak points in our work to understand the insula and, by extension, the brain. First, they allow our understanding, but, simultaneously, if we do not overcome the restrictions of these constructs or semantic categories when approaching the analysis of brain function as suggested by others [34], our work done to discover its way of functioning might go on forever. Please notice that these are merely ways of thinking and discussions based on my own experience and my approach to literature reading as I am likely applying similar methodological approaches that you follow to understand the insula and the brain. Therefore, I suggest looking rather than judging the insula's way of working as Yarkoni and Westfall [59] expressed as "being exploratory rather than confirmatory."

Conclusions

The way of naming the brain as a present prediction organ is a metaphor to highlight the need for information seeking to propose a general principle on how the brain works. In this chapter, I suggest that the brain maintains an inner pattern representation or prediction code and reacts to any violation of such patterns or unpredicted facts by information processing. Inner pattern representations or prediction codes are based on inner goals and outer goals. The brain would work as an integration of external and internal pattern representations. The insula seems to play all these roles as a central hub in connection to the rest of the brain and particularly linked to processes of perception, emotion, motivation, cognition, action, and consciousness, terms used as a way to gather all terms we use to define brain processes and functions in psychology and cognitive neuroscience. All these processes interact, and research shows that is difficult to establish and order a hierarchy in its way of functioning or interacting. As research advances, the insula is involved in interoceptive processing, emotional processing, guiding motivation, and action through its involvement in higher and lower cognitive processes. Most of the research that we conduct determines our understanding of the inner and outer stimulus processing in the insula as a whole picture in which psychological processes differentiation dissociates the insula function and structure in a probabilistic way. It is likely that future research into insula integration will reflect an unlimited definition of its structure to gain a better understanding of its pictures. However, it will involve a new paradigm in which the brain, body, and environment will likely be integrated into research that involves a rationale which summarizes all these levels but one that lacks the details that we are currently used to. Thus, the use of joint static and dynamic measures of current information processing may help us to understand the information entropy that characterizes brain processing. So interoception goes beyond the insula and beyond other brain regions, such as the orbitofrontal cortex, the ventromedial prefrontal cortex, and the anterior cingulate, which are also involved [26]. To summarize, evidence may indicate that the goal of homeostasis, explained as a free-energy system and integrates other prediction models [26] that is complementary, may serve to understand the function of both the insula and the brain as a whole. In conclusion, this chapter attempts to draw to our attention to the need to look at the brain by exploring new principles for brain functional organization. Likely, the brain function will guide our understanding of its structure, and looking at a general principle or unified theory of the brain will open up new unthinkable approaches for brain studies and applications.

References

1. Craig ADB. Significance of the insula for the evolution of human awareness of feelings from the body. Ann N Y Acad Sci. 2011;1225(1):72–82. https://doi.org/10.1111/j.1749-6632.2011.05990.x.
2. Bubic A, von Cramon DY, Schubotz RI. Prediction, cognition and the brain. Front Hum Neurosci. 2010;4:25. https://doi.org/10.3389/fnhum.2010.00025.

3. Friston K. The free-energy principle: a unified brain theory? Nat Rev Neurosci. 2010;11(2):127–38. https://doi.org/10.1038/nrn2787.
4. Gschwind M, Picard F, Craig ADB, Chang LJ, Yarkoni T, Khaw MW, et al. Building better biomarkers: brain models in translational neuroimaging. Cereb Cortex. 2013;10(3):72–82. https://doi.org/10.1002/cne.23368.
5. Seth AK, Suzuki K, Critchley HD, Kanai R, Frith C. An interoceptive predictive coding model of conscious presence. Front Psychol. 2012;2:395. https://doi.org/10.3389/fpsyg.2011.00395.
6. Singer T, Critchley HD, Preuschoff K. A common role of insula in feelings, empathy and uncertainty. Trends Cogn Sci. 2009;13(8):334–40. https://doi.org/10.1016/j.tics.2009.05.001.
7. Singer W, Lazar A. Does the cerebral cortex exploit high-dimensional, non-linear dynamics for information processing? Front Comput Neurosci. 2016;10:99. https://doi.org/10.3389/fncom.2016.00099.
8. van der Helm PA. Structural coding versus free-energy predictive coding. Psychon Bull Rev. 2016;23(3):663–77. https://doi.org/10.3758/s13423-015-0938-9.
9. Escera C, Alho K, Schröger E, Winkler IW. Involuntary attention and distractibility as evaluated with event-related brain potentials. Audiol Neurotol. 2000;5(3–4):151–66. https://doi.org/10.1159/000013877.
10. Allen M, Fardo F, Dietz MJ, Hillebrandt H, Friston KJ, Rees G, Roepstorff A. Anterior insula coordinates hierarchical processing of tactile mismatch responses. NeuroImage. 2016;127:34–43. https://doi.org/10.1016/j.neuroimage.2015.11.030.
11. Friston K, Kilner J, Harrison L. A free energy principle for the brain. J Physiol Paris. 2006;100(1–3):70–87.
12. Garfinkel SN, Critchley HD. Interoception, emotion and brain: new insights link internal physiology to social behaviour. Commentary on: "anterior insular cortex mediates bodily sensibility and social anxiety" by Terasawa et al. (2012). Soc Cogn Affect Neurosci. 2013;8(3):231–4. https://doi.org/10.1093/scan/nss140.
13. Garfinkel SN, Seth AK, Barrett AB, Suzuki K, Critchley HD. Knowing your own heart: distinguishing interoceptive accuracy from interoceptive awareness. Biol Psychol. 2015;104:65–74. https://doi.org/10.1016/j.biopsycho.2014.11.004.
14. Critchley HD, Wiens S, Rotshtein P, Öhman A, Dolan RJ. Neural systems supporting interoceptive awareness. Nat Neurosci. 2004;7(2):189–95. https://doi.org/10.1038/nn1176.
15. Terasawa Y, Shibata M, Moriguchi Y, Umeda S. Anterior insular cortex mediates bodily sensibility and social anxiety. Soc Cogn Affect Neurosci. 2013;8(3):259–66. https://doi.org/10.1093/scan/nss108.
16. Friston K. A theory of cortical responses. Philos Trans R Soc Lond B Biol Sci. 2005;360(1456):815–36. https://doi.org/10.1098/rstb.2005.1622.
17. Poldrack RA, Kittur A, Kalar D, Miller E, Seppa C, Gil Y, et al. The cognitive atlas: toward a knowledge foundation for cognitive neuroscience. Front Neuroinform. 2011;5:1–11. https://doi.org/10.3389/fninf.2011.00017.
18. Turner JA, Laird AR. The cognitive paradigm ontology: design and application. Neuroinformatics. 2012;10(1):57–66. https://doi.org/10.1007/s12021-011-9126-x.
19. Yarkoni T, Poldrack RA, Van Essen DC, Wager TD. Cognitive neuroscience 2.0: building a cumulative science of human brain function. Trends Cogn Sci. 2010;14(11):489–96. https://doi.org/10.1016/j.tics.2010.08.004.
20. Chang LJ, Yarkoni T, Khaw MW, Sanfey AG. Decoding the role of the insula in human cognition: functional parcellation and large-scale reverse inference. Cereb Cortex. 2013;23(3):739–49. https://doi.org/10.1093/cercor/bhs065.
21. Friston K., Frith C.. A duet for one; 2015. https://doi.org/10.1016/j.concog.2014.12.003.
22. Schacter DL, Addis DR, Buckner RL. Remembering the past to imagine the future: the prospective brain. Nat Rev Neurosci. 2007;8(9):657–61. https://doi.org/10.1038/nrn2213.
23. Huang Y, Rao RPN. Predictive coding. Wiley Interdiscip Rev Cogn Sci. 2011;2(5):580–93. https://doi.org/10.1002/wcs.142.
24. Friston KJ, Stephan KE. Free-energy and the brain. Synthese. 2007;159(3):417–58. https://doi.org/10.1007/s11229-007-9237-y.
25. Naqvi NH, Rudrauf D, Damasio H, Bechara A. Damage to the insula disrupts addiction to cigarette smoking. Science. 2007;315(5811):531–4. Retrieved from http://science.sciencemag.org/content/315/5811/531
26. Gallagher S, & Allen M. Active inference, enactivism and the hermeneutics of social cognition. Synthese. 2016:1–22. https://doi.org/10.1007/s11229-016-1269-8.
27. Izard CE. The many meanings/aspects of emotion: definitions, functions, activation, and regulation. Emot Rev. 2010;2(4):363–70. https://doi.org/10.1177/1754073910374661.
28. Inzlicht M, Bartholow BD, Hirsh JB. The emotional foundation of cognitive control. Trends Cogn Sci. 2015;19(3):126–32. https://doi.org/10.1016/j.tics.2015.01.004.Emotional.
29. Seeley WW, Menon V, Schatzberg AF, Keller J, Glover GH, Kenna H, et al. Dissociable intrinsic connectivity networks for salience processing and executive control. J Neurosci. 2007;27(9):2349–56.
30. Bradley MM, Codispoti M, Cuthbert BN, Lang PJ. Emotion and motivation I: defensive and appetitive reactions in picture processing. Emotion. 2001;1(3):276–98. https://doi.org/10.1037/1528-3542.1.3.276.
31. Buck R. The biological affects: a typology. Psychol Rev. 1999;106(2):301–36. https://doi.org/10.1037/0033-295X.106.2.301.
32. Pessoa L. On the relationship between emotion and cognition. Nat Rev Neurosci. 2008;9(2):148–58. https://doi.org/10.1038/nrn2317.
33. Nelson SM, Dosenbach NUF, Cohen AL, Wheeler ME, Schlaggar BL, Petersen SE. Role of the anterior insula in task-level control and focal attention. Brain Struct Funct. 2010;214(5–6):669–80. https://doi.org/10.1007/s00429-010-0260-2.

34. Yarkoni T, Poldrack RA, Nichols TE, Van Essen DC, Wager TD. Large-scale automated synthesis of human functional neuroimaging data. Nat Methods. 2011;8(8):665–70. https://doi.org/10.1038/nmeth.1635.
35. Klein TA, Endrass T, Kathmann N, Neumann J, Yves Von Cramon D, Ullsperger M. Neural correlates of error awareness. NeuroImage. 2006;34(4):1774–81. https://doi.org/10.1016/j.neuroimage.2006.11.014.
36. Bechara A, Damasio H, Tranel D, Damasio AR. Deciding advantageous before knowing the advantageous strategy. Science. 1997;275:1293–4. https://doi.org/10.1126/science.275.5304.1293.
37. Feinstein JS, Stein MB, Paulus MP. Anterior insula reactivity during certain decisions is associated with neuroticism. Soc Cogn Affect Neurosci. 2006;1(2):136–42. https://doi.org/10.1093/scan/nsl016.
38. Goldin PR, Mcrae K, Ramel W, Gross JJ. The Neural Bases of Emotion Regulation: Reappraisal and Suppression of Negative Emotion; n.d. https://www.ncbi.nlm.nih.gov/pmc/articles/PMC2483789/pdf/nihms-41619.pdf.
39. Simmons AN, Stein MB, Strigo IA, Arce E, Hitchcock C, Paulus MP. Anxiety positive subjects show altered processing in the anterior insula during anticipation of negative stimuli. Hum Brain Mapp. 2011;32(11):1836–46. https://doi.org/10.1002/hbm.21154.
40. Lazarus RS. Thoughts on the relations between emotion and cognition. Am Psychol. 1982;37(9):1019–24. https://doi.org/10.1037/0003-066X.37.9.1019.
41. Zajonc RB. Feeling and thinking: preferences need no inferences. Am Psychol. 1980;35(2):151–75. https://doi.org/10.1037/0003-066X.35.2.151.
42. Inzlicht M, Bartholow BD, Hirsh JB. Emotional foundations of cognitive control. Trends Cogn Sci. 2015;19:126–32. https://doi.org/10.1016/j.tics.2015.01.004.
43. Gray JR, Braver TS, Raichle ME Integration of emotion and cognition in the lateral prefrontal cortex; n.d. http://www.pnas.org/content/99/6/4115.full.pdf.
44. Deco G, Tagliazucchi E, Laufs H, Sanjuán A, Kringelbach ML. Novel intrinsic ignition method measuring local- global integration characterizes wakefulness and deep sleep; n.d.. https://doi.org/10.1523/ENEURO.0106-17.2017.
45. Wilkinson D, Halligan P. The relevance of behavioural measures for functional-imaging studies of cognition. Nat Rev Neurosci. 2004;5(1):67–73. https://doi.org/10.1038/nrn1302.
46. Gu X, Hof PR, Friston KJ, Fan J. Anterior insular cortex and emotional awareness. J Comp Neurol. 2013;521(15):3371–88. https://doi.org/10.1002/cne.23368.
47. Paulus MP, Stein MB. Interoception in anxiety and depression. Brain Struct Funct. 2010;214(5–6):451–63. https://doi.org/10.1007/s00429-010-0258-9.
48. Peled A, Geva AB. "Clinical brain profiling": a neuroscientific diagnostic approach for mental disorders. Med Hypotheses. 2014;83(4):450–64. https://doi.org/10.1016/j.mehy.2014.07.013.
49. Etkin A, Wager TD. Functional neuroimaging of anxiety: a meta-analysis of emotional processing in PTSD, social anxiety disorder, and specific phobia. Am J Psychiatry. 2007;164(10):1476–88. https://doi.org/10.1176/appi.ajp.2007.07030504.
50. Payne P, Levine PA, Crane-Godreau MA, Farb N, Peter Wayne CM. Somatic experiencing: using interoception and proprioception as core elements of trauma therapy. Front Psychol. 2015;6:93. https://doi.org/10.3389/fpsyg.2015.00093.
51. Pessoa L. To what extent are emotional visual stimuli processed without attention and awareness? Curr Opin Neurobiol. 2005;15(2):188–96.
52. Öhman A, Soares JJF. "Unconscious anxiety": phobic responses to masked stimuli. J Abnorm Psychol. 1994;103(2):231–40.
53. Etkin A, Prater KE, Schatzberg AF, Menon V, Greicius MD, et al. Disrupted amygdalar subregion functional connectivity and evidence of a compensatory network in generalized anxiety disorder. Arch Gen Psychiatry. 2009;66(12):1361. https://doi.org/10.1001/archgenpsychiatry.2009.104.
54. Phillips ML, Medford N, Senior C, Bullmore ET, Suckling J, Brammer MJ, et al. Depersonalization disorder: thinking without feeling. Psychiatry Res. 2001;108(3):145–60. Retrieved from http://www.ncbi.nlm.nih.gov/pubmed/11756013
55. Sepulcre J. An OP4 functional stream in the language-related neuroarchitecture. Cereb Cortex. 2015;25(3):658–66. https://doi.org/10.1093/cercor/bht256.
56. Mesulam M-M. From sensation to cognition. Brain. 1998;121:1013–52. https://doi.org/10.1093/brain/121.6.1013.
57. Chang LJ, Yarkoni T, Khaw MW, Sanfey AG. Decoding the role of the insula in human cognition: functional parcellation and large-scale reverse inference. Cereb Cortex. 2013;23:739–80.
58. Deen B, Pitskel NB, Pelphrey KA. Three systems of insular functional connectivity identified with cluster analysis. Cereb Cortex. 2011;21:1498–506. https://doi.org/10.1093/cercor/bhq186.
59. Yarkoni and Westfall. Current practice in psychology: explanation without prediction. n.d. https://doi.org/10.1177/1745691617693393.

22. The Role of the Insula in the Non-motor Symptoms of Parkinson's Disease

Braden Gardner

22.1 Function of the Insula

In order to begin to understand the insula's function in Parkinson's disease (PD), we must first describe what the insula is, as well as what it is cited to control. The insular cortex is located on the lateral wall of the cerebral hemispheres in mammals and is hidden from view just within the depths of the Sylvian fissure. Although enigmatic, this cortical region may be involved with viscerosensory, visceromotor, somatosensory, and interoceptive functions. The left and right anterior insular cortices seem to display different functions as identified by MRI testing, with the right insula participating in feelings of negative emotional valence and sympathetic activation, while the left insula was shown to activate in relation to parasympathetic function, positive affect, and approach behavior. These processes are thought to provide a homeostatic balance and unified sense of awareness [1]. Furthermore, the insula is said to be pivotal in the way we memorize procedures, in decision-making, and in risk aversion [2].

B. Gardner
Seattle University, Seattle Science Foundation,
Seattle, WA, USA
e-mail: gardne10@seattleu.edu

22.2 Non-motor Symptoms

According to Schapira et al., some of the non-motor symptoms of PD are hyposmia, sleep disorders, depression, and constipation [3]. Other symptoms include mood disorders such as apathy, anhedonia, and depression, as well as cognitive dysfunction and hallucinations, accompanied by sensory deprivation such as the registering of pain and olfactory dysfunction, disturbances in sleep-wake cycles, urogenital dysfunction, constipation, and orthostatic hypotension [4]. Furthermore, one study found that patients with PD are much more likely to develop postprandial hypotension (PPH). This paper was a literature review which identified 327 studies that looked into four different neurological disorders, one of them being PD, and evaluated whether PPH was more or less likely to exist in patients with these neurological disorders when compared next to a group of healthy controls. According to this study, 107/201 patients with PD were shown to have PPH, and only 32/136 of the control patients presented with PPH, comparatively. This makes the odds ratio 3.49 for the likelihood of getting PPH with PD versus a healthy control patient. The p-value of this finding as reported by the authors was 0.00001 [5]. In another study by Domellof et al., the authors reported on patients who at the time of diagnosis had the symptoms of olfactory dysfunction, or hyposmia, and a possible link between the existence of this hyposmia and the presence of symptoms of dementia. As identified by a brief smell test, or B-SIT, 125 patients were assessed for hyposmia from the time of diagnosis and were followed up at yearly intervals up to 10 years. The authors found that hyposmia was found in 73% of patients at diagnosis and that of these patients, 46% of them developed dementia at the 10-year check-in as compared to the 21% of those who were not diagnosed with hyposmia at the time of PD diagnosis. Using Cox proportional hazard models, it was shown that this result was statistically significant and that the presence of hyposmia at the baseline of PD indicates a higher likelihood of dementia in patients with this disease. This result goes toward showing that not only hyposmia but dementia is a non-motor symptom that someone with Parkinson's might incur [6]. In an attempt to further validate the relationship between these symptoms discussed and PD, one study attempted to determine whether early indication of these non-motor symptoms previously identified could possibly be used as an early diagnosis tool for identification of PD. This study matched sex and age with similar patients and asked them to fill out a questionnaire identifying whether or not they presented with any of these non-motor symptoms of PD. What they found was that patients who ended up having PD did indicate the presence of these symptoms (such as hyposmia, depression, constipation, and sleep disorders) 77.7% of the time, while the healthy control patients, of which there were an equal number, only indicated the presence of these symptoms 41.3% of the time. These authors concluded that these symptoms as an early indication tool for the development of PD were effective 71.2% of the time [7].

22.3 Treatments of Non-motor Symptoms

With all of these non-motor symptoms, it may leave one wondering what types of treatments patients can pursue to attempt to treat some of these symptoms. In a paper done by Schaeffer et al. in CNS drugs, they investigated the same thing. According to this study, the dopaminergic treatment of the non-motor symptoms of PD had some use, but the side effects that one has to accept as a trade-off are many. While these treatments were shown to decrease the severity of some of the symptoms of PD such as depression or anxiety, a common side effect was shown to be the worsening of psychosis and impulse control. The dopaminergic therapies not only discriminated between symptoms of different effects, but they were also shown to be selectively helpful for symptoms of the same genre as displayed by the effectiveness of treating sleep disorders. In this case, dopaminergic therapies were shown to be helpful with treating sleep disorders such as restless leg syndrome as well as sleep fragmentation, but according to this study, evidence for an improvement of rapid eye movement sleep disorder is lacking. Autonomic function as effected by these treatments was shown to increase in cases where sexual function and sweating were at a deficit; however a side effect of this treatment was constipation. It was also shown that for pain management, dopaminergic therapies were well addressed, while hyposmia and visual deficits were shown to have no change after therapy. The study concluded that personalized dopaminergic therapies have been shown to have a good response by some but not all non-motor symptoms but that the best results were obtained when therapy was individually adapted [8]. Another study cited levodopa as a possible drug that could be useful in preventing cognitive decline in patients with PD. The authors of this paper utilized a clinical study with 55 patients over the course of 3 years and were able to directly relate an increased use in levodopa to a decrease in cognitive decline using a univariate logistic regression analysis. This study began with 100 patients, but due to various factors such as death of the patient or inability to complete the instructions specified by the researchers, only 55 patients were included. It is important to state that this study did not indicate an improvement in the patients who underwent treatment with levodopa but that these patients merely showed an improvement in the sense that their cognitive decline decreased [9]. Levodopa has been found to negatively impact the prefrontal pyramidal tract-type neurons in rats [10]. So while the data from the clinical study on levodopa seems to indicate that it can be an effective treatment for PD, it fails to mention the side effects. Another treatment that has been identified for PD is the usage of EPO as therapy. Using the unified PD rating scale, this study cites the therapeutic effects of EPO on patients with PD using a 26-patient clinical study with participants randomly assigned to recombinant human EPO (rhEPO) groups and placebo groups. This study showed that after taking measurements at baseline and again at 12 months, while the motor symptoms in patients did not express any improvement, the non-motor symptoms such as cardiovascular autonomic function, sleep and fatigue conditions, mood and cognition dysfunctions, and attention and memory deficits showed significant improvement [11]. Furthermore, one study described the usefulness of EPO as a therapy for patients with PD believed to be due to the protective effects of EPO against neural apoptosis. The mechanism, although poorly understood, is thought to be that EPO activation of the EPO receptors in the brain protects neurons from apoptosis [12]. This, when coupled with the fact that EPO has been shown to be an effective treatment for PD, indicates that there may be a connection between the protection of these neurons and the hindering of advancement of the non-motor symptoms of this disease.

22.4 The Role of PD Treatments on the Insula

One study performed on newborn and juvenile pigs found that dopamine protects cerebral autoregulatory processes after fluid percussion injury in traumatic brain injury, as well as limiting neuronal cell necrosis by inhibiting the upregulation of ERK mitogen-activated protein kinase (MAPK) [13]. This ERK MAPK is an extracellular receptor that phosphorylates an oncogene, which alters the levels and activities of many transcription factors leading to cell survival [14].

In this study, the effect that levodopa seems to have on the brain that enables it to be capable of treatment of PD is that it is a neuroprotective agent that induces dopamine release. This paper cites one advantage of levodopa to other treatments for PD to be no noticeable loss in energy for the rat test subjects the cited study was conducted on. The results of the course of treatment for the non-motor symptoms in this study were measured by maze memorization tests such as the y-maze and the Morris water maze [15].

A recent article in the *International Journal of Molecular Science* cites studies done in which EPO is used to not only protect against neural apoptosis, a known NMS of PD, but was shown to also create a permissive environment for neuronal plasticity during stroke recovery and also has been shown to restore blood flow and neurovascular remodeling in cases of localized trauma [16]. While this article may just sound like it is going further to explain the idea of EPO as a treatment for PD, this is the idea behind the argument. A link can be drawn via the treatment of PD with EPO between the non-motor symptom and the effect of EPO on the brain in which the insula presides, which when paired with what the stated functions of the insula are in (Citation 1) seems to aid in drawing a convincing attachment between the NMS of PD and the insula itself. Additionally, it is said that EPO is also a protecting agent against neural apoptosis through hypoxia-related injury [17].

As it has been shown, the treatments that have been found to be effective in treating some or all of the symptoms of PD, and specifically the non-motor symptoms, have been shown in the studies displayed here and in large part have these effects because of the way they interact with tissues and pathways in the brain [14–16]. This link does not boldly say that there is a definite link between the non-motor symptoms of PD and the insula of the brain, but it leaves room for one to propose a link is possible. One could then go further to say that because of the timing of the treatments being applied coinciding with the effects shown on the non-motor symptoms of PD shown in Schaeffer et al., Ikeda et al., and Jang et al., a link between the non-motor symptoms of PD and the insula of the brain is not just a possibility but seems likely to be related when taking into account the known fact that PD is a neurodegenerative disease. It could also be further argued that regardless of all of the links drawn through the treatments of PD, a link could be drawn between the non-motor symptoms and the insula due to its purported functions in viscerosensory, visceromotor, somatosensory, and interoceptive pathways, many of which umbrella the categories of symptoms which the non-motor symptoms reside in.

References

1. Bauernfeind A, de Sousa A, Avasthi T, Dobson S, Raghanti M, Lewandowski A, Zilles K, Semendeferi K, Allman J, Craig A, Hof P, Sherwood C. A volumetric comparison of the insular cortex and its subregions in primates. J Hum Evol. 2013;64(4):263–79.
2. Naqvi N, Gaznick N, Tranel D, Bechara A. The insula: a critical neural substrate for craving and drug seeking under conflict and risk. Ann N Y Acad Sci. 2014; 1316:53–70.
3. Schapira AHV, Chaudhuri KR, Jenner P. Non-motor features of Parkinson disease. Nat Rev Neurosci. 2017;18(7):435–50.
4. Poewe W. Non-motor symptoms in Parkinson's disease. Eur J Neurol. 2008;15(Suppl 1):14–20.
5. Pavelić A, Krbot Skorić M, Crnošija L, Habek M. Postprandial hypotension in neurological disorders: systematic review and meta-analysis. Clin Auton Res. 2017;27(4):263–71.
6. Domellöf ME, Lundin KF, Edström M, Forsgren L. Olfactory dysfunction and dementia in newly diagnosed patients with Parkinson's disease. Parkinsonism Relat Disord. 2017;38:41–7.
7. Rodríguez-Violante M, de Saráchaga AJ, Cervantes-Arriaga A, Davila-Avila NM, Carreón-Bautista E, Estrada-Bellmann I, Parra-López G, Cruz-Fino D, Pascasio-Astudillo F. Premotor symptoms and the risk of Parkinson's disease: a case-control study in Mexican population. Clin Neurol Neurosurg. 2017;160:46–9.
8. Schaeffer E, Berg D. Dopaminergic therapies for non-motor symptoms in Parkinson's disease. CNS Drugs. 2017;31(7):551–70.
9. Ikeda M, Kataoka H, Ueno S. Can levodopa prevent cognitive decline in patients with Parkinson's disease? Am J Neurodegener Dis. 2017;6(2):9–14.
10. Nishijima H, Ueno T, Ueno S, Mori F, Miki Y, Tomiyama M. Levodopa-induced morphologic changes of prefrontal pyramidal tract-type neurons in a rat model of Parkinson's disease. Neurosci Res. 2017;115:54–8.
11. Jang W, Park J, Shin KJ, Kim JS, Kim JS, Youn J, Cho JW, Oh E, Ahn JY, Oh KW, Kim HT. Safety and efficacy of recombinant human erythropoietin treatment of non-motor symptoms in Parkinson's disease. J Neurol Sci. 2014;337(1–2):47–54.
12. Zhang DX, Zhang LM, Zhao XC, Sun W. Neuroprotective effects of erythropoietin against sevoflurane-induced neuronal apoptosis in primary rat cortical neurons involving the EPOR-Erk1/2-Nrf2/Bach1 signal pathway. Biomed Pharmacother. 2017;87:332–41.
13. Curvello V, Hekierski H, Pastor P, Vavilala MS, Armstead WM. Dopamine protects cerebral autoregulation and prevents hippocampal necrosis after traumatic brain injury via block of ERK MAPK in juvenile pigs. Brain Res. 2017;1670:118–24.
14. Setia S, Nehru B, Nath Sanyal S. Upregulation of MAPK/Erk and PI3K/Akt pathways in ulcerative colitis-associated colon cancer. Biomed Pharmacother. 2014;68(8):1023–9.
15. Yadav SK, Pandey S, Singh B. Role of estrogen and levodopa in 1-methyl-4-pheny-l-1, 2, 3, 6-tetrahydropyridine (mptp)-induced cognitive deficit in Parkinsonian ovariectomized mice model: a comparative study. J Chem Neuroanat. 2017;85:50–9.
16. Kimáková P, Solár S, Solárová Z, Komel R, Debeljak N. Erythropoietin and its angiogenic activity. Int J Mol Sci. 2017;18(7):1519.
17. Jeong JE, Park JH, Kim CS, Lee SL, Chung HL, Kim WT, Lee EJ. Neuroprotective effects of erythropoietin against hypoxic injury via modulation of the mitogen-activated protein kinase pathway and apoptosis. Korean J Pediatr. 2017;60(6):181–8.

Part III
Clinical Disorders Related with Insula

Insular Cortex Epilepsy

Manish Jaiswal

Abbreviations

ECoG	Electrocorticography
EEG	Electroencephalography
FLAIR	Fluid-attenuated inversion recovery
ICE	Insular cortex epilepsy
MCDs	Malformations of cortical development
MRI	Magnetic resonance imaging
MTLE	Medial temporal lobe epilepsy
PET	Positron-emission tomography
SEEG	Stereo-electroencephalography
SISCOM	Subtraction ictal single-photon emission computed tomography coregistered to MRI
SPECT	Single-photon emission computed tomography
TLE	Temporal lobe epilepsy

23.1 Introduction

Insular cortex epilepsy (ICE) is an uncommon form of complex partial seizures. Since first described by Penfield and Faulk, seizures arising from the insular cortex have become ever more documented, although they are infrequent [1–5]. The insula has long been concerned in the 30% failure rate after temporal lobe resections for medial temporal lobe epilepsy (MTLE). Penfield illustrated the resemblance between many of the symptoms of MTLE and those he established with insular stimulation, signifying that, in theory, insular seizures could be confused with MTLE seizures [3, 5]. Initial surgeons time and again found residual post excision spikes on their electrocorticograms (ECoG) originating from the insula after temporal lobe resection, confirming its epileptic potential [4, 5]. Semiology, electroencephalography, and even surface electrocorticography recordings may falsely localize other cortical foci, leading to wrong diagnosis and treatment.

Studies of patients with recognized insular epileptic foci have revealed that conventional video-EEG monitoring cannot be trusted to distinguish between temporal and insular discharges [6, 7]. Invasive monitoring, such as grid ECoG and subdural strip electrodes, has been used to effectively localize seizure foci in intractable epilepsy when conventional EEG is ambiguous. However, seizures originating from deeper arrangements (such as the insula) can blow out to the cortical surface. In such cases, direct recording by depth electrodes aids in localization [8–11]. The use of insular depth electrodes permits more accurate localization of seizure foci. When patients preselected using clinical seizure characteristics, scalp EEG recordings (with or without video correlation), MRI, SPECT, and PET

M. Jaiswal, M.S., M.Ch.
Department of Neurosurgery, King George's
Medical University, Lucknow, Uttar Pradesh, India
e-mail: manishjaiswal@kgmcindia.edu

undergo insular electrode recording, a seizure-onset zone specifically within the insula may be found in approximately 10–20% of cases [12, 13]. These findings, as well as the functional connectivity of the insula to the orbitofrontal cortex, cingulate cortex, and temporo-limbic structures, mandate the thought of the insula's role (and potential intracranial EEG recording of the insula) not only in insular epilepsy but also in so-called frontal or temporal lobe epilepsy [13–18].

23.2 Semiological Characterization of Insular Epilepsy

Insular epilepsy is classically accompanying a sensation of laryngeal constriction and paresthesias affecting perioral or large cutaneous regions with well-preserved consciousness, after that dysarthric speech and/or elementary auditory hallucinations or motor signs. Later observations have also exhibited that insular epilepsy could feature hypermotor symptoms similar to frontal lobe seizures, early visceral signs or dysphasia with consequently altered consciousness suggestive of temporal lobe seizures, and early somatosensory symptoms in the absence of laryngeal constriction simulating parietal lobe seizures [12, 15, 17–21]. Penfield wrote in 1954: "On the insula there appears to be representation of the gastrointestinal tract as judged by the fact that stimulation here produces various types of abdominal sensation such as nausea, umbilical sensation, borborygmi, belching, and the desire to defecate" [3, 5].

Insular epilepsy diagnosed by stereo-EEG is generally simple and partial in nature, with common symptoms being laryngeal discomfort, dysphonia, paresthesias, and somatomotor symptoms [22–28]. They may additionally include hypermotor features mimicking frontal lobe seizures, visceral symptoms, or dysphasia mimicking temporal lobe seizures, and early somatosensory symptoms in the absence of laryngeal constriction mimicking parietal lobe seizures [24–30].

23.3 Surgical Anatomy of the Insula

The anatomical features of the insula present unique challenges in surgical exposure for electrode coverage. The insula covers the lateral surface of the hemispheric core and has a triangular shape with its apex directed anteriorly and inferiorly toward the limen insula. The insula is encircled and separated from the frontal, parietal, and temporal opercula by a shallow, limiting circular sulcus, which has superior, inferior, and anterior borders [31]. The insula also has radially projecting sulci and gyri (directed superiorly and posteriorly) from the insular apex. The central sulcus is the deepest of these sulci and extends superoposteriorly, dividing the insula into anterior and posterior parts. Accessing the insula, therefore, requires dissection of the Sylvian fissure, retraction of potentially functional opercular cortex, and further dissection through the M2 middle cerebral artery branches on the surface of the insula. Human cadaveric and primate studies have demonstrated that the insula receives main afferents from the amygdala, the dorsal thalamus, and different cortical regions, particularly the sensory cortices and the auditory cortex [7, 23, 31–36]. Most of these afferents terminate in the posterior granular part of the insula, whereas the ventral anterior agranular insula receives predominantly afferents from the limbic cortex, e.g., the entorhinal, perirhinal, posterior, and orbitofrontal cortices and the cingulate gyrus. In addition, the efferents of the ventral anterior insula reciprocate the afferents of the anterior insula, although this is not the case in the posterior insula. Relatively little is understood about the function of the insula, although several investigators have suggested it may play a role in secondary sensory processing, language and motor control, or higher autonomic control and as a component of the limbic system [37–39]. The anatomical connectivity described above and the seizure characteristics seen in documented insular epilepsy are consistent with this concept.

23.4 Intracranial EEG Investigation of the Insula

The use of intracranial EEG to investigate seizure onset in patients with medically intractable epilepsy is well established, and the role of the insula in seizure onset has received increasing interest over the past decade [3, 5, 15, 19, 23, 34, 40]. Several groups have published reports of intracranial monitoring electrodes implanted into the insula using a variety of methods. Broadly speaking, the electrodes may be intracerebral depth electrodes (located within the insula) or subdural strip electrodes (located on the insula surface) and may be placed stereotactically, with or without the use of a stereotactic frame, or under direct visualization. The techniques of electrode placement within or onto the insula may be categorized as follows:

1. Craniotomy and direct visualization method, with or without frameless stereotactic neuronavigation
2. Stereotactic orthogonal method
3. Stereotactic posterior oblique electrode method
4. Stereotactic anterior oblique electrode method
5. Combined stereotactic anterior and posterior oblique electrode method

Each method has potential advantages and disadvantages and should be chosen accordingly. These approaches have been extensively described, and each has potential advantages and disadvantages.

23.4.1 Depth Electrode for Seizure Foci Localization

Intracranial electroencephalography monitoring of the insula is an important tool in the investigation of the insula in medically intractable epilepsy and has been shown to be safe and reliable [4, 6, 8, 12–18, 25, 28, 36, 41–46]. Several methods of placing electrodes for insular coverage have been reported and include open craniotomy as well as stereotactic orthogonal and stereotactic anterior and posterior oblique trajectories. The spread of epileptogenic activity from the insula to the adjacent cortex may be falsely localized to the cortical surface by scalp EEG and even ECoG, leading to persistence of seizures after topectomy. Lobar corticography alone would have incorrectly localized these seizures and incorrectly classified the seizures as multifocal. Only by interpretation of depth electrode data can insular epilepsy be localized and treated [26, 34, 39]. Insular electrode implantation is a safe method of determining primary versus secondary insular involvement prior to surgery. Previous work has shown that, although combining conventional temporal lobectomy with insulectomy did not reduce seizures in patients with temporal lobe epilepsy, some patients in whom the previous temporal lobectomy failed experienced a significant seizure reduction after undergoing reoperation and insulectomy [23, 27, 28, 31, 39, 41, 44]. That study, done without depth electrodes, implied that patients ultimately shown to have insular onset may appear to have temporal lobe origin.

23.5 MRI Brain for Insular Epilepsy Evaluation

Structural abnormalities underlying temporal and frontal lobe epilepsies have been well described. Little is known about the type and frequency of structural brain abnormalities that give rise to insular/peri-insular cortex epilepsy (ICE) [3, 5, 11, 15, 27, 31, 37, 46, 47].

There are six common radio pathologic categories responsible for insular epilepsy:

1. Neoplastic
2. Malformations of cortical development [MCDs]
3. Vascular malformations
4. Atrophy/gliosis from acquired insults
5. Other
6. Normal

Neoplastic lesions are found in about 27% of the patients in the form of low-grade glioma and high-grade glioma like astrocytomas, oligoastrocytoma, and glioblastoma multiforme. MR imaging in these patients demonstrates infiltrative masses centered over the insula with extension to the frontal and temporal lobes. MCDs are detected in 20% of the patients as cortical dysplasia involving the insula and polymicrogyria implicating the insula with perisylvian extension. Some time bilateral perisylvian polymicrogyria and tuberous sclerosis presenting multiple cortical tubers are also found. Vascular malformations are found in 19% of cases as arteriovenous malformation with a nidus in the insula or inducing a FLAIR signal abnormality in the insula, Sylvian bifurcation aneurysm with mass effect on the insular cortex, and cavernoma centered on the insula. Atrophy/gliosis from acquired insults has been observed in 17% of cases responsible for structural insular epilepsy in the form of encephalomalacia and atrophy after trauma, old ischemic infarct exclusively in the insular cortex, cystic encephalomalacia after herpes encephalitis affecting the insula and temporal lobes, encephalomalacia and atrophy after a congenital insult, encephalomalacia and atrophy of indeterminate cause involving the insula, and the frontal or temporal opercula and pial enhancement over the insula and frontal operculum of an unknown cause. Serial MR imaging in a patient with Rasmussen encephalitis reveals an initial insular fluid-attenuated inversion recovery (FLAIR) signal abnormality followed by unilateral fronto-temporo-insular atrophy and gliosis. A millimetric, innocuous-looking T2 signal abnormality over the posterior insula is found in some patients. A neuroepithelial cyst vs. a signal abnormality in the insula is seen in some patients [32, 36]. Few patients have no insular lesions (despite depth electrode-proven insular seizures) but rather hippocampal sclerosis (and concomitant hippocampal seizures). MR imaging results are sometimes normal in patients.

23.6 Other Investigations for Insular Epilepsy

Data from noninvasive imaging studies would be helpful. Positron-emission tomography with 18[F] fluorodeoxyglucose and [11C] flumazenil-PET scans show insular hypometabolism and decreased benzodiazepine-receptor binding in the insula in the majority of patients with MTLE, but this finding likely indicates frequent spread, rather than initiation. Subtraction ictal single-photon emission computed tomography coregistered to MRI (SISCOM), which should be more sensitive to ictal onsets, often demonstrates insular as well as medial and lateral temporal lobe hyperperfusion during MTLE. However, the technique also has poor temporal resolution and cannot always differentiate onsets from early spread [48]. It would be interesting to perform SISCOM on a patient with proven insular onsets to see if this technique can identify these patients noninvasively.

23.7 Refractoriness to Antiepileptic Drugs

Refractoriness to antiepileptic drug treatment is usually 100% for the patients with MCD, 50.0% in the presence of atrophy/gliosis from acquired insults, 39% for neoplastic lesions, and 22.2% for vascular malformations [22, 26, 28, 41, 48–50]. An estimated 75% of patients with normal MR imaging results have refractory epilepsy. The patient with Rasmussen encephalitis has refractory epilepsia partialis continua with occasionally complex partial seizures most often.

23.8 Surgical Safety in Insular Cortex Epilepsy

The early reports of 20% morbidity are clearly overestimates of current morbidity rates. In the last few years, several reports of surgery in the insula have appeared in the literature, mainly for low-grade gliomas. In the era of frameless stereotactic navigation and intraoperative MRI, surgery in the insula is becoming increasingly common and safe [23, 26, 28, 31–36, 41–46, 49–52]. One recent report of seizure outcome after removal of insular tumors demonstrated an 82% Engel I outcome with 45% transient and no long-term morbidity in a series of 11 patients [37].

Conclusion

- Seizure activity in insular epilepsy may simulate temporal, parietal, or other cortical areas.
- Identification of insular cortex epilepsy by semiology is sometimes effective for making a diagnosis.
- The electrographic study of the insula is challenging.
- Epileptologists and epilepsy surgeons should be more aggressive about localization with depth electrodes and even surgical resection within the insula.

References

1. Guillaume MMJ, Mazars G. Cinq cas de foyers épileptogènes insulaires opérés. Soc Française Neurol. 1949;a:766–9.
2. Guillaume MMJ, Mazars G. Technique de resection de l'insula dans les epilepsies insulaires. Rev Neurol. 1949;81:900–3.
3. Penfield W, Jasper WW, editors. Epilepsy and the functional anatomy of the human brain. Boston: Little, Brown; 1954.
4. Guillaume MMJ, Mazars G, Mazars Y. Indications chirurgicales dans les epilepsies dites "temporalis". Rev Neurol. 1953;88:461–501.
5. Penfield W, Faulk ME. The insula: further observations on its function. Brain. 1955;78:445–70.
6. Silfvenius H, Gloor P, Rasmussen T. Evaluation of insular ablation in surgical treatment of temporal lobe epilepsy. Epilepsia. 1964;5:307–20.
7. Hatashita S, Sakakibara T, Ishii S. Lipoma of the insula: case report. J Neurosurg. 1983;58(2):300.
8. Roper SN, Levesque MF, Sutherling WW. Surgical treatment of partial epilepsy arising from the insular cortex: report of two cases. J Neurosurg. 1993;79:266–9.
9. Duffau H, Capelle L, Lopes M. Medically intractable epilepsy from insular low-grade gliomas: improvement after an extended lesionectomy. Acta Neurochir. 2002;144:563–72.
10. Isnard J, Guenot M, Ostrowsky K. The role of the insular cortex in temporal lobe epilepsy. Ann Neurol. 2000;48:614–23.
11. Williamson D, Engel J, Munari C. Anatomic classification of localization-related epilepsy. In: Engel J, Pedley T, editors. Epilepsy. Philadelphia: Lippincott-Raven; 1997. p. 2405–26.
12. Prats JM, Garaizar C, Garcia-Nieto ML. Opercular epileptic syndrome: an unusual form of benign partial epilepsy in childhood. Rev Neurol. 1999;29:375–80.
13. Bancaud J. Clinical symptomatology of epileptic seizures of temporal origin. Rev Neurol. 1987;143:392–400.
14. Guenot M, Isnard J, Ryvlin P. Neurophysiological monitoring for epilepsy surgery: the Talairach SEEG method: StereoElectroEncephaloGraphy: indications, results, complications and therapeutic applications in a series of 100 consecutive cases. Stereotact Funct Neurosurg. 2001;77:29–32.
15. Talairach J, Tournoux P. Co-planar stereotactic atlas of the human brain. Stuttgart: Thieme Verlag; 1988.
16. Kahane P, Tassi L, Francione S. Electroclinical manifestations elicited by intracerebral electric stimulation "shocks" in temporal lobe epilepsy. Neurophysiol Clin. 1993;23:305–26.
17. Ostrowsky K, Magnin M, Ryvlin P. Representation of pain and somatic sensation in the human insula: a study of responses to direct electrical cortical stimulation. Cereb Cortex. 2002;12:376–85.
18. French JA, Williamson PD, Thadani VM. Characteristics of medial temporal lobe epilepsy, I: results of history and physical examination. Ann Neurol. 1993;34:774–80.
19. Commission on Classification and Terminology of the International League Against Epilepsy. Proposal for revised clinical and electroencephalographic classification of epileptic seizures. Epilepsia. 1981;22:489–501.
20. Mauguière F, Courjon J. Somatosensory epilepsy: a review of 127 cases. Brain. 1978;101:307–32.
21. Salanova V, Andermann F, Rasmussen T. Parietal lobe epilepsy: clinical manifestations and outcome in 82 patients treated surgically between 1929 and 1988. Brain. 1995;118:607–27.
22. Blume WT, Jones DC, Young GB. Seizures involving secondary sensory and related areas. Brain. 1992;115:1509–20.
23. Picard F, Baulac S, Kahane P. Dominant partial epilepsies: a clinical, electrophysiological and genetic study of 19 European families. Brain. 2000;123:1247–62.
24. Bartolomei F, Wendling F, Vignal JP. Neural networks underlying epileptic humming. Epilepsia. 2002;43:1001–12.
25. Ostrowsky K, Isnard J, Ryvlin P. Functional mapping of the insular cortex: clinical implication in temporal lobe epilepsy. Epilepsia. 2000;41(6):681.
26. Augustine JR. The insular lobe in primates including humans. Neurol Res. 1985;7:2–10.
27. Augustine JR. Circuitry and functional aspects of the insular lobe in primates including humans. Brain Res Rev. 1996;22:229–44.
28. Cereda C, Ghika J, Maeder P. Strokes restricted to the insular cortex. Neurology. 2002;59:1950–5.
29. Robinson CJ, Burton H. Organization of somatosensory receptive fields in cortical areas 7b, retroinsula, postauditory and granular insula of M. fascicularis. J Comp Neurol. 1980;192:69–92.
30. Schneider RJ, Friedman DP, Mishkin M. A modality-specific somatosensory area within the insula of the rhesus monkey. Brain Res. 1993;621:116–20.

31. Burton H, Videen TO, Raichle ME. Tactile-vibration-activated foci in insular and parietal-opercular cortex studied with positron emission tomography: mapping the second somatosensory area in humans. Somatosens Mot Res. 1993;10:297–308.
32. Bornhovd K, Quante M, Glauche V. Painful stimuli evoke different stimulus-response functions in the amygdala, prefrontal, insula and somatosensory cortex: a single-trial fMRI study. Brain. 2002;125:1326–36.
33. Frot M, Mauguiere F. Dual representation of pain in the operculo-insular cortex in humans. Brain. 2003;126:438–50.
34. Aziz Q, Furlong PL, Barlow J. Topographic mapping of cortical potentials evoked by distention of the human proximal and distal esophagus. Electroencephalogr Clin Neurophysiol. 1995;96:219–28.
35. Daniels SK, Foundas AL. The role of the insular cortex in dysphagia. Dysphagia. 1997;12:146–56.
36. Krolak-Salmon P, Henaff MA, Isnard J. An attention modulated response to disgust in human ventral anterior insula. Ann Neurol. 2003;53:446–53.
37. Carmant L, Carrazana E, Kramer U. Pharyngeal dysesthesia as an aura in temporal lobe epilepsy. Epilepsia. 1996;37:911–3.
38. Garganis K, Papadimitriou C, Gymnopoulos K. Pharyngeal dysesthesias as an aura in temporal lobe epilepsy associated with amygdalar pathology. Epilepsia. 2001;42:565–71.
39. Ferro JM, Martins IP, Pinto F. Aphasia following right striato-insular infarction in a left-handed child: a clinico-radiological study. Dev Med Child Neurol. 1982;24:173–82.
40. Habib M, Daquin G, Milandre L. Mutism and auditory agnosia due to bilateral insular damage: role of the insula in human communication. Neuropsychologia. 1995;33:327–39.
41. Duffau H, Bauchet L, Lehericy S. Functional compensation of the left dominant insula for language. Neuroreport. 2001;12:2159–63.
42. Munari C, Talairach J, Bonnis A. Differential diagnosis between temporal and perisylvian epilepsy in a surgical perspective. Acta Neurochir. 1980;30: S97–100.
43. Bossi L, Munari C, Stoffels C. Somatomotor manifestations in temporal lobe seizures. Epilepsia. 1984;25:70–6.
44. Babkin BP, Van Buren JM. Mechanism and cortical representation of the feeding pattern. Arch Neurol Psychiatr. 1950;66:1–19.
45. Hauser-Hauw C, Banncaud J. Gustatory hallucinations in epileptic seizures. Brain. 1987;101:339–59.
46. Cascino GD, Karnes WE. Gustatory and second sensory seizures associated with lesions in the insular cortex seen on magnetic resonance imaging. J Epilepsy. 1990;3:185–7.
47. Smith-Swintosky VL, Plata-Salaman CR, Scott TR. Gustatory neural coding in the monkey cortex: stimulus quality. J Neurophysiol. 1991;66:1156–65.
48. Fiol ME, Leppik IE, Mireles R. Ictus emeticus and the insular cortex. Epilepsy Res. 1988;2:127–31.
49. Kahane P, Hoffmann D, Minotti L. Reappraisal of the human vestibular cortex by cortical electrical stimulation study. Ann Neurol. 2003;54:615–24.
50. Engel J, Williamson PD, Wieser HG. Mesial temporal lobe epilepsies. In: Engel J, Pedley TA, editors. Epilepsy. Philadelphia: Lippincott-Raven; 1997. p. 2417–26.
51. Wieser HG. Electroclinical features of the psychomotor seizure. Stuttgart: Gustav Fisher; 1983.
52. Doyle WK, Spencer DD. Anterior temporal resections. In: Engel J, Pedley TA, editors. Epilepsy. Philadelphia: Lippincott-Raven; 1997. p. 1807–17.

Insular Ischemic Stroke

Bing Yu Chen, Olivier Boucher, Christian Dugas, Dang Khoa Nguyen, and Laura Gioia

24.1 Introduction

Ischemic strokes including the insula commonly occur following a proximal occlusion of the middle cerebral artery (MCA) and are associated with the classic clinical presentation of a MCA stroke [1]. Isolated insular infarcts are less common but may surprisingly give rise to a variety of neurological deficits [2, 3]. The purpose of this chapter is to provide an overview of insular ischemic strokes, with a focus on *isolated* insular infarcts. In this chapter, the insular vascular anatomy of the insula, the acute clinical presentations of insular strokes as well as the cardiovascular complications associated with insular strokes, and the underlying anatomical correlates will be detailed. Long-term outcomes following insular strokes will also be discussed.

B. Y. Chen
Faculty of Medicine, McGill University, Montréal, QC, Canada

O. Boucher
Centre de recherche du Centre Hospitalier de l'Université de Montréal, Montréal, QC, Canada

C. Dugas
Centre de recherche du Centre Hospitalier de l'Université de Montréal, Montréal, QC, Canada

Département de Neurosciences, Université de Montréal, Montréal, QC, Canada

D. K. Nguyen, M.D., Ph.D.
Centre de recherche du Centre Hospitalier de l'Université de Montréal, Montréal, QC, Canada

Département de Neurosciences, Université de Montréal, Montréal, QC, Canada

Service de Neurologie, Centre Hospitalier de l'Université de Montréal, Montréal, QC, Canada

L. Gioia, M.D. (✉)
Département de Neurosciences, Université de Montréal, Montréal, QC, Canada

Service de Neurologie, Centre Hospitalier de l'Université de Montréal, Montréal, QC, Canada

Division of Neurology, CHUM Centre Hospitalier de l'Université de Montréal, Montréal, QC, Canada

24.2 Overview of Insular Structural Anatomy

The insular lobe is located deep inside the Sylvian fissure and is covered by the frontal, parietal, and temporal opercula [4]. In the shape of a pyramid, it has an apex (the limen insulae) and a base, with the apex oriented downward toward the Sylvian fissure [4]. The insula is bound by three sulci (anterior, superior, and inferior limiting sulci) with its inferior pole bound by the lateral stem of the Sylvian fissure. Its average size is 5.2 cm (width) × 2.9 cm (height), with the left insula being generally slightly larger than the right, which may be partially explained by a more frequent left hemispheric dominance.

The insular cortex is divided into an anterior and a posterior portion by the central insular sulcus [4]. The anterior insula is composed of three or four short gyri, and the posterior insula contains two long gyri. The subinsular area is defined as the subcortical region deep to the insular cortical ribbon which extends parallel and subjacent to the insular cortex and is continuous with the paraventricular component [5].

24.3 Vascular Anatomy

Cadaveric and surgical studies have suggested that the insular cortex receives its blood supply from the MCA [4]. The superior division of the second branch (M2) of the MCA predominantly supplies the anterior insula while the posterior insula is supplied by the inferior M2 branch [4, 6]. Variants of insular vascularization have been described in which the superior M2 branch of the MCA occasionally supplies part of or the entire posterior insular region [6]. However, it is not uncommon that a small vascular contribution arises from the insular branches of the main MCA [1], supplying the limen insulae [6]. Similarly, a few branches could arise from the M3 division of the MCA to supply region of either the superior or inferior peri-insular sulcus [6]. Currently, there is no evidence supporting collateral vascularization from the anterior or posterior cerebral arteries [1]. The subinsular area lies in a vascularization border zone between small insular penetrating arteries and branches of the lenticulostriate arteries from the proximal branch of the MCA [5].

It is noteworthy that arteries supplying the insula often arise from branches of the MCA that have a further destination, [4] which includes the extreme capsule, the claustrum, the external capsule, the corona radiata, and the medial surface of the operculum [6]. Consequently, a proximal MCA occlusion could lead to basal ganglia infarction (territory of the lenticulostriate arteries) as well as infarction in both the anterior and posterior portions of the insula [1] and damage to the cortical and subcortical frontal, parietal, and temporal lobes [7]. In fact, evidence of insular infarction on brain imaging is often an early sign of a proximal MCA stroke [1].

The insula contains three overlapping venous drainage zones: the subapical region, the anterior lobe, and the posterior lobe [4]. The anterior insula is drained by the superficial Sylvian veins [4]. The subapical region and the posterior insula are drained by the deep middle cerebral vein, which subsequently drains into the basal vein proper [4]. There is some evidence that the deep middle cerebral vein drains primarily the lateral lenticular veins and secondarily the insula [4].

24.4 Overview of Ischemic Stroke Involving the Insular Cortex

The most common form of insular stroke is a non-isolated insular infarct occurring in the context of a partial or total anterior circulation stroke of the MCA. In a prospective study of 150 patients with acute MCA stroke, insular involvement occurred in 48% with concomitant damage to the lenticulostriate region in 46% of cases. The underlying stroke etiology is often thromboembolic (non-lacunar) [2] with nearly 40% of cases being cardioembolic in origin. Major insular infarction is associated with proximal vascular occlusions of the MCA, a more severe stroke severity and larger infarct size.

Isolated insular stroke, on the other hand, is a very rare entity, occurring in less than 0.1% of patients with a first-ever ischemic stroke, according to a consecutive stroke registry of 4800 patients from Switzerland [2]. In another study, Manes et al. [8] found 12 patients with isolated insular infarcts (3.6%) from among 330 patients admitted for treatment after an acute cerebral infarction. In a prospective acute stroke cohort of 605 patients evaluated for a possible ischemic cerebrovascular events and after excluding 501 patients for multiple reasons (transient ischemic attack, onset of symptoms past 72 h, misdiagnosis, lack of informed consent, a history of congestive heart failure, atrial fibrillation, no MRI data, or no cardiological work-up), Frontzek et al. [9] identified in the remaining 104 patients 18 subjects (17%) who had an ischemic stroke which involved the insula and 86 (84%) who had a stroke which spared it. Only seven patients (8%)

had an isolated insular stroke. In a series of 23 isolated insular strokes (7 newly identified cases and 22 cases from the literature), Lemieux et al. [3] found that 61% of cases were in the dominant insula, 17% affected the anterior insula, 52% the posterior insula, and 30% both the anterior and posterior insulae. Isolated insular strokes are most likely due to embolic sources, occluding the M2 branches of the MCA, with presumably good collateral flow protecting the more distal cortical structures [2, 3].

Isolated stroke of the subinsular region is also rare, occurring in 0.4% of all ischemic strokes [5]. Due to its border zone vascularization, the pathogenesis of isolated subinsular infarcts may be due to either a cerebral hypoperfusion in context of internal carotid artery stenosis and/or occlusion or microembolism from either a carotid or cardiac source [5].

The most common clinical presentation of an acute ischemic stroke involving (but not limited to) the insular cortex is the constellation of neurological symptoms associated with a partial anterior circulation cerebral infarct of the MCA (contralateral hemiparesis, hemisensory loss, aphasia in the dominant hemisphere, or visuospatial deficits in the non-dominant hemisphere). On the other hand, even when the ischemic damage is limited to the insula, a variety of clinical manifestations have been reported. In a series of 21 cases of isolated insular ischemic strokes, the most common clinical neurological deficits associated with insular damage included aphasia (43%), dysarthria (43%), somatosensory manifestations (43%), a vestibular-like syndrome (35%), and motor symptoms (26%) [3]. Each category will be detailed below along with the most likely anatomical correlates.

24.4.1 Speech and Language Deficits

Heterogeneous speech and language deficits have been reported in association with insular infarcts although upon close examination, many patients had lesions extending beyond the insula limiting precise anatomico-clinical correlations. For example, Ferro et al. [10] reported the case of a 6-year-old left-handed child with transient (2 weeks) nonfluent aphasia due to a cerebrovascular accident involving not only the right insula but also the internal capsule and lenticular nucleus. Hyman and Tranel [11] reported a 61-year-old man with sudden dense right hemianesthesia and conduction aphasia (which resolved nearly completely after several months) due to an infarct involving the posterior insula but also the white matter subjacent to the posterior temporal and inferior parietal cortices. Shuren [12] reported a 59-year-old woman with impaired speech initiation and aberrant linguistic prosody as the result of a left anterior insular infarct with pressure on the extreme capsule and claustrum and bilateral putaminal lacunar infarcts. Pronunciation and prosody eventually improved within weeks, but speech initiation remained impaired. Habib et al. [13] reported a 44-year-old patient with transient (1 month) mutism and persistent auditory agnosia due to two successive ischemic infarcts mainly involving the insula on both hemispheres; however, these insular infarcts also extended to the opercular central region, Heschl's gyrus, and the temporal pole on the right side and the opercular central region and the striatum on the left side. Marshall et al. [7] reported a patient with a subtype of conduction aphasia characterized by a predominance of semantic paraphasias in association with an infarct in the posterior insula also extending to the intrasylvian parietal opercular cortex. Marien et al. [14] reported an 83-year-old stroke patient with severe speech disorder that evolved into mutism within a few hours; rapid recovery from mutism ensued with persistent severe apraxia of speech and phonological agraphia receding within a year. Here, the ischemic infarct involved the left anterior insula and the adjacent part of the intrasylvian frontal opercular cortex.

There are however a few cases of speech and language disturbances where the infarct appears limited to the insula. Nagao et al. [15] reported a case of speech apraxia due to an infarct in the left precentral gyrus of the insula. In the four patients reported in the case series of Cereda et al. [2], two patients with a left posterior insular infarct had language difficulties: the first presented a transient jargonaphasic fluent aphasia characterized by numerous phonemic paraphasias and dysarthria and altered comprehension and repetition, which fortunately regressed after a few days and the second had nonfluent aphasia with several phonemic paraphasias, anomia, phonemic distortion, and dysarthria (but preserved comprehension and repetition) [2]. Lemieux et al. [3] also identified four new patients with aphasia with dominant hemisphere restricted insular infarcts. More recently, Julayanont et al. [16] reported a case of left posterior insular stroke manifesting transient right facial weakness, dysarthria and impairment of semantic fluency, phonemic fluency, and repetition of complex sentences.

Combined with evidence from cortical stimulation [17, 18], computerized lesion reconstruction [19], and functional MRI studies [20, 21], these case reports of insular or opercular insular infarcts support the notion that the insula is part of the brain network for speech and language. Of note, the role of the insula in the motor aspects of speech is not confined to the left hemisphere. While aphasia has been described only in cases of insular injury in the dominant hemisphere, dysarthria may occur following either left or right posterior insular damage [22]. Indeed, in their study of ten patients with isolated insular stroke, Baier et al. [22] reported that only those with left hemisphere damage showed aphasia during the acute period, while damage to either of the hemispheres led to dysarthria. In addition to the damage to the insula per se in the dominant hemisphere, interruption of the arcuate fasciculus interconnecting Wernicke's area and Broca's area can explain some of the observed language deficits when the infarct extends subcortically to the extreme capsule.

24.4.2 Somatosensory Deficits

An isolated insular stroke may present clinically with a contralateral somatosensory deficit [3, 23, 24]. Sensory deficits may involve large cutaneous territories such as the whole hemibody with or without facial sparing or be limited to one or two limbs; they can affect all elementary modalities (touch, pain, temperature, vibration, and position sense) and discriminative modalities or may be limited to specific modalities. For instance, in the case series of Cereda et al. [2], one had moderate right hemisensory deficit in all elementary modalities for face, arm, trunk, and leg with dramatic impairment in the right leg; another had a right hemisensory deficit involving the face, arm, trunk, and leg for all elementary sensory modalities as well as discriminative sensory modalities (two-point discrimination, graphesthesia, stereognosia); and a third patient had hypesthesia for touch and pain of the left upper extremity, alteration of graphesthesia, and stereognosis (vibration sense was intact). Birklein et al. [25] reported a case of acute ischemic infarct restricted to the middle/posterior left insular cortex producing a contralateral deficit in cold, cold pain, and pinprick sensation (with no other sensory deficits). Similarly, Cattaneo et al. [23] described a patient with a posterior dorsal insular infarct who presented with isolated thermal anesthesia for both cold and warm sensations (without impairment of light touch, superficial pain, vibration, and position sense nor discriminative modalities) on the left side of the body, sparing the face. Finally, out of a series of 270 patients investigated for somatosensory abnormalities, Garcia-Larrea and Perchet [26] identified five subjects presenting with central pain and pure thermoalgesic sensory loss contralateral to cortical stroke involving the posterior insula and the inner parietal operculum.

These deficits are in line with the important role of the insula in the processing of somatosensory stimuli, especially the posterior insula ([27–29]). The dorsal posterior insula at the junction of S2 is particularly important for thermal nociception [30–32]. Because receptive fields are larger and the somatotopy is not as refined in the insula compared to the primary sensory cortex, sensory deficits tend to involve large cutaneous territories despite a limited infarct size (similar to a pure lacunar sensory stroke) [3, 23, 24].

24.4.3 Neglect

At least one case study has reported somesthetic, auditory, and visual neglect following damage to the right insula [8]. Neglect resulting from a right insular infarct may be due to disruption in the connections between the insula and the right parietal lobe [4] involved in arousal, attention, and activation [8]. Furthermore, body schema disorders, anosognosia, visuospatial neglect, and tactile extinction have been suggested in patients with subinsular strokes, particularly posterior lesions [5].

24.4.4 Motor Deficits

Contralateral weakness of various patterns (brachial, brachiofacial, or limited to the lower face) has been reported following insular stroke [3]. These deficits are possibly explained by the presence of connections between the anterior insula and the primary motor, premotor, and supplementary motor areas [5, 7]. Other studies have also suggested damage to connections between the posterior insula and the motor association area as well [7]. Subinsular stroke is also associated with motor deficits of various distributions as well (faciobrachiocrural, faciobrachial, or brachiocrural), presumably through the involvement of corticocapsular motor pathways [5].

24.4.5 Disorders of Swallowing

Dysphagia has been reported following insular stroke affecting either the anterior lobe or the subinsular area or following bilateral insular involvement [2]. The anterior insula is believed to have an important role in volitional swallowing due to its proximity to the frontal opercula as well as connections with the primary and supplementary motor cortices [5].

24.4.6 Deficits in Taste and Olfaction

Although rarely reported, isolated insular infarcts are associated with impairment of taste perception, notably increased taste sensitivity, and/or an unpleasant taste in the mouth [2, 33, 34]. Indeed, the insula is the primary cortical site devoted to taste processing: insular neurons respond to gustatory stimulation in animals [35]; the insula is activated in functional imaging studies with olfactory paradigms [36]; and electrical cortical stimulation of the insula can elicit gustatory responses [18].

In addition to gustatory inputs, the insula also has a role in the processing of olfactory inputs. The insula receives projections from the primary olfactory cortex [37], is activated by odors in functional imaging studies [38], and can generate olfactory symptoms when electrically stimulated particularly in the mid-dorsal part [18]. Mak et al. [33] reported a 70-year-old patient who, following ischemic damage to the posterior two-thirds of the insula extending into the supramarginal gyrus, developed heightened taste intensity, alteration of hot and cold sensations, mild right neglect, language difficulties, increased odor sensitivity in the contralateral nostril, and decreased intensity ratings in the ipsilateral nostril.

24.4.7 Vestibular Deficits

A vestibular-like syndrome has been reported in several studies following insular stroke [2, 3, 39, 40]. Specifically, infarcts in the posterior insular or subinsular region may produce an acute pseudovestibular syndrome including dizziness, gait instability, difficulty remaining upright, and a nonlateralized tendency to fall without nystagmus or cerebellar dysfunction, as well as transient rotational vertigo [2, 3, 5, 39, 40]. The posterior insula is indeed part of the widely distributed vestibular cortical system, as shown by activation studies of the vestibular system using functional neuroimaging [41] and cortical stimulation [42, 43].

24.4.8 Auditory Deficits

Impairments in central auditory function have been reported following insular stroke, albeit with some extension to a variety of cortical and subcortical areas [3, 44]. For instance, Spreen et al. [45] reported a case of auditory agnosia without aphasia associated with a stroke in the right Sylvian fissure involving (based on necropsy findings) the long and short gyri of the insula but also the superior temporal, angular, inferior, and middle frontal gyri and a large portion of the inferior parietal lobe. Fifer [46] described a patient with a stroke involving the right insula and adjacent white matter who presented a unilateral auditory processing disorder when presenting speech materials to the left ear. Finally, Bamiou et al. [44] assessed central auditory function in a series of eight patients with stroke affecting the insula and adjacent areas (the lesion spared the adjacent auditory areas in three patients and included other auditory structures in five cases). Compared to healthy controls, results of the gaps in noise test were abnormal contralateral to the lesion in three and bilateral in five cases [44].

24.4.9 Cardiovascular Effects of Insular Stroke

Beyond the multitude of neurological deficits that are associated with isolated insular stroke, the effects of insular stroke on the cardiovascular system are among the most intriguing. The insular cortex is known to be involved in the regulation of autonomic functions, notably cardiac rhythms [47–52]. Several studies have uncovered some of the pathways providing an anatomical substrate for the modulatory role of the insula. Animal studies have suggested that the rostral posterior insula could be responsible for tachycardia, while bradycardia may be provoked by stimulation of the caudal-posterior insula [53]. Since most studies used rodent, cat, and non-human primate models, generalization of the results to humans should be made with caution. Furthermore, it is important to keep in mind that insular lesions are often associated with large infarcts caused by occlusion of the MCA [1]. An overwhelming majority of case series described cardiovascular effects of strokes *not limited* to the insular regions. Unless otherwise specified, insular infarcts reported in this section are not isolated lesions and must be interpreted cautiously.

Afferent sensory axons from cardiovascular structures terminate within the dorsomedial part of the nucleus tractus solitarii (NTS). Information from the NTS is then relayed to the posterior insula via the parabrachial nucleus (PBN) and ventroposterior, ventral basal, and medial thalamic nuclei [54]. There are also direct reciprocal projections from the PBN and NTS to the posterior insula [55]. Efferent pathways are established between the posterior insula and infralimbic cortex [56], contralateral insula [57], central nuclei of the amygdala, medial dorsal and ventroposterior nuclei of the thalamus, lateral hypothalamus [58], and ventrolateral, medial, and rostral parts of the NTS [56].

Several case studies have reported that strokes with insular involvement may lead to autonomic abnormalities, such as lower or higher blood pressure, [1, 3, 52] tachycardia, [59] lipothymia and syncope, [3] absence of circadian cardiovascular variation [52], and increases in sympathetic autonomic function [60] leading to cardiac myocytolysis [48]. Regarding cardiac rhythms, a wide variety of ECG abnormalities and cardiac arrhythmias have been documented following insular stroke [61] likely secondary to sympathetic activation [47, 50, 61, 62]. These may include abnormal repolarization [58], T-wave inversion [3], QTc interval prolongation [52], prominent U waves [52], ST elevation [59], ectopic beats [59], ventricular arrhythmias [52], left anterior hemiblock [59], and atrial fibrillation [47]. A recent study in 83 patients has shown that isolated ischemic lesions of the insula ($n = 7$) are associated with a ten-fold increased risk for newly diagnosed AF compared to lesions sparing the insula ($n = 76$) [9]. It has been hypothesized that ultimately, these neurogenic changes could result in sudden cardiac death following insular stroke through various mechanisms.

Several studies have suggested that autonomic functions and roles in cardiac rhythm regulation are lateralized, such that the left and right insulae produce opposing effects in the autonomic system [52]. However, this laterality hypothesis is still controversial in humans, and conflicting results have been reported in the literature. According to human studies involving carotid infusion of amobarbital as well as insular stimulation during surgery for intractable epilepsy, the left anterior insula has a parasympathetic cardiovascular predominance and could be responsible for bradycardia, hypotension, baroreceptor reflex sensitivity regulation, and depressor responses [63]. On the other hand, the right anterior insula appears to have a sympathetic predominance and likely plays a role in tachycardia and hypertension, leading to myocytolysis with arrhythmogenic potential [63]. It has also been suggested that subcortical regions adjacent to the insula may play a role in insular autonomic function regulation, such that these regions inhibit sympathetic functions from the right insula or activate parasympathetic functions in opposition to the latter.

Despite the possible lateralization in insular autonomic functions, it appears that insular stroke may be associated with sympathetic activation rather than parasympathetic predominance, either by direct damage to the left insula or through release of inhibition on the right insula on account of damage to adjacent regions. Infarct in the left insula may produce decreased parasympathetic tone [51] and increased sympathetic tone [59], decrease in heart rate variability [59] by disrupting interactions of central oscillators regulating cardiac rhythmicity [52], transient ST depression [52], global ventricular wall motion dysfunction [51], and excess cardiac mortality [51]. Infarct in the left insula is also associated with impaired baroreceptor reflex sensitivity [64]. Right insular lesions are associated with hypertension and tachycardia [60], reduced respiratory heart rate variability [60], more complex dysrhythmias, including atrial fibrillation, AV block, and ectopic beats [50]. Furthermore, infarct in the right insula, sparing the anterior insula, may lead to myocardial infarction due to possible disinhibition of the anterior insula, leading to enhanced cardiac sympathetic activity and myocardial injury, as evidenced by elevated troponin levels [58].

24.5 Long-Term Deficits and Prognosis Following Insular Stroke

Overall, isolated insular stroke is associated with a good to excellent prognosis, with the majority of patients asymptomatic or with minimal deficits within a week from stroke onset [3]. Lemieux et al. [3] identified seven patients from two centers with isolated insular strokes and 16 previously published cases. At 6 month follow-up, the modified Rankin Scale was 0 in 8/23 (35%) patients, 1–2 in 7/23 (30%), and unknown in 8/23 (35%). Although residual deficits have been reported, they are generally mild in intensity [2]. For instance, Boucher et al. [65] described a case of insular stroke who still had hyperacusis and attention impairments 4 years after a right posterior insular stroke [65]. This rapid recovery is likely explained by the small volume of isolated insular strokes and the large structural/functional connectivity of this multimodal structure allowing functional compensation. The scenario is different in the presence of larger MCA strokes with ischemic damage involving not only the insula but several other structures. Furthermore, because of the negative effects on the cardiovascular system, insular damage may adversely affect cardiac prognosis [52] and could thus be associated with poor functional outcome from a mortality perspective [1].

Conclusion
While large MCA strokes damaging the insula are frequent, isolated insular infarcts are relatively rare. Although limited in volume, isolated insular infarcts may result in multimodal deficits (combining somatosensory, motor, speech/language, vestibular, gustatory and olfactory symptoms) in line with the multiple functions of the insula. Prognosis is generally good with rapid recovery within days albeit

mild permanent deficits are occasionally encountered. While there is cumulative evidence to suggest that MCA strokes with insular involvement increase the risk of autonomic abnormalities and cardiovascular complications, there is only limited information for isolated insular strokes. Considering the low prevalence of isolated insular infarcts, larger multicenter prospective studies would be required to better define the clinical presentation, cardiovascular risks, and outcome.

References

1. Fink JN, Selim MH, Kumar S, Voetsch B, Fong WC, Caplan LR. Insular cortex infarction in acute middle cerebral artery territory stroke: predictor of stroke severity and vascular lesion. Arch Neurol. 2005;62(7):1081–5. https://doi.org/10.1001/archneur.62.7.1081.
2. Cereda C, Ghika J, Maeder P, Bogousslavsky J. Strokes restricted to the insular cortex. Neurology. 2002;59(12):1950–5.
3. Lemieux F, Lanthier S, Chevrier MC, Gioia L, Rouleau I, Cereda C, et al. Insular ischemic stroke: clinical presentation and outcome. Cerebrovasc Dis Extra. 2012;2(1):80–7. https://doi.org/10.1159/000343177.
4. Varnavas GG, Grand W. The insular cortex: morphological and vascular anatomic characteristics. Neurosurgery. 1999;44(1):127–36. discussion 136-8
5. Kumral E, Ozdemirkiran T, Alper Y. Strokes in the subinsular territory: clinical, topographical, and etiological patterns. Neurology. 2004;63(12):2429–32.
6. Ture U, Yasargil MG, Al-Mefty O, Yasargil DC. Arteries of the insula. J Neurosurg. 2000;92(4):676–87. https://doi.org/10.3171/jns.2000.92.4.0676.
7. Marshall RS, Lazar RM, Mohr JP, Van Heertum RL, Mast H. "Semantic" conduction aphasia from a posterior insular cortex infarction. J Neuroimaging. 1996; 6(3):189–91.
8. Manes F, Springer J, Jorge R, Robinson RG. Verbal memory impairment after left insular cortex infarction. J Neurol Neurosurg Psychiatry. 1999;67(4):532–4.
9. Frontzek K, Fluri F, Siemerkus J, Muller B, Gass A, Christ-Crain M, et al. Isolated insular strokes and plasma MR-proANP levels are associated with newly diagnosed atrial fibrillation: a pilot study. PLoS One. 2014;9(3):e92421. https://doi.org/10.1371/journal.pone.0092421.
10. Ferro JM, Martins IP, Pinto F, Castro-Caldas A. Aphasia following right striato-insular infarction in a left-handed child: a clinico-radiological study. Dev Med Child Neurol. 1982;24(2):173–82.
11. Hyman BT, Tranel D. Hemianesthesia and aphasia. An anatomical and behavioral study. Arch Neurol. 1989;46(7):816–9.
12. Shuren J. Insula and aphasia. J Neurol. 1993;240(4): 216–8.
13. Habib M, Daquin G, Milandre L, Royere ML, Rey M, Lanteri A, et al. Mutism and auditory agnosia due to bilateral insular damage—role of the insula in human communication. Neuropsychologia. 1995;33(3):327–39.
14. Marien P, Pickut BA, Engelborghs S, Martin JJ, De Deyn PP. Phonological agraphia following a focal anterior insulo-opercular infarction. Neuropsychologia. 2001;39(8):845–55.
15. Nagao M, Takeda K, Komori T, Isozaki E, Hirai S. Apraxia of speech associated with an infarct in the precentral gyrus of the insula. Neuroradiology. 1999;41(5):356–7.
16. Julayanont P, Ruthirago D, JC DT. Isolated left posterior insular infarction and convergent roles in verbal fluency, language, memory, and executive function. Proc (Bayl Univ Med Cent). 2016;29(3):295–7.
17. Afif A, Minotti L, Kahane P, Hoffmann D. Anatomo-functional organization of the insular cortex: a study using intracerebral electrical stimulation in epileptic patients. Epilepsia. 2010;51(11):2305–15. https://doi.org/10.1111/j.1528-1167.2010.02755.x.
18. Mazzola L, Mauguiere F, Isnard J. Electrical stimulations of the human insula: their contribution to the ictal semiology of insular seizures. J Clin Neurophysiol. 2017;34(4):307–14. https://doi.org/10.1097/WNP.0000000000000382.
19. Dronkers NF. A new brain region for coordinating speech articulation. Nature. 1996;384(6605):159–61. https://doi.org/10.1038/384159a0.
20. Bohland JW, Guenther FH. An fMRI investigation of syllable sequence production. NeuroImage. 2006;32(2):821–41. https://doi.org/10.1016/j.neuroimage.2006.04.173.
21. Oh A, Duerden EG, Pang EW. The role of the insula in speech and language processing. Brain Lang. 2014;135:96–103. https://doi.org/10.1016/j.bandl.2014.06.003.
22. Baier B, zu Eulenburg P, Glassl O, Dieterich M. Lesions to the posterior insular cortex cause dysarthria. Eur J Neurol. 2011;18(12):1429–31. https://doi.org/10.1111/j.1468-1331.2011.03473.x.
23. Cattaneo L, Chierici E, Cucurachi L, Cobelli R, Pavesi G. Posterior insular stroke causing selective loss of contralateral nonpainful thermal sensation. Neurology. 2007;68(3):237. https://doi.org/10.1212/01.wnl.0000251310.71452.83.
24. Peskine A, Galland A, Chounlamountry AW, Pradat-Diehl P. Sensory syndrome and aphasia after left insular infarct. Rev Neurol. 2008;164(5):459–62.
25. Birklein F, Rolke R, Muller-Forell W. Isolated insular infarction eliminates contralateral cold, cold pain, and pinprick perception. Neurology. 2005;65(9):1381. https://doi.org/10.1212/01.wnl.0000181351.82772.b3.

26. Garcia-Larrea L, Perchet C, Creach C, Çonvers P, Peyron R, Laurent B, et al. Operculo-insular pain (parasylvian pain): a distinct central pain syndrome. Brain. 2010;133(9):2528–39. https://doi.org/10.1093/brain/awq220.
27. Augustine JR. Circuitry and functional aspects of the insular lobe in primates including humans. Brain research. Brain Res Brain Res Rev. 1996;22(3):229–44.
28. Baier B, zu Eulenburg P, Geber C, Rohde F, Rolke R, Maihofner C, et al. Insula and sensory insular cortex and somatosensory control in patients with insular stroke. Eur J Pain. 2014;18(10):1385–93. https://doi.org/10.1002/j.1532-2149.2014.501.x.
29. Nieuwenhuys R. The insular cortex: a review. Prog Brain Res. 2012;195:123–63. https://doi.org/10.1016/B978-0-444-53860-4.00007-6.
30. Craig AD. Topographically organized projection to posterior insular cortex from the posterior portion of the ventral medial nucleus in the long-tailed macaque monkey. J Comp Neurol. 2014;522(1):36–63. https://doi.org/10.1002/cne.23425.
31. Denis DJ, Marouf R, Rainville P, Bouthillier A, Nguyen DK. Effects of insular stimulation on thermal nociception. Eur J Pain. 2016;20(5):800–10. https://doi.org/10.1002/ejp.806.
32. Dum RP, Levinthal DJ, Strick PL. The spinothalamic system targets motor and sensory areas in the cerebral cortex of monkeys. J Neurosci. 2009;29(45):14223–35. https://doi.org/10.1523/JNEUROSCI.3398-09.2009.
33. Mak YE, Simmons KB, Gitelman DR, Small DM. Taste and olfactory intensity perception changes following left insular stroke. Behav Neurosci. 2005;119(6):1693–700. https://doi.org/10.1037/0735-7044.119.6.1693.
34. Metin B, Melda B, Birsen I. Unusual clinical manifestation of a cerebral infarction restricted to the insulate cortex. Neurocase. 2007;13(2):94–6. https://doi.org/10.1080/13554790701316100.
35. Maffei A, Haley M, Fontanini A. Neural processing of gustatory information in insular circuits. Curr Opin Neurobiol. 2012;22(4):709–16. https://doi.org/10.1016/j.conb.2012.04.001.
36. Small DM, Zald DH, Jones-Gotman M, Zatorre RJ, Pardo JV, Frey S, et al. Human cortical gustatory areas: a review of functional neuroimaging data. Neuroreport. 1999;10(1):7–14.
37. Carmichael ST, Clugnet MC, Price JL. Central olfactory connections in the macaque monkey. J Comp Neurol. 1994;346(3):403–34. https://doi.org/10.1002/cne.903460306.
38. Small DM, Voss J, Mak YE, Simmons KB, Parrish T, Gitelman D. Experience-dependent neural integration of taste and smell in the human brain. J Neurophysiol. 2004;92(3):1892–903. https://doi.org/10.1152/jn.00050.2004.
39. Brandt T, Botzel K, Yousry T, Dieterich M, Schulze S. Rotational vertigo in embolic stroke of the vestibular and auditory cortices. Neurology. 1995;45(1):42–4.
40. Liou LM, Guo YC, Lai CL, Tsai CL, Khor GT. Isolated ataxia after pure left insular cortex infarction. Neurol Sci. 2010;31(1):89–91. https://doi.org/10.1007/s10072-009-0164-1.
41. Fasold O, von Brevern M, Kuhberg M, Ploner CJ, Villringer A, Lempert T, et al. Human vestibular cortex as identified with caloric stimulation in functional magnetic resonance imaging. NeuroImage. 2002;17(3):1384–93.
42. Kahane P, Hoffmann D, Minotti L, Berthoz A. Reappraisal of the human vestibular cortex by cortical electrical stimulation study. Ann Neurol. 2003;54(5):615–24. https://doi.org/10.1002/ana.10726.
43. Mazzola L, Lopez C, Faillenot I, Chouchou F, Mauguiere F, Isnard J. Vestibular responses to direct stimulation of the human insular cortex. Ann Neurol. 2014;76(4):609–19. https://doi.org/10.1002/ana.24252.
44. Bamiou DE, Musiek FE, Stow I, Stevens J, Cipolotti L, Brown MM, et al. Auditory temporal processing deficits in patients with insular stroke. Neurology. 2006;67(4):614–9. https://doi.org/10.1212/01.wnl.0000230197.40410.db.
45. Spreen O, Benton AL, Fincham RW. Auditory agnosia without aphasia. Arch Neurol. 1965;13:84–92.
46. Fifer RC. Insular stroke causing unilateral auditory processing disorder: case report. J Am Acad Audiol. 1993;4(6):364–9.
47. Abboud H, Berroir S, Labreuche J, Orjuela K, Amarenco P, Investigators G. Insular involvement in brain infarction increases risk for cardiac arrhythmia and death. Ann Neurol. 2006;59(4):691–9. https://doi.org/10.1002/ana.20806.
48. Cheshire WP Jr, Saper CB. The insular cortex and cardiac response to stroke. Neurology. 2006;66(9):1296–7. https://doi.org/10.1212/01.wnl.0000219563.87204.7d.
49. Christensen H, Boysen G, Christensen AF, Johannesen HH. Insular lesions, ECG abnormalities, and outcome in acute stroke. J Neurol Neurosurg Psychiatry. 2005;76(2):269–71. https://doi.org/10.1136/jnnp.2004.037531.
50. Colivicchi F, Bassi A, Santini M, Caltagirone C. Cardiac autonomic derangement and arrhythmias in right-sided stroke with insular involvement. Stroke. 2004;35(9):2094–8. https://doi.org/10.1161/01.STR.0000138452.81003.4c.
51. Laowattana S, Zeger SL, Lima JA, Goodman SN, Wittstein IS, Oppenheimer SM. Left insular stroke is associated with adverse cardiac outcome. Neurology. 2006;66(4):477–83.; discussion 63. https://doi.org/10.1212/01.wnl.0000202684.29640.60.
52. Oppenheimer S. Cerebrogenic cardiac arrhythmias: cortical lateralization and clinical significance.

Clin Auton Res. 2006;16(1):6–11. https://doi.org/10.1007/s10286-006-0276-0.
53. Oppenheimer SM, Cechetto DF. Cardiac chronotropic organization of the rat insular cortex. Brain Res. 1990;533(1):66–72.
54. Allen GV, Saper CB, Hurley KM, Cechetto DF. Organization of visceral and limbic connections in the insular cortex of the rat. J Comp Neurol. 1991;311(1):1–16. https://doi.org/10.1002/cne.903110102.
55. Saper CB, Loewy AD. Efferent connections of the parabrachial nucleus in the rat. Brain Res. 1980;197(2):291–317.
56. Yasui Y, Breder CD, Saper CB, Cechetto DF. Autonomic responses and efferent pathways from the insular cortex in the rat. J Comp Neurol. 1991;303(3):355–74. https://doi.org/10.1002/cne.903030303.
57. Zhang Z, Oppenheimer SM. Electrophysiological evidence for reciprocal insulo-insular connectivity of baroreceptor-related neurons. Brain Res. 2000;863(1–2):25–41.
58. Ay H, Koroshetz WJ, Benner T, Vangel MG, Melinosky C, Arsava EM, et al. Neuroanatomic correlates of stroke-related myocardial injury. Neurology. 2006;66(9):1325–9. https://doi.org/10.1212/01.wnl.0000206077.13705.6d.
59. Mandrioli J, Zini A, Cavazzuti M, Panzetti P. Neurogenic T wave inversion in pure left insular stroke associated with hyperhomocysteinaemia. J Neurol Neurosurg Psychiatry. 2004;75(12):1788–9. https://doi.org/10.1136/jnnp.2003.035295.
60. Meyer S, Strittmatter M, Fischer C, Georg T, Schmitz B. Lateralization in autonomic dysfunction in ischemic stroke involving the insular cortex. Neuroreport. 2004;15(2):357–61.
61. Pasquini M, Laurent C, Kroumova M, Masse I, Deplanque D, Leclerc X, et al. Insular infarcts and electrocardiographic changes at admission: results of the PRognostic of Insular CErebral infarctS Study (PRINCESS). J Neurol. 2006;253(5):618–24. https://doi.org/10.1007/s00415-006-0070-x.
62. Rey V, Cereda C, Michel P. Neurological picture. Recurrent asystolia in right middle cerebral artery infarct with predominant insular involvement. J Neurol Neurosurg Psychiatry. 2008;79(6):618. https://doi.org/10.1136/jnnp.2007.127357.
63. Zamrini EY, Meador KJ, Loring DW, Nichols FT, Lee GP, Figueroa RE, et al. Unilateral cerebral inactivation produces differential left/right heart rate responses. Neurology. 1990;40(9):1408–11.
64. Sykora M, Diedler J, Rupp A, Turcani P, Steiner T. Impaired baroreceptor reflex sensitivity in acute stroke is associated with insular involvement, but not with carotid atherosclerosis. Stroke. 2009;40(3):737–42. https://doi.org/10.1161/STROKEAHA.108.519967.
65. Boucher O, Turgeon C, Champoux S, Menard L, Rouleau I, Lassonde M, et al. Hyperacusis following unilateral damage to the insular cortex: a three-case report. Brain Res. 2015;1606:102–12. https://doi.org/10.1016/j.brainres.2015.02.030.

Attention, Salience, and Self-Awareness: The Role of Insula in Meditation

Jordi Manuello, Andrea Nani, and Franco Cauda

25.1 The Multifaceted Function of the Insular Cortex

The insular cortex is undoubtedly among the most investigated brain areas in neuroscientific literature [1]. During the last years, researchers have delved into its role within a variety of domains, from more general functions, such as awareness [2, 3], emotion [4], pain perception [5] saliency, and attention [6] to more specific processes, such as addiction [7], taste [8], hearing [9], and time perception [10]. The multifaceted profile of the insula reflects its rich connectivity, both structural and functional, to many other brain regions, including the temporal, cingulate, and sensorimotor cortices [11, 12].

Given its involvement in so many cognitive, emotional, and sensory processes, the insula is at the interface between two worlds, one which is the inner milieu of our feelings and sensations and the other which is the external environment of our perceptions and stimuli. This privileged position makes the insula an ideal candidate for being a key component of the neural correlates of meditation, a practice that aims to blend harmoniously our inner and external worlds [13].

In the last 20 years, the Western scientific community has started to show a great interest in meditation, whose origins can be traced back to different ancient cultures, such as Indian, Chinese, and Arab worlds. The scientific study of meditation has led to several therapeutic approaches involving this practice, in particular the ones based on the concept of mindfulness meditation, of which the most renowned is mindfulness-based stress reduction (MBSR) [14].

Although meditation is commonly thought of as something esoteric or religious, many of its aspects can be investigated by neuroscientific methods as well as described in neuroscientific terms. For instance, the concept of *sati*, a key feature of meditation that is often translated as *mindfulness* [15, 16], has been defined by Jon Kabat-Zinn as "the awareness that emerges through paying attention on purpose, in the present moment, and nonjudgmentally to the unfolding of experience moment to moment" [14]. Interestingly, a similar definition can also be found in the words of the Theravada monk Nyanaponika Thera [17].

J. Manuello · A. Nani (✉)
GCS fMRI, Koelliker Hospital and University of Turin, Turin, Italy

FOCUS Lab, Department of Psychology, University of Turin, Turin, Italy
e-mail: andrea.nani@unito.it

F. Cauda
GCS fMRI, Koelliker Hospital and University of Turin, Turin, Italy

FOCUS Lab, Department of Psychology, University of Turin, Turin, Italy

Neuroscience Institute of Turin, Turin, Italy

© Springer International Publishing AG, part of Springer Nature 2018
M. Turgut et al. (eds.), *Island of Reil (Insula) in the Human Brain*,
https://doi.org/10.1007/978-3-319-75468-0_25

This important connection between meditation, attention, and awareness becomes even more evident if we evaluate the guidelines for practicing meditation: meditators are usually asked to focus on their breath or alternatively to try to increase their awareness of whatever emerges into consciousness during the ongoing experience [16]. The control of attention appears therefore to be the gate of a deeper level of consciousness. Accordingly, attention regulation and body awareness are among the cognitive domains that are more often investigated in order to understand how meditation works [18].

In the next paragraphs, we will illustrate the role of the insula in meditation under both the perspectives of its involvement in attention and awareness. We will also discuss the cognitive domain of salience, which is strictly related to attention and awareness. As we will see, these important cognitive functions are supported by brain networks in which the insula is thought to play an essential role.

25.2 The Insular Cortex and Attention

A classic definition of attention was proposed by William James in his masterpiece *The Principles of Psychology*. According to James "Everyone knows what attention is. It is the taking possession by the mind, in clear and vivid form, of one out of what seem several simultaneously possible objects or trains of thought. Focalization, concentration, of consciousness are of its essence. It implies withdrawal from some things in order to deal effectively with others" [19]. James' definition remains interesting and can still intuitively grasp what we broadly consider to be both the nature and function of attention: a fundamental cognitive process that helps any human being to orient his or her conduct within a world full of stimuli and incoming information [20]. However, since James' times many scientific experiments addressing attention have been carried out, and numerous theoretical models have been proposed to account for this complex function. Although a comprehensive scientific theory of attention is far to come, there is good agreement on the idea that attention is a cognitive process that selects relevant information from our sense data. In other words, "the concept of attention refers to one of the basic characteristics of cognition, namely the capacity to voluntary and involuntary give priority to some parts of the information that is available at a given moment" [21]. The voluntary or involuntary character of attention is of crucial importance, as it allows to distinguish between a top-down attention and a bottom-up attention. Top-down attention is based on endogenous factors and results in the capacity of the mind to point at a particular feature (feature-based attention), object (object-based attention), or region in space (focal attention). In turn, bottom-up attention is elicited by exogenous factors and results in the quality of the mind to be captured by stimuli beyond a certain degree of intensity that continuously attract one person's focus.

Activity of the insular cortex has been found in both types of attention. Touroutoglou et al. [22] have proved that the right insula, and specifically its dorsal anterior segment, is a key component of a brain network generally involved in attention processing. Furthermore, the higher degree of resting-state connectivity between the dorsal anterior insula and dorsal anterior cingulate cortex has been proportionally associated with a good performance on the *trail making test*, a neuropsychological test of visual attention and task switching. [23]. In an original experiment, Veldhuizen and Small [24] have found that the anterior insula is activated when attention is focused to detect the taste of a solution or a smell in the air.

Many experimental designs are conceived to investigate the neural correlates of attention by using external cue signals to elicit bottom-up attention [25]. Bengson et al. [26], on the contrary, tried to examine volitional or top-down attention and found out that the anterior insula, along with the anterior cingulate cortex and the left middle frontal gyrus, participates to attentional processes even in the absence of cue constraints.

Further evidence has been gathered from studies on pathological conditions. Boehme et al. [27] studied people suffering from social anxiety, who are characterized by the tendency to exceed self-focused attention while experiencing unpleasant social situations. According to the authors, this pathological trait is associated with an increase in the activity of the insula, thus confirming the role of this brain region in attentional processes. Furthermore, in a recent study by Odriozola et al. [28], it was found that children with autism exhibit an atypical functioning of the dorsal anterior insula while they are engaged in a task requiring attention to social stimuli. And reduction in the volume of the insular cortex was also found in adolescents diagnosed with attention deficit and hyperactivity disorder compared with healthy controls [29].

The insula, therefore, is strongly involved in supporting attentional processes, both voluntary and involuntary, and its functional or structural disruption has been associated with attentional deficits in several pathological conditions.

25.3 The Insular Cortex and Self-Awareness

Awareness is an extremely wide concept, which escapes a univocal and precise definition. Its meaning and features can vary according to the topic or the field of inquiry. Sometimes it refers to the conscious perception of something (e.g., a stimulus), and in this case it is considered to be tantamount to consciousness. In this chapter we focus on a particular type of awareness, that is, the awareness of the self, which emerges through the sense of interoception (or body awareness). Self-awareness occurs when an "organism becomes aware that it is awake and actually experiencing specific mental events, emitting behaviors, and possessing unique characteristics" [30]. Such a highly complex phenomenon is supposed to derive from the activity of a network of brain areas [31], within which the insula could be an essential component [32]. In fact, after traumatic brain injuries, patients that experience disrupted self-awareness typically exhibit abnormal activity in the anterior insula [31].

A fundamental ingredient of self-awareness is interoception, a process which refers to the elaboration of bodily sensations due to breathing, digestion, blood flow, and visceral motility [13, 33–35]. The insular cortex, with its rich amount of both afferent and efferent fibers, is a key brain area for supporting interoception [36]. On the one hand, the insula is the arrival point of spinothalamo-cortical pathways, which make it the primary interoceptive cortex—especially its posterior part [37]. On the other hand, the insula is the origin of descending projections that reach motor and sensory areas of the brainstem [10]. Of note, Craig [32] reviewed a large number of studies reporting the activation of the anterior insula induced with a variety of interoceptive cues (e.g., heartbeat, visceral distension, thirst, warmth, coolness, itch, dyspnea, sexual arousal, wine tasting, and the so-called air hunger).

The role of the insula in interoception and self-awareness is also confirmed by neurotransmitters' activity. For instance, it has been demonstrated that during a heartbeat perception task, there is a strong coupling between GABA/NAA concentration in the insula and the neural activity of this region [3]. Furthermore, Ernst et al. [38] have suggested that the positive correlation between GABAergic activity and insula engagement could reflect the increase of interoceptive awareness in alexithymic people.

25.4 The Insular Cortex and Salience

Salience is the feature of objects, stimuli, sensations, etc., that determines their likelihood of becoming the focus of attentional processes [39]. This concept is strictly linked to the concepts of attention and awareness, since it constitutes an essential condition of what we become aware through attention. A number of studies, carried out to analyze brain activity when we evaluate salience, have highlighted the engagement of a set of brain areas, the so-called salience network, which comprises the insula and the dorsal anterior cingulate cortex [40]. The pivotal role of the insula within the salience network has been repeatedly confirmed, even when we try to dissociate the effect of the valuation of the experimental stimuli from their saliency, which may act as a confounding factor [41].

Accordingly, the altered activity of the insular cortex has been found to be the case in a wide range of different pathological conditions characterized, to some extent, by abnormal salience processing. For instance, both functional and structural alterations of the insula have been associated with aberrant evaluation of salience in patients with schizophrenia [42]. Moreover, compared with healthy controls, people suffering from insomnia exhibit a greater activation of the insula [43]. This hyperactivity prevents not only sleep, but also causes a misperception of sleep duration and quality [44]. The level of insular activity and related capacity of salience processing have also been inversely correlated with the body mass index in obese people [45]. More generally, dysfunction of the insula can be found in other addictive behaviors, such as Internet gaming disorder [46], nicotine dependence [47], and drug addiction [48]. However, it must be said that, despite the specific goals of different experimental paradigms, salience seems to be such a pervasive feature of the way we perceive the world that is likely to be evaluated almost every time. So, it is not surprising that the insula is often found active in functional magnetic resonance imaging (fMRI) task experiments [4, 49].

Before drawing our attention to meditation, we want to briefly review an interesting model that illustrates the role of the insula in the cognitive domains mentioned above. This theoretical model is based on the evidence that the insula, in addition to being part of the salience network, also appears to play a central role in switching brain activity from the default mode network to the central executive network, and vice versa [50]. In other words, the insula can trigger the transition from the state in which the brain is not directly engaged in a specific task (i.e., resting state) to the state in which the brain needs to process relevant information and perform goal-oriented actions. According to Menon and Uddin [6], this transition occurs in four steps: (1) external stimuli, which are filtered and processed by sensory specific brain structures, reach the insula; (2) the insula promotes the access to attentional resources and allows the processing of the more salient stimuli; (3) the autonomic system is modulated to properly react to the salient stimuli and induce a state of body awareness; (4) the insula sends information to the anterior cingulate cortex, which promotes the motor response. This model, which includes the insula as a key area in supporting salience, attention, and interoceptive awareness, provides a parsimonious account of how bottom-up inputs, such as environmental stimuli, are detected and processed and how pertinent top-down motor outputs are consequently generated.

25.5 The Role of the Insular Cortex in Meditation

As already said, during the last decade, meditation has stirred a lot of interest in neuroscience. According to PubMed, more than 4400 peer-reviewed studies about this practice have been published to date, with a pace of at least 100 a year from 2005 onward. In addition to experimental trials aimed to discover the neural correlates of meditation and the effects of this practice over the brain, great efforts have been done to construct a theoretical framework capable of accounting for its different features. One of the most comprehensive theories has been proposed by Hölzel et al. [18]. According to the authors, "mindfulness meditation practice comprises a process of enhanced self-regulation that can be differentiated into distinct but interrelated components, namely, attention regulation, body awareness, emotion regulation (reappraisal and extinction), and the change in perspective on the self." In this theoretical model, we find again the cognitive and emotional processes that have been repeatedly associated with the activation of the insular cortex. In particular with regard to the first two elements of the model (attention and awareness), there is increasing evidence that meditation improves the ability to focus attention as well as to distinguish among different bodily states. For instance, meditation practice is related to an improvement of both attention orientation and executive control [51]. Furthermore, meditators performed better than controls in the *attention network test*, probably in virtue of a better allocation of attentional resources [52].

Interestingly, the scientific investigation of the relationship between body awareness and meditation has been driven not only by a theoretical purpose, but also by the possible benefits that meditation practice may exert on mental health [53]. In line with this research, it has been suggested that meditation could increase the confidence in the signals coming from one's own body, on the basis of a bottom-up process [53]. Preliminary findings indicate that meditators exhibit a subtler interoceptive capacity of regulating the breath, compared with non-meditators [54]. Moreover, even a brief meditation training was found to improve performances on the *somatic signal detection task* [55].

A great variety of techniques and traditions fall under the term "meditation", and it has been argued whether their effects should be considered as a whole or independently of each other [15]. The discussion of this point goes beyond the scope of this chapter, which focuses only on the role of the insular cortex in meditation under the features of attention, salience, and self-awareness. It is worth noting, however, that every variant of meditation brings about effects on these cognitive domains. What is more, a recent meta-analysis has found that different meditation techniques promote the recruitment in brain activity of the insula [56].

Meditation has been investigated by means of many different neuroimaging methods. However, here we will briefly take into account only some of the contributions of structural and functional MRI.

An intriguing line of research in neuroimaging aims to investigate how the brain morphologically changes with the practice of meditation. In accordance with the studies of gray matter plasticity, findings suggest that a prolonged practice of meditation could induce cortical thickness in specific brain regions. Indeed, Lazar et al. [57] have found that practitioners with a long experience in Buddhist insight meditation show an increase of cortical thickness in the right anterior insula. This is consistent with what we discussed in the previous paragraphs, as this particular tradition is based on focusing the attention on internal and external stimuli, the processing of which requires the involvement of the insula [6, 37].

Another study conducted on mindfulness meditation practitioners has found similar results [58]. Significantly, authors interpreted these findings in terms of the involvement of the insular cortex in supporting interoceptive awareness. Cortical changes were also analyzed by Luders et al. [59] with the help of the morphological parameter of "gyrification," which is the process that shapes the characteristic folds of the cerebral cortex. The study has showed that long-lasting meditators have higher values of this parameter in the right anterior dorsal insula than non-meditators. Moreover, the level of gyrification of this brain area was positively correlated with the years of practice, thus suggesting that the effect depends on how long meditation is performed. Appreciable effects of cortical thickening in the right insula, however, have been found after only 8 weeks of MBSR in a sample of meditation-naïve subjects [60]. And a meta-analysis of 21 studies about the structural effects of meditation provides evidence that the insula is among the eight brain areas that appear to differ between meditators and control subjects [61]. Morphological differences between meditators and non-meditators are so relevant that it is possible to distinguish between the former and the latter on the basis of their brain structure [62]. Specifically, the use of a support vector machine, which is a statistical approach capable of assigning elements to one out of multiple categories, allowed authors to identify correctly the membership of 37 out of 39 participants. This important result further confirms the effects of brain plasticity due to meditation practice.

Although most of the studies have focused on gray matter changes, there is also evidence that practicing meditation can alter the white matter. For instance, Laneri et al. [63] examined changes in the fractional anisotropy (FA) of diffusion tensor imaging data. FA is a parameter that measures the motility of water molecules within a tissue. Authors found that meditators have higher FA values than control subjects in the proximity of the insula. A higher level of FA is commonly interpreted as evidence of increased structural connectivity, which could reflect an enhanced process of myelination as well as an increased density of axons. This finding suggests that meditation may promote the bidirectional exchange of information between the insula and the other connected brain regions.

The study of the structural changes related to meditation has been paralleled by the investigation of the functional modifications due to this practice. Farb et al. [35] have interestingly showed that meditation alters interoception by acting on the insula. Compared with controls, meditators have demonstrated a higher level of functional connectivity between the middle and posterior insula, which is thought to be, as we have already said, the primary interoceptive cortex. This result, in line with the purpose of MBSR, can be seen as evidence that practicing meditation can improve the moment-to-moment body awareness [14]. Farb et al. [35] also found a positive correlation between the amount of practice and the degree of attentional modulation of the activity in the posterior insula. Increased activity in the left anterior insula, on the contrary, has been described after only 8 weeks of mindfulness-oriented training [64]. Finally, Tang et al. [65] have recently demonstrated that just after 5 h of mindfulness meditation training, there is an effective bilateral increase in the resting state connectivity of the insulae. Intriguingly, this suggests that meditation can exert its effects even on the basal activity of the brain.

It could be argued that similar effects might be found in relation with any behavioral practice comparable to meditation. To resolve the point, Tang et al. [66] compared a short training based on meditation with an alternative approach based on a relaxation technique, finding that meditation caused different patterns of activity in the insula.

Intriguingly, another study has found that different functional profiles of the insula in meditation-naïve subjects can predict the success of engagement with meditation training [67]. This interesting result, which deserves further investigation, suggests that people could have individual neurobiological predispositions or inclinations to engage with different meditation techniques.

Conclusions

The insular cortex has been associated with a variety of cognitive functions, and in this chapter we reviewed its role in supporting three of those (attention, salience, and self-awareness) in connection with the ancient practice of meditation. Notably, these important mental processes are significantly involved in any type of meditative practice, and their modulation can determine relevant changes both in the structure and in the functional activity of the insular cortex. In a sense, the processes of attention, salience, and self-awareness are like bridges that unite brain activity (of which the insula is an essential component) and meditation (Fig. 25.1).

As a final consideration, we would like to point out the similarity between the role of the insula in the practice of meditation described in this chapter and the abovementioned model proposed by Menon and Uddin. Both in this model and in the review that we have discussed, the insula lies at the crossroads of two worlds, the inner and the outer, thus receiving inputs related to the processing of both internal and external

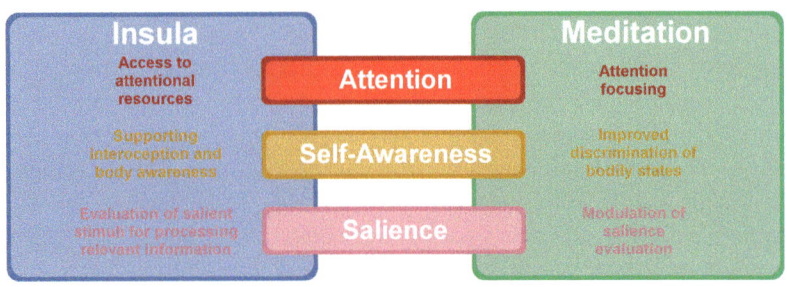

Fig. 25.1 Schematic representation of the relationship between the insular cortex and meditation

stimuli. This unique position allows the insula to promote the allocation of attentional resources as well as the evaluation of the surroundings' saliency. In turn, this helps to construct a coherent and meaningful picture of self-awareness. In the particular case of meditation, all these three processes are enhanced and modulated to achieve an improved state of being. Attention, for instance, can be focused on a specific target (e.g., the breath) or, more widely, on whatever comes into the ongoing experience, while salient features can be better detected and recognized. And a deeper sense of self-awareness can be achieved through an improved discrimination of bodily states. Therefore the insular cortex, by virtue of its great involvement in supporting attention, self-awareness, and salience, appears to be an essential part of the neural correlates of meditation.

References

1. Gogolla N. The insular cortex. Curr Biol. 2017;27: R580–r586.
2. Craig AD. Significance of the insula for the evolution of human awareness of feelings from the body. Ann N Y Acad Sci. 2011;1225:72–82.
3. Wiebking C, Duncan NW, Tiret B, Hayes DJ, Marjanska M, Doyon J, Bajbouj M, Northoff G. GABA in the insula—a predictor of the neural response to interoceptive awareness. NeuroImage. 2014; 86:10–8.
4. Gasquoine PG. Contributions of the insula to cognition and emotion. Neuropsychol Rev. 2014;24:77–87.
5. Garcia-Larrea L, Peyron R. Pain matrices and neuropathic pain matrices: a review. Pain. 2013;154 (Suppl 1):S29–43.
6. Menon V, Uddin LQ. Saliency, switching, attention and control: a network model of insula function. Brain Struct Funct. 2010;214:655–67.
7. Droutman V, Read SJ, Bechara A. Revisiting the role of the insula in addiction. Trends Cogn Sci. 2015; 19:414–20.
8. Small DM. Taste representation in the human insula. Brain Struct Funct. 2010;214:551–61.
9. Bamiou DE, Musiek FE, Luxon LM. The insula (island of Reil) and its role in auditory processing. Literature review. Brain Res Brain Res Rev. 2003; 42:143–54.
10. Craig AD. Emotional moments across time: a possible neural basis for time perception in the anterior insula. Philos Trans R Soc Lond Ser B Biol Sci. 2009a;364:1933–42.
11. Cauda F, D'Agata F, Sacco K, Duca S, Geminiani G, Vercelli A. Functional connectivity of the insula in the resting brain. NeuroImage. 2011;55:8–23.
12. Cauda F, Costa T, Torta DM, Sacco K, D'Agata F, Duca S, Geminiani G, Fox PT, Vercelli A. Meta-analytic clustering of the insular cortex: characterizing the meta-analytic connectivity of the insula when involved in active tasks. NeuroImage. 2012a; 62:343–55.
13. Manuello J, Vercelli U, Nani A, Costa T, Cauda F. Mindfulness meditation and consciousness: an integrative neuroscientific perspective. Conscious Cogn. 2016;40:67–78.
14. Kabat-Zinn J. Mindfulness-based interventions in context: past, present, and future. Clin Psychol Sci Pract. 2003;10:144–56.
15. Awasthi B. Issues and perspectives in meditation research: in search for a definition. Front Psychol. 2012;3:613.
16. Siegel RD, Germer CK, Olendzki A. Mindfulness: what is it? Where did it come from? In: Didonna F, editor. Clinical handbook of mindfulness. New York: Springer; 2008.
17. Thera N. Heart of Buddhist meditation. Kandy: Buddhist Publication Society; 1962.
18. Hölzel BK, Lazar SW, Gard T, Schuman-Olivier Z, Vago DR, Ott U. How does mindfulness meditation work? Proposing mechanisms of action from a conceptual and neural perspective. Perspect Psychol Sci. 2011;6:22.
19. James W. The principles of psychology. New York: Dover Publications; 1890.
20. Knudsen EI. Fundamental components of attention. Annu Rev Neurosci. 2007;30:57–78.
21. Naghavi HR, Nyberg L. Common fronto-parietal activity in attention, memory, and consciousness: shared demands on integration? Conscious Cogn. 2005; 14:390–425.
22. Touroutoglou A, Hollenbeck M, Dickerson BC, Feldman Barrett L. Dissociable large-scale networks anchored in the right anterior insula subserve affective experience and attention. NeuroImage. 2012; 60:1947–58.
23. Reitan RM. Validity of the trail making test as an indicator of organic brain damage. Percept Mot Skills. 1958;8:271–6.
24. Veldhuizen MG, Small DM. Modality-specific neural effects of selective attention to taste and odor. Chem Senses. 2011;36:747–60.
25. Taylor PCJ, Rushworth MFS, Nobre AC. Choosing where to attend and the medial frontal cortex: an fMRI study. J Neurophysiol. 2008;100:1397–406.
26. Bengson JJ, Kelley TA, Mangun GR. The neural correlates of volitional attention: a combined fMRI and ERP study. Hum Brain Mapp. 2015;36:2443–54.
27. Boehme S, Miltner WH, Straube T. Neural correlates of self-focused attention in social anxiety. Soc Cogn Affect Neurosci. 2015;10:856–62.
28. Odriozola P, Uddin LQ, Lynch CJ, Kochalka J, Chen T, Menon V. Insula response and connectivity during social and non-social attention in children with autism. Soc Cogn Affect Neurosci. 2016;11:433–44.

29. Lopez-Larson MP, King JB, Terry J, McGlade EC, Yurgelun-Todd D. Reduced insular volume in attention deficit hyperactivity disorder. Psychiatry Res. 2012;204:32–9.
30. Morin A. Levels of consciousness and self-awareness: a comparison and integration of various neurocognitive views. Conscious Cogn. 2006;15:358–71.
31. Ham TE, Bonnelle V, Hellyer P, Jilka S, Robertson IH, Leech R, Sharp DJ. The neural basis of impaired self-awareness after traumatic brain injury. Brain. 2014;137:586–97.
32. Craig AD. How do you feel—now? The anterior insula and human awareness. Nat Rev Neurosci. 2009b;10:59–70.
33. Cauda F, Torta DME, Sacco K, Geda E, D'Agata F, Costa T, Duca S, Geminiani G, Amanzio M. Shared "core" areas between the pain and other task-related networks. PLoS One. 2012b;7:e41929.
34. Craig AD. How do you feel? Interoception: the sense of the physiological condition of the body. Nat Rev Neurosci. 2002;3:655–66.
35. Farb NAS, Segal ZV, Anderson AK. Mindfulness meditation training alters cortical representations of interoceptive attention. Soc Cogn Affect Neurosci. 2013;8:15–26.
36. Cauda F, Torta DE, Sacco K, D'Agata F, Geda E, Duca S, Geminiani G, Vercelli A. Functional anatomy of cortical areas characterized by Von Economo neurons. Brain Struct Funct. 2013;218:1–20.
37. Flynn FG. Anatomy of the insula functional and clinical correlates. Aphasiology. 1999;13:55–78.
38. Ernst J, Boker H, Hattenschwiler J, Schupbach D, Northoff G, Seifritz E, Grimm S. The association of interoceptive awareness and alexithymia with neurotransmitter concentrations in insula and anterior cingulate. Soc Cogn Affect Neurosci. 2014;9:857–63.
39. Uddin LQ. Salience processing and insular cortical function and dysfunction. Nat Rev Neurosci. 2015;16:55–61.
40. Seeley WW, Menon V, Schatzberg AF, Keller J, Glover GH, Kenna H, Reiss AL, Greicius MD. Dissociable intrinsic connectivity networks for salience processing and executive control. J Neurosci. 2007;27:2349–56.
41. Litt A, Plassmann H, Shiv B, Rangel A. Dissociating valuation and saliency signals during decision-making. Cereb Cortex. 2011;21:95–102.
42. Palaniyappan L, Liddle PF. Does the salience network play a cardinal role in psychosis? An emerging hypothesis of insular dysfunction. J Psychiatry Neurosci. 2012;37:17–27.
43. Chen MC, Chang C, Glover GH, Gotlib IH. Increased insula coactivation with salience networks in insomnia. Biol Psychol. 2014;97:1–8.
44. Hairston IS, Talbot LS, Eidelman P, Gruber J, Harvey AG. Sensory gating in primary insomnia. Eur J Neurosci. 2010;31:2112–21.
45. Hendrick OM, Luo X, Zhang S, Li CS. Saliency processing and obesity: a preliminary imaging study of the stop signal task. Obesity (Silver Spring). 2012;20:1796–802.
46. Zhang JT, Yao YW, Li CS, Zang YF, Shen ZJ, Liu L, Wang LJ, Liu B, Fang XY. Altered resting-state functional connectivity of the insula in young adults with Internet gaming disorder. Addict Biol. 2016;21:743–51.
47. Janes AC, Farmer S, Peechatka AL, Frederick Bde B, Lukas SE. Insula-dorsal anterior cingulate cortex coupling is associated with enhanced brain reactivity to smoking cues. Neuropsychopharmacology. 2015;40:1561–8.
48. Koob GF, Volkow ND. Neurobiology of addiction: a neurocircuitry analysis. Lancet Psychiatry. 2016;3:760–73.
49. Sterzer P, Kleinschmidt A. Anterior insula activations in perceptual paradigms: often observed but barely understood. Brain Struct Funct. 2010;214:611–22.
50. Sridharan D, Levitin DJ, Menon V. A critical role for the right fronto-insular cortex in switching between central-executive and default-mode networks. Proc Natl Acad Sci U S A. 2008;105:12569–74.
51. Tsai MH, Chou WL. Attentional orienting and executive control are affected by different types of meditation practice. Conscious Cogn. 2016;46:110–26.
52. Jo HG, Schmidt S, Inacker E, Markowiak M, Hinterberger T. Meditation and attention: a controlled study on long-term meditators in behavioral performance and event-related potentials of attentional control. Int J Psychophysiol. 2016;99:33–9.
53. Farb N, Mehling WE. Editorial: interoception, contemplative practice, and health. Front Psychol. 2016;7:1898.
54. Daubenmier J, Sze J, Kerr CE, Kemeny ME, Mehling W. Follow your breath: respiratory interoceptive accuracy in experienced meditators. Psychophysiology. 2013;50:777–89.
55. Mirams L, Poliakoff E, Brown RJ, Lloyd DM. Brief body-scan meditation practice improves somatosensory perceptual decision making. Conscious Cogn. 2013;22:348–59.
56. Fox KC, Dixon ML, Nijeboer S, Girn M, Floman JL, Lifshitz M, Ellamil M, Sedlmeier P, Christoff K. Functional neuroanatomy of meditation: a review and meta-analysis of 78 functional neuroimaging investigations. Neurosci Biobehav Rev. 2016;65:208–28.
57. Lazar SW, Kerr CE, Wasserman RH, Gray JR, Greve DN, Treadway MT, McGarvey M, Quinn BT, Dusek JA, Benson H, et al. Meditation experience is associated with increased cortical thickness. Neuroreport. 2005;16:1893–7.
58. Holzel BK, Ott U, Gard T, Hempel H, Weygandt M, Morgen K, Vaitl D. Investigation of mindfulness meditation practitioners with voxel-based morphometry. Soc Cogn Affect Neurosci. 2008;3:55–61.
59. Luders E, Kurth F, Mayer EA, Toga AW, Narr KL, Gaser C. The unique brain anatomy of meditation

practitioners: alterations in cortical gyrification. Front Hum Neurosci. 2012;6:34.
60. Santarnecchi E, D'Arista S, Egiziano E, Gardi C, Petrosino R, Vatti G, Reda M, Rossi A. Interaction between neuroanatomical and psychological changes after mindfulness-based training. PLoS One. 2014; 9:e108359.
61. Fox KC, Nijeboer S, Dixon ML, Floman JL, Ellamil M, Rumak SP, Sedlmeier P, Christoff K. Is meditation associated with altered brain structure? A systematic review and meta-analysis of morphometric neuroimaging in meditation practitioners. Neurosci Biobehav Rev. 2014;43:48–73.
62. Sato JR, Kozasa EH, Russell TA, Radvany J, Mello LE, Lacerda SS, Amaro E Jr. Brain imaging analysis can identify participants under regular mental training. PLoS One. 2012;7:e39832.
63. Laneri D, Schuster V, Dietsche B, Jansen A, Ott U, Sommer J. Effects of long-term mindfulness meditation on brain's white matter microstructure and its aging. Front Aging Neurosci. 2015;7:254.
64. Tomasino B, Fabbro F. Increases in the right dorsolateral prefrontal cortex and decreases the rostral prefrontal cortex activation after 8 weeks of focused attention based mindfulness meditation. Brain Cogn. 2016;102:46–54.
65. Tang YY, Tang Y, Tang R, Lewis-Peacock JA. Brief mental training reorganizes large-scale brain networks. Front Syst Neurosci. 2017;11:6.
66. Tang YY, Ma Y, Fan Y, Feng H, Wang J, Feng S, Lu Q, Hu B, Lin Y, Li J, et al. Central and autonomic nervous system interaction is altered by short-term meditation. Proc Natl Acad Sci U S A. 2009;106: 8865–70.
67. Mascaro JS, Rilling JK, Negi LT, Raison CL. Pre-existing brain function predicts subsequent practice of mindfulness and compassion meditation. NeuroImage. 2013;69:35–42.

Neuropsychological Deficits Due to Insular Damage

26

Olivier Boucher, Daphné Citherlet, Benjamin Hébert-Seropian, and Dang Khoa Nguyen

26.1 Introduction

The insula is the fifth lobe of the brain and is also the least understood for its role in neuropsychological function. It is considered a multimodal structure involved in several aspects of sensory processing, cognitive function, and emotion [1]. Neuroimaging studies have revealed a pattern of functional specialization within the insula, with its anterior part being activated by cognitive (anterior-dorsal insula) and socio-emotional (anterior-ventral insula) stimuli and tasks, whereas sensorimotor experiments activate its posterior portion [2]. The responses elicited by the electrical stimulation of the insular cortex also differ according to the specific site that is stimulated [3]. Although the advent of functional neuroimaging and deep cortical stimulation techniques has led to important advances in our comprehension of insular functions, they do not allow a conclusion to be expressed about its essential roles and, thereby, of the possible neuropsychological consequences of insular damage. Lesions of the insular cortex may occur in the context of a stroke, tumor, neurosurgery, or traumatic brain injury. While each pathology may be associated with neuropsychological impairments, comparisons across studies should be made with caution, as each possible cause of insular damage is intrinsically associated with certain characteristics that will influence the observed deficits [4].

The insula is supplied by about 100 small arteries originating from the middle cerebral artery (MCA), mainly the M2 segment, but also from the M1 and M3 segments [5]. Besides the insular cortex, insular arteries also supply the extreme capsule, and a small proportion also supplies the claustrum and the external capsule and may extend to the corona radiata. Cerebral infarction due to occlusion of the MCA is a common cause of insular injury but usually also damages neighboring regions such as the basal ganglia. MCA strokes resulting in isolated insular damage are thus extremely uncommon. As an example,

O. Boucher · B. Hébert-Seropian
Département de Psychologie, Université de Montréal, Montreal, QC, Canada

Centre de recherche du CHUM, Montreal, QC, Canada

D. Citherlet
Centre de recherche du CHUM, Montreal, QC, Canada

Département de Neurosciences, Université de Montréal, Montreal, QC, Canada

D. K. Nguyen, M.D., Ph.D. (✉)
Centre de recherche du CHUM, Montreal, QC, Canada

Département de Neurosciences, Université de Montréal, Montreal, QC, Canada

Division of Neurology, CHUM, Montreal, QC, Canada
e-mail: d.nguyen@umontreal.ca

Cereda and colleagues identified only four patients with selective insular stroke among a consecutive sample of 4800 patients with first ever acute stroke [6]. This accounts for the very small number of group studies involving patients with isolated insular stroke and also accounts for the inclusion of subjects whose damage extends to neighboring regions in several studies to increase the sample size.

The low incidence of strokes restricted to the insula is a major obstacle to characterizing the neuropsychological deficits due to insular damage. Still more limited is our ability to understand the differences according to lateralization and to the specific insular subregion affected. To overcome this limitation, studies have used lesion overlap and voxel-based lesion-symptom mapping (VLSM) techniques to relate insular lesions to post-stroke deficits [7–10]. However, it has been shown that infarcts including the insula are typically larger than infarcts that exclude it and that the larger strokes due to MCA occlusion usually involve the insula [11]. Thus, these techniques may overestimate the involvement of insular damage in post-stroke deficits, as the insula is a likely candidate for any deficit due to MCA stroke and because frequent involvement of the insula in strokes increases its statistical power to detect an association with a given deficit, in comparison to other structures. Some of the reported associations between insular damage and post-stroke deficits may reflect the fact that insular lesions are a marker of larger strokes, rather than result from the insular damage per se.

Insular damage may also occur as a result of glioma or after repeated epileptic activity due to a congenital malformation in the insula such as focal cortical dysplasia [12]. However, since these types of lesion typically develop over a long period of time, their impact on neuropsychological function may be reduced due to functional plasticity, and their study is further complicated by the distribution of tumoral cells to adjacent structures and the propagation of epileptic activity to other connected areas [13]. More acute deficits may occur following neurosurgical ablation of the insular cortex due to the presence of a tumor or in the context of epilepsy surgery [14]. Again, however, because most of the resected tissue is damaged by pathology, the consequences of insular cortex resection may be reduced as a result of presurgical functional reorganization. Moreover, neurosurgery in the insula is complex due to its deep location and rich vascularization, and thus deficits may occur because of a stroke in underlying subcortical structures resulting from the surgical procedure [15, 16]. Finally, its location makes the insula an unlikely candidate for specific injury due to head trauma.

In this chapter, we perform a selective review of the literature on the neuropsychological consequences of insular damage in humans, with a focus on the long-term impairments in the sensory, cognitive, and emotion domains. More consideration was given to isolated ischemic insular lesions because of the limitations inherent to the other types of pathology. However, results from lesion-mapping techniques and postsurgical assessments were also provided, as they offer complementary information about the possible consequences of insula lesions.

26.2 Sensory Processing

26.2.1 Somatosensory Deficits

The insular cortex is crucial for the processing of a wide range of somatosensory signals. Nociception is one of the modalities that is most often associated with the insula, with afferent nociceptive information relayed rostrally from the second somatosensory cortex (SII) to the posterior insula and subsequently to its anterior portion [17]. Although a wealth of evidence from fMRI and iEEG studies has convincingly demonstrated the involvement of the insula in the representation of pain signals, studies involving patients with insular lesions have provided mixed findings. For example, a lesion study involving three patients with posterior insula damage showed elevated pain thresholds on the hand contralateral to the lesion, but not in two other subjects with substantial damage to the anterior insula [18]. On the other hand, a more recent study by Starr et al. did not replicate these results in two individuals with extensive lesions to the insula, as these patients exhibited significantly higher pain sensitivity than healthy controls [19]. Also of interest is the lack of insular activity in the fMRI data during painful stimulation in these patients, whereas their SI cortices showed strong activation when compared to controls. The authors concluded that SI and its associated networks could be compensating for the insular damage in the processing of nociceptive information, thus suggesting that the experience of pain does not require the insula. More complex alterations of the experience of pain were also reported in a study of six patients with extensive unilateral hemispheric damage following ischemic lesions in five and a traumatic hematoma in another. In all six patients, damage to the insula was extensive, and they presented a severe case of pain asymbolia, a condition in which individuals exhibit inappropriate affective responses to noxious stimuli and do not withdraw when faced with threatening gestures or when painful stimuli is applied over their body [20].

In several cases, following insular damage, central pain is also accompanied by thermoalgesic sensory loss. For example, a study by Garcia-Larrea and colleagues found that out of 270 patients with somatosensory abnormalities, five subjects were identified as presenting both central pain and pure thermoalgesic sensory loss contralateral to their stroke, and all five patients' lesions had the particularity of involving the posterior insula and inner parietal operculum [21]. All patients presented profoundly altered thermal and pain sensations. Similarly, in a study with epileptic patients, contralateral loss of thermal sensibility, decreased thermal nociception, and central pain syndrome were observed in two out of three subjects with supero-posterior-operculo insulectomy as part of epilepsy surgery [22]. When administering a noxious thermal stimulus to the hand contralateral to their lesion, they could feel an unpleasant sensation without being able to distinguish cold from warm. In comparison to the other two subjects, the insulectomy undergone by the individual that did not report changes in thermal sensations was much smaller.

In all cases, pressure pain was not affected by the lesion, a finding that is in accordance with existing studies showing that lemniscal sensory modalities are usually preserved when an ischemic lesion is restricted to the operculo-insular region [21]. In a case study of a 64-year-old man with acute middle-posterior left insular infarct, it was reported that he suffered from contralateral loss of cold, cold pain, and pinprick pain, while other sensory modalities remained normal. [23]

Despite these findings, it is still unclear whether nonpainful and painful thermal sensations are mainly represented in the insular cortex, and whether the two modalities are segregated. For example, in a VLSM study comprised of 24 patients with acute unilateral cortical damage, lesions of the posterior insula were associated with deficits in temperature perception, while not showing other deficits such as thermoalgesic pain [24]. In addition, a case study reported pure thermal anesthesia contralateral to a small insular infarct in the posterior dorsal region of 59-year-old man [25]. The deficits experienced by the patient were restricted to nonpainful thermal sensations. As exemplified in the studies cited above, the posterior insula appears to be central for the processing of nonpainful and painful thermal and nociceptive stimuli, while not being essential for the processing of mechanic signals.

26.2.2 Audition

Considering the wide array of efferent projections received from the auditory cortices, it is not surprising that insular damage may lead to central auditory processing deficits [26]. Fifer reported the case of a 60-year-old patient who presented left ear speech identification deficits 3 weeks to 2 months after stroke involving the right insula and adjacent white matter [27]. Identification of pure tone was normal, as revealed by audiologic evaluations. Four months later, performance was back to normal. Conversely, Habib and colleagues documented a case of nonverbal auditory agnosia in a 44-year-old woman who suffered from a bilateral insular lesion caused by two successive ischemic infarcts [28]. Other regions affected included a small part of the left striatum and right temporal pole. In the acute stage, the patient also showed a severe mutism, which disappeared within 1 month. An auditory agnosia battery revealed impairments in recognizing nonverbal sounds (e.g., pigeon), nursery rhymes, and famous voices. Moreover, the patient was impaired in her ability to match intonations with the corresponding emotion, reflecting a deficit in emotional prosody recognition. Additionally, she was unable to play piano pieces she had known before the stroke. These impairments were still present 2 years after the infarct.

Amusia was described in other case studies of stroke patients whose brain damage involved the insula. Griffiths and colleagues described the case of a 75-year-old male patient, who showed difficulties recognizing tunes and piano music he had been familiar with, following a right hemisphere infarction involving the insula [29]. Assessment conducted 1 year after stroke revealed impairments in tune recognition, suggesting a deficit in the processing of rapid temporal sequences of notes. By contrast, performance was normal on tasks aimed to assess nonverbal and speech sound processing. Ayotte and colleagues reported two other cases of amusia following right insula and right superior temporal lobe damage after brain surgery for clipping of an aneurysm of the MCA [30]. Both patients failed in music memory tasks and discrimination tests involving musical perception. In another case report of a 52-year-old patient who suffered from an infarction involving the left insula and amygdala, music and prosody perception were not impaired, but the patient reported having lost the emotional experience associated with music listening [31].

In a rare group comparison study on the effects of insular damage on auditory function, Bamiou and colleagues assessed central auditory function in eight patients with a stroke affecting the insula (five in the right hemisphere, three in the left)—three of whom had their adjacent auditory areas spared—and compared their performance to that of eight matched controls [32]. Participants were assessed during the acute stage on an auditory test battery comprising dichotic speech test, frequency and duration pattern recognition, and a temporal resolution test. All patients were impaired on the duration pattern and gaps in noise tests, and all but one were impaired on frequency pattern recognition. On most occasions, impairments were contralesional or bilateral. These results suggested that insular lesions affected temporal resolution and sequencing more specifically.

Hyperacusis is another possible consequence of insular cortex damage. Boucher et al. reported three cases of increased sensitivity to loud sounds—in two cases following isolated insular stroke and in one following right operculo-insular resection as part of epilepsy surgery [33]. One patient with a right insular stroke reported being easily annoyed by sounds, such as television, clocks, and children's voice, was easily disturbed by surrounding sounds, and had difficulty following conversations involving several people talking simultaneously. The other stroke patient, whose lesion involved the left posterior insula, reported that her ear became "overdeveloped," feeling annoyed by high-pitched sounds, which caught were attention more easily. Patients were assessed at least a full year after brain injury on a battery of central auditory function. Objective assessment of loudness discomfort levels confirmed hyperacusis at both ears in each patient, in comparison to matched controls and to normative data. Deficits were also observed on pattern recognition tasks. Increased sensitivity to sounds had also been reported in a patient who underwent right insular resection, in a series of neurosurgical patients [34].

26.2.3 Chemosensation

Effects of insular damage on chemosensation have been documented in a few case reports and include either increased or decreased sensitivity to taste and smell and unpleasant taste or smell sensations. Mak and colleagues reported the case of a 70-year-old patient who showed deficits on taste and olfactory intensity perception following a stroke involving left posterior insula and extending into the supramarginal gyrus [35]. During the acute stage, the patient reported that previously familiar and appreciated food tasted intensely unpleasant and unfamiliar. One week after the stroke, his subjective experience of food returned to normal, but his appetite had declined, and he had lost weight. Gustatory and olfactory tests were administered, and performance was compared with a control group. Results revealed an increased sensitivity to taste and odors located to the side opposite to the lesion. This increase was more pronounced for unpleasant odors and for strong tastes, either unpleasant or pleasant. By contrast, Ribas and Duffau reported a case of complete anosmia associated with ageusia following a resection of a low-grade glioma involving the left insula, anterior temporal lobe, uncus, and amygdala which persisted 3 years after surgery [36].

Metin and colleagues described the impairment of a 66-year-old man who suffered a left anterior insular damage due to a stroke [37]. The patient complained of diminished pleasure in taste due to an annoying taste of rotten melon in his mouth during the acute phase. His ability to distinguish different tastes (i.e., bitter, sour, and sweet) was preserved. About 2 weeks after his admission to the hospital, these complaints completely disappeared. In another case report, an infarction involving the right posterior insula and the pre- and postcentral gyri was associated with olfactory hallucinations in a 55-year-old woman [38]. The day after the stroke, the patient showed a mild hyposmia with a displeasing smell of burned hair, and 6 months after, she still reported hyposmia, intermittent dysosmia, and a mild loss of taste.

Pritchard and colleagues assessed taste quality and intensity perception in a group of six patients with unilateral insular damage extending to neighboring brain structures due to strokes or tumor resections and compared them to a group of 3 patients with brain damage outside the insula (head trauma group) as well as to 11 healthy controls [39]. Prior to tumor resection, one patient reported unusual predilection for sweet foods and chocolate milk and also reported food-induced vomiting. After a stroke, one patient reported mild paresthesia of the lips. Although patients did not complain of any problem with taste, objective assessments revealed that patients with damage to the right insula displayed ipsilateral taste recognition and intensity deficits, whereas damage to the left insula caused an ipsilateral deficit in taste intensity but a bilateral deficit in taste recognition. Differences between the healthy controls and the head trauma group were not significantly different.

In another group comparison study, 14 patients with unilateral insular lesions were compared to 42 healthy controls participants [40]. Participants were assessed on two tasks. In the first task (taste discrimination), they had to judge whether two tastes presented in each side of the tongue tip were the same or different. In the second, they were assessed on judgments of quality, intensity, hedonism, and name recognition for stimulus sampling through 8-point rating scales (e.g., pleasant/unpleasant, salty, bitter, sweet, sour, and hedonic rating). Patients with unilateral insular lesions to either the left or the right side presented bilateral impairments in discrimination, quality of judgment, and name recognition in comparison to controls, and patients with left insular lesion showed poorer performance for salt-related information compared to controls and to patients with right insula lesion. They also had difficulties judging the salty taste quality and to name it bilaterally. On the other hand, they presented a reduced taste intensity on the lesion side and a greater taste intensity on the contralateral side.

26.2.4 Vestibular Function

In agreement with stimulation and recent activation likelihood estimation meta-analyses supporting a role of the parietal operculum-insular region in vestibular processing [41–43], a few studies have reported vestibular symptoms following isolated insular damage, although the evidence remains scarce. Papathanasiou and colleagues described a case of recurrent episodes of vertigo and imbalance due to insular damage; surprisingly however, the lesion was localized to the anterior insula (right hemisphere) [44]. Normal results on evoked potential studies and pure tone audiometry ruled out the possibility of peripheral or brainstem causes for vestibular symptoms. Vestibular-like symptoms, including dizziness, nonrotatory vertigo, and unsteadiness leading to gait difficulty, were also reported in three of the four cases described by Cereda et al., this time in relation to posterior insular injury [6]. By contrast, in a study of ten consecutive cases of acute unilateral stroke restricted to the insula, Baier et al. found no case of vestibular otolith dysfunction nor vertigo [45].

26.2.5 Interoceptive Function, Body Representation, and Self-Awareness

Partly due to the rare nature of isolated insular lesions, studies attempting to verify the link between insular damage, altered interoceptive function, and dysfunctional body representation are very few. One of the only known cases of bilateral insular damage has been extensively studied. The patient, known as Roger, suffered bilateral insular, anterior cingulate cortex, and medial prefrontal cortex damage following a severe episode of herpes simplex encephalitis. When administered isoproterenol, a sympathetic agonist with effects similar to adrenaline, it was shown that this patient's awareness of his own heartbeat was dose-dependent, whereby higher doses of intravenous isoproterenol injections were associated with higher heartbeat and breathing intensity as reported by the participants using a dial. [46] This trend was similar in the control group, except for the fact that Roger's ratings of his heartbeat intensity were consistently delayed, as if additional cognitive efforts were required to properly assess the interoceptive sensations. In a VLSM study where left and right focal brain-damaged patients were asked to answer questions pertaining to their interoceptive feelings, the authors found that insular damage was linked to lower scores on their Interoceptive Awareness Questionnaire [47]. They posited that damage to short connections between the insular cortex and the amygdala contributed to the observed defective interoception.

In terms of body representation, another VLSM study in 27 stroke patients showed that the right posterior insula was commonly damaged in patients with anosognosia for hemiplegia/hemiparesis [8]. However, such lesions were significantly less involved in hemiplegic/hemiparetic patients without anosognosia, pointing to a role in awareness of one's own body rather than motor control. Concordantly, an interesting case study found that a 43-year-old male with extensive damage to the right insula caused by a tumor exhibited highly unusual experiences during a full-body illusion paradigm where body self-consciousness is manipulated using cardio-visual stimuli [48]. After showing a virtual body surrounded by a glow flashing in synchrony with the subject's heartbeat, the patient reported strong self-identification with the virtual body while also experiencing an illusory duplication of his body, as if he was at two places at the same time. These experiences were not reported by the 16 age-matched controls. The authors also assessed heartbeat awareness pre- and postoperatively using an established heartbeat detection task [49]. While the patient's heartbeat awareness was similar to the control group before surgery, 3 months after the near-to-complete removal of the patient's right insula to resect his tumor, his heartbeat awareness diminished considerably and was significantly lower than that of the control group.

Thanks to its rich connectivity with both afferent somatic signals and the cortex as well as the limbic system, it was proposed that the insula could be crucial for the emergence of a global meta-representation of the state of the body via its pivotal role in interoception and socio-emotional processing [50]. Craig posits that the insula is essential to the emergence of a "material self" a continuous but fluctuating stream of physical awareness [51]. However, the latter hypothesis is controversial; going back to Roger's extremely rare example of bilateral insular damage, Philippi et al. demonstrated that this patient exhibited largely preserved concept of self, at the primordial (feelings of the living body, personal agency, ownership of the body), extended (awareness of autobiographical self), and introspective levels (metacognitive and higher-order executive functions) [52]. Based on these findings, the authors proposed that information derived from the body form the basis of feeling states and emotional experience, without requiring the insula to be the source of awareness itself.

26.2.6 Neglect

Neglect occurs when a patient fails to respond to and report or orient to meaningful stimuli appearing in the space contralateral to the brain lesion, in the absence of sensory or motor deficits that could account for this failure. Spatial neglect occurs more frequently with right hemispheric damage (left hemispatial neglect) and has typically been associated with parietal lesions, but some studies suggest that the insula might also be involved [53]. Berthier, Starkstein, and Leiguarda have first reported a case of severe multimodal neglect following right hemispheric infarction involving the insula, adjacent white matter, and the inner cortical surface of the fronto-temporo-parietal operculum [54]. Manes and colleagues assessed neglect 4–8 weeks after stroke in nine patients with isolated insular infarctions—four with right insular lesions and five patients with left insular lesions—using tasks in the visual, tactile, and auditory modalities [55]. Assessments, conducted 4–8 weeks after stroke, comprised the double simultaneous test (visual, auditory, and somatic stimulation) and the line bisection task. Compared to patients with left insular damage, those with right insular lesions showed significant neglect in all three modalities. Studies using lesion-symptom mapping among patients with right hemisphere stroke have also identified the insula, among other structures, as more often damaged among patients with visual neglect compared than in those without [53, 56]. Other studies suggest that right insular lesions are more specifically associated with egocentric neglect, i.e., inattention to stimuli presented on the contralesional side of the body, than with allocentric neglect, i.e., inattention to elements on the contralesional side of individual objects [57, 58]. Other studies, however, have failed to find a relation between damage to the insula and spatial neglect, which puts into question the specific role of the insula in this syndrome and might suggest that damage to other neighboring structures account for the rare cases of neglect following insular injury [59, 60].

26.3 Cognition

26.3.1 Language

Although quite heterogeneous, speech impairments are among the most common clinical manifestations of insular injury. Shuren reported the case of a 59-year-old woman who, following a left infero-anterior insular infarct, presented difficulty initiating speech and pronouncing words, along with aberrant prosody [61]. Although pronunciation and prosody improved during the following weeks, speech initiation, as assessed with word fluency testing, remained impaired. The role of the left anterior insula in speech initiation has been further supported by a VLSM study involving 101 aphasic stroke patients, which found that insular and deep parietal white matter lesions are the best predictors of poor verbal fluency performance [9]. Borovsky and colleagues rated conversational speech of 50 aphasic patients with left-hemisphere lesions on measures of amount of speech, grammatical complexity, and semantic variation and found that damage to the anterior insula predicted reduced amount of speech and poorer grammatical complexity. These findings suggest that the verbal fluency impairments seen on formal neuropsychological exams can also be observable in day-to-day conversation [62].

Impairments in speech repetition, or conduction aphasia, have also been reported following damage to the left insula. Marshall et al. reported a case of a 54-year-old woman presenting with what they called "semantic" conduction aphasia 1 week after infarction limited to the gray matter of the left posterior insula and the intrasylvian parietal opercular cortex [63]. This manifested as fluent speech along with impaired verbal fluency and picture naming, repetition difficulties, impaired rhythm reproduction, dysgraphia, and frequent semantic paraphasias. One month later, most deficits were substantially improved, but there were still numerous semantic paraphasias during the naming task. Single-photon emission computed tomography (SPECT) showed hypoperfusion localized in the inferomedial and lateral regions of the left temporal lobe. Altered repetition was also described in a 76-year-old patient who suffered from insular stroke [64]. After a first subcortical stroke in the left hemisphere, this patient showed mixed transcortical aphasia with intact repetition and reading, and 10 days later, a second stroke involving the anterior insula and adjacent white matter bundles led to a marked deterioration of repetition, which was characterized by numerous phonological paraphasias, whereas reading remained intact. These cases are consistent with a lesion-symptom mapping study using imaging data from 107 stroke patients which related insular/external capsule lesions to repetition disorder [65].

Speech articulation coordination deficit is another possible consequence of insular cortex damage. Using lesion-symptom mapping in 25 stroke patients who displayed chronic apraxia of speech and in 19 patients without this deficit, Dronkers found that all patients with articulatory planning deficits had lesions that included the superior part of the left precentral gyrus of the insula, whereas this area was spared in all patients without apraxia of speech [7]. Nagao and colleagues also reported the case of a 67-year-old Japanese woman with an acute infarct limited to the left precentral insular gyrus who displayed transient speech and oral apraxia [66]. The role of the left precentral gyrus of the insula for coordination of speech articulation has also been supported by a recent VLSM study involving 33 patients with left hemisphere stroke [67].

A case of pure dysarthria without aphasia in a 72-year-old patient who suffered from a cortical infarction restricted to the left posterior insula suggested that the insula is involved in the motor aspects of speech [68]. The role of the insula in the motor aspects of speech may not be confined to the left hemisphere. Indeed, in their study of ten patients with isolated insular stroke, Baier et al. reported that only those with left hemisphere damage showed aphasia during the acute period, while damage to either of the hemispheres led to dysarthria [45]. In a group of 18 patients who underwent insulectomy as part of an epilepsy surgery, the only significant deterioration found at the group level 6 months after surgery was observed on the color naming condition of the Stroop task, which requires oro-motor speed [69]. Interestingly, this deterioration was found for patients with left and right insulectomy, separately.

The impacts of insular lesions on speech thus include speech initiation and naming deficits, conduction aphasia, speech apraxia, dysarthria, or a combination of those. Among their four cases of isolated insular injury, Cereda et al. reported speech impairments in both patients with left hemispheric stroke: one showed a transient jargonophasic fluent aphasia, characterized by phonemic paraphasia and dysarthria, with altered comprehension and repetition, and the other patient showed phonemic paraphasia and dysarthria, along with anomia and phonemic distortion, with preserved comprehension and repetition [6]. However, the fact that language impairments following insular injury are usually transient, as further evidenced by neurosurgical experience, indicates that the insula may not be essential for all aforementioned aspects of language [70–72].

26.3.2 Memory

The evidence relating isolated insular damage to memory impairments is sparse. Manes and colleagues compared verbal and visuospatial memory performance of six patients with left hemispheric infarction restricted to the insula and adjacent white matter to that of four patients with right hemispheric infarction in the same region [73]. Patients were assessed 4–8 weeks after stroke with the Logical Memory and Verbal Paired Associates subtests from the Wechsler Memory Scale, the CERAD Word List Memory subtest, and the Benton Visual Retention Test. Patients with left insular damage had significantly poorer Logical Memory (both immediate and delayed recall) and CERAD Word List Memory (delayed recall) compared to those with right insular damage. These results are consistent with data from a neurosurgical series involving 33 patients with Grade II or III insular gliomas who were assessed on a full standard neuropsychological test battery (including tasks of attention, executive function, learning and memory, language, and visuo-construction) before and 3 months after resection [74]. Although rates of decline did not statistically differ from those of a control group comprised of patients in whom gliomas had been resected in nearby brain regions, patients with left hemispheric insular gliomas showed a greater decline in the domain of learning and memory and lexical fluency following surgery, whereas those with right hemispheric insular gliomas presented a greater decline in the domain of visuo-construction. However, it should be noted that the effects on memory and fluency decreased as the time between surgery and postsurgical assessment increased. By contrast, a neurosurgical series of 18 epileptic patients who underwent insular resection (5 left hemisphere, 13 right hemisphere) for control of drug-resistant seizures found no deterioration in memory, attention, or executive function performance ≥5 months after surgery at the group level, but the small sample size limited further comparisons according to the hemisphere of resection [69]. Taken together, these results suggest that episodic memory impairments may occur following left insular damage, but these deficits may be transitory and probably reflect general left hemisphere dysfunction (e.g., due to insular connections with memory-related areas in the mesial temporal lobe and prefrontal cortex) rather than specific insular-related memory impairments.

26.3.3 Attention and Executive Functions

Despite extensive functional neuroimaging evidence supporting a central role of the anterior insula in the saliency network [75], there is little clinical evidence supporting this assertion. One recent study reported the case of a 45-year-old woman with left anterior insular stroke who showed a specific deficit in executive functioning in the acute stroke period [76]. Neuropsychological assessment, conducted 3 days post-stroke, revealed poor (<10th percentile) performance on word (oral phonemic and written) and design fluency tasks, on attention and inhibition control (Stroop Color-Word), and on the Wisconsin Card Sorting Test in which she could not even complete the first category, suggesting impaired conceptual thinking. By contrast, she did not show impairments in language, visuospatial perception, attention, memory, and working memory. This study was limited by the lack of a follow-up assessment. In their case study of a patient with personality changes following a left posterior insula lesion, Borg and colleagues reported only minor cognitive impairments mainly affecting verbal working memory (Digit Span) and task planning (Six Elements Test), whereas other attention and executive functions were relatively preserved [77]. Attention and executive function, as assessed using Wechsler's Digit-Symbol Coding subtest, the Trail Making Test, and the Stroop Word-Color Interference test, showed no decline at the group level following insular surgery in neurosurgical series of patients with insular tumors and epilepsy [69, 74].

26.4 Emotion and Empathy

26.4.1 Emotional Experience

Lesion studies provide clinical support for a contribution of the insular cortex to emotional processing. In a seminal case study, Calder and colleagues reported the case of a patient with left hemisphere infarction involving the insula, internal capsule, putamen, and globus pallidus, who was impaired in his ability to recognize expressions of disgust through multiple modalities, including nonverbal emotional sounds and visual emotional pictures [78]. In self-administered questionnaires aimed to investigate his experience of disgust, fear, and anger, the patient evaluated disgust-provoking scenarios as less disgusting than controls, in accordance with the view that the insula plays a central role in processing the emotion of disgust [79–81]. However, other studies did not find such specific deficits in disgust processing. For instance, Straube and colleagues found no impairment in disgust recognition or experience in a patient who suffered from a complete right insula and right basal ganglia damage due to stroke [82]. Hemispheric differences in disgust processing have been proposed to account for such discordant findings [83].

Emotional disturbances following insular cortex damage may not be limited to the emotion of disgust. Manes and colleagues assessed neuropsychiatric symptoms 4–8 weeks after acute stroke in patients with left ($n = 6$) and right ($n = 6$) insular stroke and in patients with non-insular stroke [84]. They found that patients with right insular damage were more likely to display tiredness, anergia, and underactivity in comparison to patients with non-insular and left insular stroke. More recently, Borg and colleagues reported the fascinating case of a young woman who suffered from a stroke in the left posterior insula, including SII, who presented a selective reduction of disgust identification for facial pictures, and evaluated expressions of disgust, fear, and happiness as less intense than a control group did [77]. This patient complained of emotional and behavioral changes following her stroke. She reported being able to identify emotions, but their intensity was less intense than before. For instance, 3 months after her stroke, when her family members cried tears of joy upon seeing her at the airport, she understood and identified their emotion of happiness but was unable to feel happy. Changes in emotional processing were evaluated by the Iowa Scales of Personality Changes. The most striking changes were loss of sensitivity, lack of resistance, impassiveness, mood changes, irritability, and inadequate emotion. The patient also reported a decrease in libido, and though she was never interested in art before, she developed a compulsive need to paint 6 months after the stroke. She was able to paint for several hours without pausing, and during these phases of creative activity, she felt better and experienced new sensations of physical pleasures though she couldn't describe her sensory experience [77, 85]. In the same view, Hébert-Seropian and colleagues used the Iowa Scales of Personality Changes to assess personality changes in 19 patients who underwent epilepsy surgery involving insular cortex resection and compared them with 19 patients who underwent standard temporal lobe epilepsy surgery [86]. Although postoperative changes did not differ significantly between these groups, insular resection was associated with a mild increase in irritability, emotional lability, frugality, and anxiety, while these changes were not seen in the temporal lobe epilepsy group, with the exception of anxiety.

Cho and colleagues also reported the case of a 66-year-old woman who exhibited an isolated right anterior insula infarction, resulting in an abrupt change in social behavior immediately after her stroke [87]. The patient began to act impulsively and impatiently, she made inappropriate remarks, and she presented new food habits (new preference for sweets) and ate compulsively. Her behavior was improved 45 days post-stroke.

In a group comparison study, a small group of seven patients with unilateral insular damage that included more than 50% of the insula (anterior and posterior parts) due to an ischemic stroke were compared to a group of patients with amygdala lesions following anterior temporal lobectomies and to another lesion-control group of patients who had unilateral middle cerebral artery strokes that spared the insula or posterior cerebral artery strokes [88]. All patients were asked to rate a set of pleasant, unpleasant, and neutral emotional pictures for their emotional valence and arousal on a Likert-type scale. Patients with insular lesions reported a reduction of arousal ratings for both pleasant and unpleasant pictures and attenuation of valence ratings, in comparison to the lesion-control group. This suggests that insular lesions may lead to a reduction in the intensity of the subjective emotional experience induced by pleasant and unpleasant information.

VLSM techniques also suggest that insular lesions lead to emotional disturbances. A series of studies using VLSM studies in a large group of Vietnam veterans found that insular lesions were associated with higher anxiety and apathy symptoms and that the degree of anterior insula damage predicts increased alexithymia (impaired awareness of one's emotional state) over anterior cingulate cortex and total lesion volume, providing evidence that this structure supports emotional awareness [89–91].

26.4.2 Social Cognition

In recent years, neuroimaging paradigms have convincingly established that the insula is involved extensively in functions of social cognition. Indeed, in tasks assessing empathy, others' pain, or recognition of emotional expressions in faces, the insula is one of the brain areas most often reported as being activated. [92] Lesion studies tend to corroborate these findings. For example, poor emotional sensitivity across all emotions was reported in a study assessing emotion recognition in three patients with damage to the right anterior insula, caused, respectively, by cerebral infarct, aneurysmal subarachnoid hemorrhage, and brain edema [93]. Notably, the patients misidentified anger as disgust or sadness and misinterpreted disgust as sadness. The authors posited that these errors could be the result of an acquired inability to distinguish low (e.g., sadness) from high (e.g., anger) arousal levels in emotions while still retaining the ability to identify the general valence of the stimuli. In another case study, emotion recognition was assessed in a patient with extensive bilateral insular damage and to the dorsolateral prefrontal cortex and found that the subject couldn't recognize emotions apart from happiness [94]. However, if dynamic images pertaining to the emotions being illustrated were shown instead of static stimuli, the patient could properly identify all emotions, with the notable exception of disgust. They concluded that information about actions is primarily assessed in occipito-parietal and dorsal frontal cortices, regions that are intact in the subject's brain. Similarly, a large study using VLSM among 180 veterans with traumatic brain injury related insular lesions to deficits in emotion recognition regardless of the valence of the stimuli, though these deficits were also associated with damage to the medial prefrontal cortex and anterior cingulate cortex [95].

Emotion recognition underlies more complex social cognition processes allowing individuals to perceive, share, predict, and react to another person's affective state. Multiple lesion studies corroborate the idea that the insula is primordial for empathy, the lens through which we view others' emotion expressions. For example, in another VLSM study from the research group stated above, a self-reported emotional empathy questionnaire was administered to 192 veterans with TBIs and revealed that damage to the insula was associated with diminished emotional empathy [96]. This correlation was also associated with damage to the ventrolateral prefrontal cortex, as well as to the left and right posterior temporal lobe. Both emotion recognition and empathy were assessed in 15 patients who underwent unilateral insulectomy for the treatment of pharmacoresistant epilepsy and were compared to a group of patients who underwent temporal lobe epilepsy surgery and to a healthy control group [97]. Whereas both groups of patients were poorer than healthy controls to recognize expressions of fear, only those with insular epilepsy surgery were impaired in recognizing happiness and surprise. However, in contradiction to previous accounts from neuroimaging studies linking insular activity to the processing of disgust, no deficits in disgust recognition were observed. Moreover, when compared to healthy controls, insular patients displayed lower scores on the perspective taking scale on the self-administered empathy questionnaire. In another study, Gu and colleagues examined the ability to perceive others' pain in six cases divided equally in two groups consisting of patients with anterior insula and anterior cingulate cortex damage [98]. Interestingly, they found that empathetic pain was impaired in the insular group only, suggesting that while the processing of empathetic pain involves multiple brain regions, the insula may be primordial for its correct functioning. In a final example, it was reported that in 17 patients who suffered insular damage due to cerebral low-grade gliomas, emotion recognition, empathic concern, perspective taking, and the ability to perceive others' pain were all impaired [99]. Being able to interpret others' feelings or to put ourselves in their shoes is a crucial social skill, and as the data from lesion studies illustrates, it is one that can be severely impaired in individuals with insular damage, especially when the anterior insula is involved.

26.4.3 Decision-Making

As for other "hot" cognitive processes, risky decision-making is seldom assessed as part of standard clinical neuropsychological exams. Because of a growing interest in the neural basis of risk-taking behavior and neuroeconomics, decision-making deficits have been assessed in a few studies involving groups of patients with cerebral damage affecting the insular cortex. However, these studies are limited by the extent of insular damage, which varied across patients and often exceeded the insula. Clark and colleagues assessed risky decision-making in 13 patients with focal insular lesion, 20 with lesion to the ventromedial prefrontal cortex, 12 with a lesion in other regions, and 41 healthy control participants on the Cambridge Gamble Task [100]. The insular group was selectively poorer than the other groups in their ability to adjust their bets by the odds of winning. This group showed poor adjustment to risk, finished the task with a lower point score, and experienced more bankruptcies. Weller and colleagues then used the Cups Task to assess risk adjustment under gain and loss conditions separately, in a group of ten patients with unilateral insular damage due to a middle cerebral artery stroke [101]. They found that insular patients were severely impaired in adjustment to risk in both the gain and loss conditions when compared to healthy volunteers. By contrast, a study using the same task in 13 epileptic patients who underwent insular resection revealed a specific impairment in the loss condition [102]. These patients did not differ from healthy controls when assessed with the Iowa Gambling Task, a more established risky decision-making task involving implicit learning.

Clark et al. assessed the gambler's fallacy and near misses distortion effects on decision-making in patients with focal lesions to the ventromedial prefrontal cortex, insula, or amygdala and in healthy controls [103]. While other groups displayed these cognitive distortions, insular patients did not show such effects. Another lesion study assessed temporal discounting—a phenomenon by which the subjective value attributed to a reward decreases as the delay until it is given increases—in 12 patients with lesions involving the insular cortex and the surrounding gray and white matter, in 13 lesion-control patients, and in 64 healthy controls [104]. In comparison to both other groups, insular patients showed reduced temporal discounting of future rewards, as they behaved more patiently than control participants and were less sensitive to sooner rewards. Taken together, these studies suggest that insular lesions affect the emotional aspects of decision-making, making patients less sensitive to emotion-related "cognitive biases" when making risky decisions.

Conclusion

The neuropsychological deficits following damage to the insular cortex are numerous and heterogeneous and include sensory, cognitive, and/or emotional disturbances. These multiple manifestations of insular lesions are consistent with data from functional neuroimaging and cortical stimulation studies. Although specific location-deficit relationships are difficult to establish due to the variability in insular lesions and in their clinical presentations, the reviewed data suggest that posterior insular damage is more frequently associated with sensory (especially somatosensory/pain and auditory) impairments, whereas anterior insular damage in the left hemisphere has often been associated with speech initiation impairments. Many of the reviewed deficits were transient or attenuated with time, suggesting compensating mechanisms due to cerebral plasticity.

References

1. Uddin LQ, Nomi JS, Hébert-Seropian B, Ghaziri J, Boucher O. Structure and function of the human insula. J Clin Neurophysiol. 2017;34:300–6.
2. Kurth F, Zilles K, Fox PT, Laird AR, Eickhoff SB. A link between the systems: functional differentiation and integration within the human insula revealed by meta-analysis. Brain Struct Fucnt. 2010;214:519–34.
3. Mazzola L, Mauguière F, Isnard J. Electrical stimulations of the human insula: their contribution to the ictal semiology of insular seizures. J Clin Neurophysiol. 2017;34:307–14.
4. Ibañez A, Gleichgerrcht E, Manes F. Clinical effects of insular damage in humans. Brain Struct Funct. 2010;214:397–410.
5. Türe U, Yaşargil MG, Al-Mefty O, Yaşargil DC. Arteries of the insula. J Neurosurg. 2000;92:676–87.
6. Cereda C, Ghika J, Maeder P, Bogousslavsky J. Strokes restricted to the insular cortex. Neurology. 2002;59:1950–5.
7. Dronkers NF. A new brain region for coordinating speech articulation. Nature. 1996;384:159–61.
8. Karnath HO, Baier B, Nagele T. Awareness of the functioning of one's own limbs mediated by the insular cortex? J Neurosci. 2005;25:7134–8.
9. Bates E, Wilson SM, Saygin AP, Dick F, Sereno MI, Knight RT, et al. Voxel-based lesion-symptom mapping. Nat Neurosci. 2003;6:448–50.
10. Moon HI, Pyun SB, Tae WS, Kwon HK. Neural substrates of lower extremity motor, balance, and gait function after supratentorial stroke using voxel-based lesion symptom mapping. Neuroradiology. 2016;58:723–31.
11. Kodumuri N, Sebastian R, Davis C, Posner J, Kim EH, Tippett DC, et al. The association of insular stroke with lesion volume. Neuroimage Clin. 2016;11:41–5.
12. Gras-Combe G, Minotti L, Hoffmann D, Krainik A, Kahane P, Chabardes S. Surgery for nontumoral insular epilepsy explored by stereoelectroencephalography. Neurosurgery. 2016;79:578–88.
13. Kalani MY, Kalani MA, Gwinn R, Keogh B, Tse VC. Embryological development of the human insula and its implications for the spread and resection of insular gliomas. Neurosurg Focus. 2009;27:E2.
14. Skrap M, Mondani M, Tomasino B, Weis L, Budai R, Pauletto G, et al. Surgery of insular nonenhancing gliomas: volumetric analysis of tumoral resection, clinical outcome, and survival in a consecutive series of 66 cases. Neurosurgery. 2012;70:1081–93.
15. Finet P, Nguyen DK, Bouthillier A. Vascular consequences of operculoinsular corticectomy for refractory epilepsy. J Neurosurg. 2015;122:1293–8.
16. Simon M, Neuloh G, von Lehe M, Meyer B, Schramm J. Insular gliomas: the case for surgical management. J Neurosurg. 2009;110:685–95.
17. Mesulam MM, Mufson EJ. Insula of the old world monkey. III: efferent cortical output and comments on function. J Comput Neurosci. 1982;212:38–52.
18. Greenspan JD, Lee RR, Lenz FA. Pain Sensitivity alterations as a function of lesion location in the parasylvian cortex. Pain. 1999;81:273–82.
19. Starr CJ, Sawaki L, Wittenberg GF, Burdette JH, Oshiro Y, Quevedo AS, Coghill RC. Roles of the insular cortex in the modulation of pain: insights from brain lesions. J Neurosci. 2009;29:2684–94.
20. Berthier M, Starkstein S, Leiguarda R. Asymbolia for pain: a sensory-limbic disconnection syndrome. Ann Neurol. 1988;24:41–9.
21. Garcia-Larrea L, Perchet C, Creac'h C, Convers P, Peyron R, Laurent B, et al. Operculo-insular pain parasylvian pain: a distinct central pain syndrome. Brain. 2010;133:2528–39.
22. Denis DJ, Marouf R, Rainville P, Bouthillier A, Nguyen DK. Effects of insular stimulation on thermal nociception. Eur J Pain. 2016;20:800–10.
23. Birklein F, Rolke R, Müller-Forell W. Isolated insular infarction eliminates contralateral cold, cold pain, and pinprick perception. Neurology. 2005;65:1381.
24. Baier B, zu Eulenburg P, Geber C, Rohde F, Rolke R, Maihöfner C, et al. Insula and sensory insular cortex and somatosensory control in patients with insular stroke. Eur J Pain. 2014;18:1385–93.
25. Cattaneo L, Chierici E, Cucurachi L, Cobelli R, Pavesi G. Posterior insular stroke causing selective loss of contralateral nonpainful thermal sensation. Neurology. 2007;68:237.
26. Augustine JR. Circuitry and functional aspects of he insular lobe in primates including humans. Brain Res Brain Res Rev. 1996;22:229–44.
27. Fifer RC. Insular stroke causing unilateral auditory processing disorder: case report. J Am Acad Audiol. 1993;4:364–9.
28. Habib M, Daguin G, Milandre L, Royere ML, Rey M, Lanteri A, et al. Mutism and auditory agnosia due to bilateral insular damage – role of the insula in human communication. Neuropsychologia. 1995;33:327–39.
29. Griffiths TD, Rees A, Witton C, Cross PM, Shakir RA, Green GG. Spatial and temporal auditory processing deficits following right hemisphere infarction. A psychophysical study. Brain. 1997;120:785–94.
30. Ayotte J, Peretz I, Rousseau I, Bard C, Bojanowski M. Patterns of music agnosia associated with middle cerebral artery infarcts. Brain. 2000;123:1926–38.
31. Griffiths TD, Warren JD, Dean JL, Howard D. "When the feeling's gone": a selective loss of musical emotion. J Neurol Neurosurg Psychiatry. 2004;75:344–5.
32. Bamiou DE, Musiek FE, Stow I, Stevens J, Cipolotti L, Brown MM, et al. Auditory temporal processing

deficits in patients with insular stroke. Neurology. 2006;67:614–9.
33. Boucher O, Turgeon C, Chapoux S, Ménard L, Rouleau I, Lassonde M, et al. Hyperacusis following unilateral damage to the insular cortex: a three-case report. Brain Res. 2015;1606:102–12.
34. Tomasino B, Marin D, Canderan C, Maieron M, Skrap M, Rumiati RI. Neuropsychological patterns following lesions of the anterior insula in a series of forty neurosurgical patients. AIMS Neurosci. 2014; 1:225–44.
35. Mak YE, Simmons KB, Gitelman DR, Small DM. Taste and olfactory intensity perception changes following left insular stroke. Behav Neurosci. 2005; 119:1693–700.
36. Ribas ES, Duffau H. Permanent anosmia and ageusia after resection of a left temporoinsular low-grade glioma: anatomofunctional considerations. J Neurosurg. 2012;116:1007–13.
37. Metin B, Melda B, Birsen I. Unusual clinical manifestation of a cerebral infarction restricted to the insulate cortex. Neurocase. 2007;13:94–6.
38. Beume LA, Klingler A, Reinhard M, Niesen WD. Olfactory hallucinations as primary symptom for ischemia in the right posterior insula. J Neurol Sci. 2015;354:138–9.
39. Pritchard TC, Macaluso DA, Eslinger PJ. Taste perception in patients with insular cortex lesions. Behav Neurosci. 1999;113:663–71.
40. Stevenson RJ, Miller LA, McGrillen K. The lateralization of gustatory function and the flow of information from tongue to cortex. Neuropsychologia. 2013; 51:1408–16.
41. Mazzola L, Lopez C, Faillenot I, Chouchou F, Mauguière F, Isnard J. Vestibular responses to direct stimulation of the human insular cortex. Ann Neurol. 2014;76:609–19.
42. Lopez C, Blanke O, Mast FW. The human vestibular cortex revealed by coordinate-based activation likelihood estimation meta-analysis. Neuroscience. 2012;212:159–79.
43. zu Eulenburg P, Caspers S, Roski C, Eickhoff SB. Meta-analytical definition and functional connectivity of the human vestibular cortex. NeuroImage. 2012; 60:162–9.
44. Papathanasiou ES, Papacostas SS, Charalambous M, Eracleous E, Thodi C, Pantzaris M. Vertigo and imbalance caused by a small lesion in the anterior insula. Electromyogr Clin Neurophysiol. 2006;46: 185–92.
45. Baier B, Conrad J, Zu Eulenburg P, Best C, Müller-Forell W, Birklein F, et al. Insular strokes cause no vestibular deficits. Stroke. 2013;44: 2604–6.
46. Khalsa SS, Rudrauf D, Feinstein JS, Tranel D. The pathways of interoceptive awareness. Nat Neurosci. 2009;12:1494–6.
47. Grossi D, De Vita A, Palermo L, Sabatini U, Trojano L, Guariglia C. The brain network for self-feeling: a symptom-lesion mapping study. Neuropsychologia. 2014;63:92–8.
48. Ronchi R, Bello-Ruiz J, Lukowska M, Herbelin B, Cabrilo I, Schaller K, Blanke O. Right insular damage decreases heartbeat awareness and alters cardio-visual effects on bodily self-consciousness. Neuropsychologia. 2015;70:11–20.
49. Schandry R. Heart beat perception and emotional experience. Psychophysiology. 1981;18:483–8.
50. Damasio A. Descartes' error: emotion, reason and the human brain. Grosset/Putnam: New York, NY; 1994.
51. Craig AD. How do you feel—now? The anterior insula and human awareness. Nat Rev Neurosci. 2009;10:59–70.
52. Philippi CL, Feinstein JS, Khalsa SS, Damasio A, Tranel D, Landini G, et al. Preserved self-awareness following extensive bilateral brain damage to the insula, anterior cingulate, and medial prefrontal cortices. PLoS One. 2012;7:e38413.
53. Karnath HO, Berger MF, Küker W, Rorden C. The anatomy of spatial neglect based on voxelwise statistical analysis: a study of 140 patients. Cereb Cortex. 2004;14:1164–72.
54. Berthier M, Starkstein S, Leiguarda R. Behavioral effects of damage to the right insula and surrounding regions. Cortex. 1987;23:673–8.
55. Manes F, Paradiso S, Springer JA, Lamberty G, Robinson RG. Neglect after right insular cortex infarction. Stroke. 1999;30:946–8.
56. Golay L, Schnider A, Ptak R. Cortical and subcortical anatomy of chronic spatial neglect following vascular damage. Behav Brain Funct. 2008;4:43.
57. Chechlacz M, Rotshtein P, Roberts KL, Bickerton WL, Lau JKL, Humphreys GW. The prognosis of allocentric and egocentric neglect: evidence from clinical scans. PLoS One. 2012;7:e47821.
58. Kenzie JM, Girgulis K, Semrau JA, Findlater SE, Desai JA, Dukelow SP. Lesion sites associated with allocentric and egocentric visuospatial neglect in acute stroke. Brain Connect. 2015;5:413–22.
59. Pedrazzini E, Schnider A, Ptak R. A neuroanatomical model of space-based and object-centered processing in spatial neglect. Brain Struct Funct. 2017;222(8):3605–13. https://doi.org/10.1007/s00429-017-1420-4.
60. Smith DV, Clithero JA, Rorden C, Karnath HO. Decoding the anatomical network of spatial attention. Proc Natl Acad Sci U S A. 2013;110:1518–23.
61. Shuren J. Insula and aphasia. J Neurol. 1993;240: 216–8.
62. Borovsky A, Saygin AP, Bates E, Dronkers N. Lesion correlates of conversational speech production deficits. Neuropsychologia. 2007;45:2525–33.
63. Marshall RS, Laza RM, Mohr JP, Van Heertum RL, Mast H. "Semantic" conduction aphasia from a posterior insular cortex infarction. J Neuroimaging. 1996; 6:189–91.
64. Carota A, Annoni JM, Marangolo P. Repeating through the insula: evidence from two consecutive strokes. Neuroreport. 2007;18:1367–70.
65. Kreisler A, Godefroy O, Delmaire C, Debachy B, Leclercq M, Pruvo JP, et al. The anatomy of aphasia revisited. Neurology. 2000;54:1117–23.

66. Nagao M, Takeda K, Komori T, Isozaki E, Hirai S. Apraxia of speech associated with an infarct in the precentral gyrus of the insula. Neuroradiology. 1999;41:356–7.
67. Baldo JV, Wilkins DP, Ogar J, Willock S, Dronkers NF. Role of the precentral gyrus of the insula in complex articulation. Cortex. 2011;47:800–7.
68. Hiraga A, Tanaka S, Kamitsukasa I. Pure dysarthria due to an insular infarction. J Clin Neurosci. 2010;17:812–3.
69. Boucher O, Rouleau I, Escudier F, Malenfant A, Denault C, Charbonneau S, et al. Neuropsychological performance before and after partial or complete insulectomy in patients with epilepsy. Epilepsy Behav. 2015;43:53–60.
70. Lang FF, Olansen NE, DeMonte F, Gokaslan ZL, Holland EC, Kalhorn C, et al. Surgical resection of intrinsic insular tumors: complication avoidance. J Neurosurg. 2001;95:638–50.
71. Duffau H, Taillandier L, Gatignol P, Capelle L. The insular lobe and brain plasticity: lessons from tumor surgery. Clin Neurol Neurosurg. 2006;108:543–8.
72. Sanai N, Polley MY, Berger MS. Insular gioma resection: assessment of patient morbidity, survival, and tumor progression. J Neurosurg. 2010;112:1–9.
73. Manes F, Springer J, Jorge R, Robinson RG. Verbal memory impairment after left insular cortex infarction. J Neurol Neurosurg Psychiatry. 1999;67(4):532.
74. Wu AS, Witgert ME, Lang FF, Xiao L, Bekele BN, Meyers CA, et al. Neurocognitive function before and after surgery for insular gliomas. J Neurosurg. 2011;115:1115–25.
75. Menon V, Uddin LQ. Saliency, switching, attention and control: a network model of insula function. Brain Struct Funct. 2010;214:655–67.
76. Markostamou I, Rudolf J, Tsiptsious I, Kosmidis MH. Impaired executive functioning after left anterior insular stroke: a case report. Neurocase. 2014;21:148–53.
77. Borg C, Bedoin N, Peyron R, Bogey S, Laurent B, Thomas-Antérion C. Impaired emotional processing in a patient with a left posterior insula-SII lesion. Neurocase. 2013;19:592–603.
78. Calder AJ, Keane J, Manes F, Antoun N, Young AW. Impaired recognition and experience of disgust following brain injury. Nat Neurosci. 2000;3:1077–8.
79. Brown S, Gao X, Tisdelle L, Eickhoff SB, Liotti M. Naturalizing aesthetics: brain areas for aesthetic appraisal across sensory modalities. NeuroImage. 2011;58:250–8.
80. Phillips ML, Young AW, Senior C, Brammer M, Andrew C, Calder AJ, et al. A specific neural substrate for perceiving facial expressions of disgust. Nature. 1997;389:495–8.
81. Stark R, Zimmermann M, Kagerer S, Schienle A, Walter B, Weygandt M, et al. Hemodynamic brain correlates of disgust and fear ratings. NeuroImage. 2007;37:663–73.
82. Straube T, Weisbrod A, Schmidt S, Raschdorf C, Preul C, Mentzel HJ, et al. No impairment of recognition and experience of disgust in a patient with a right-hemispheric lesion of the insula and basal ganglia. Neuropsychologia. 2010;48:1735–41.
83. Wicker B, Keysers C, Plailly J, Royet JP, Gallese V, Rizzolatti G. Both of us disgusted in my insula: the common neural basis of seeing and feeling disgust. Neuron. 2003;40:655–64.
84. Manes F, Paradiso S, Robinson RG. Neuropsychiatric effects of insular stroke. J Nerv Ment Dis. 1999;187(12):707.
85. Thomas-Antérion C, Creac'h C, Dionet E, Borg C, Extier C, Faillenot I, et al. De novo artistic activity following insular-SII ischemia. Pain. 2010;150:121–7.
86. Hébert-Seropian B, Boucher O, Sénéchal C, Rouleau I, Bouthillier A, Lepore F, et al. Does unilateral insular resection disturb personality? A study with epileptic patients. J Clin Neurosci. 2017;43:121–5. https://doi.org/10.1016/j.jocn.2017.04.001.
87. Cho HJ, Kim SJ, Hwang SJ, Jo MK, Kim HJ, Seeley WW, et al. Social-emotional dysfunction after isolated right anterior insular infarction. J Neurol. 2012;259:764–7.
88. Berntson GG, Norman GJ, Bechara A, Bruss J, Tranel D, Cacioppo JT. The insula and evaluative processes. Psychol Sci. 2011;22:80–6.
89. Knutson KM, Rakowsky ST, Solomon J, Krueger F, Raymont V, Tierney MC, et al. Injured brain regions associated with anxiety in Vietnam veterans. Neuropsychologia. 2013;51:686–94.
90. Knutson KM, Monte OD, Raymont V, Wassermann EM, Krueger F, Grafman J. Neural correlates of apathy revealed by lesion mapping in participants with traumatic brain injuries. Hum Brain Mapp. 2014;35:943–53.
91. Hogeveen J, Bird G, Chau A, Krueger F, Grafman J. Acquired alexithymia following damage to the anterior insula. Neuropsychologia. 2016;82:142–8.
92. Bernhardt BC, Singer T. The neural basis of empathy. Ann Rev Neurosci. 2012;35:1–23.
93. Terasawa Y, Kurosaki Y, Ibata Y, Moriguchi Y, Umeda S. Attenuated sensitivity to the emotions of others by insular lesion. Front Psychol. 2015;6:1314.
94. Adolphs R, Tranel D, Damasio A. Dissociable neural systems for recognizing emotions. Brain Cogn. 2003;52:61–9.
95. Dal Monte O, Krueger F, Solomon JM, Schintu S, Knutson KM, Strenziok M, et al. A voxel-based lesion study on facial emotion recognition after penetrating brain injury. Soc Cogn Affect Neurosci. 2013;8:632–9.
96. Driscoll DM, Dal Monte O, Solomon J, Krueger F, Grafman J. Empathic deficits in combat veterans with traumatic brain injury: a voxel-based lesion-symptom mapping study. Cogn Behav Neurol. 2012;25:160–6.
97. Boucher O, Rouleau I, Lassonde M, Lepore F, Bouthillier A, Nguyen DK. Social information processing following resection of the insular cortex. Neuropsychologia. 2015;71:1–10.

98. Gu X, Gao Z, Wang X, Liu X, Knight RT, Hof PR, Fan J. Anterior insular cortex is necessary for empathetic pain perception. Brain. 2012;135:2726–35.
99. Chen P, Wang G, Ma R, Jing F, Zhang Y, Wang Y, et al. Multidimensional assessment of empathic abilities in patients with insular glioma. Cogn Affect Behav Neurosci. 2016;16:962–75.
100. Clark L, Bechara A, Damasio H, Aitken MR, Sahakian BJ, Robbins TW. Differential effects of insular and ventromedial prefrontal cortex lesions on risky decision-making. Brain. 2008;131:1311–22.
101. Weller JA, Levin IP, Shiv B, Bechara A. The effects of insula damage on decision-making for risky gains and losses. Soc Neurosci. 2009;4:347–58.
102. Von Siebenthal Z, Boucher O, Rouleau I, Lassonde M, Lepore F, Nguyen DK. Decision-making impairments following insular and medial temporal lobe resection for drug-resistant epilepsy. Soc Cogn Affect Neurosci. 2017;12:128–37.
103. Clark L, Studer B, Bruss J, Tranel D, Bechara A. Damage to insula abolishes cognitive distortions during simulated gambling. Proc Natl Acad Sci U S A. 2014;111:6098–103.
104. Sellitto M, Ciaramelli E, Mattioli F, di Pellegrino G. Reduced sensitivity to sooner reward during intertemporal decision-making following insula damage in humans. Front Behav Neurosci. 2016; 9:367.

The Role of the Insula in Schizophrenia

27

Cameron Schmidt

27.1 Introduction

Affecting approximately 1% of the population worldwide [1], schizophrenia (SZ) is a rare but devastating neuropsychiatric disorder characterized by broad cognitive and functional impairments. The symptoms of SZ are typically categorized into positive and negative symptoms. Positive symptoms include sensory hallucinations, delusional thinking, grandiosity, paranoia, disorganized thinking, and hostility [2]. Negative symptoms, by contrast, include blunted affect (affective flattening), amotivation, social withdrawal, and a poverty of speech [2, 3]. The broad cognitive impairments related to SZ yield problems in social cognition, sensory processing, verbal semantic processing, and executive attention, to name a few [4–6].

Cognitive deficits are present at the earliest stages of SZ [6, 7] but may manifest long before disease onset. Motor and neurological deficits, including deficits in attention, verbal memory, and motor skills [8, 9], and deficits in social competence [10, 11] are prospective indicators of SZ. A 48-year longitudinal study showed that teacher ratings of interpersonal deficits within school-age children were significantly related to the later development of SZ [12]. This suggests that rather than interpersonal deficits arising as a product of disease onset, such deficits may instead represent the earliest manifestation of neurodevelopmental alterations.

Cognitive functioning in SZ patients is associated with negative symptom severity [6, 13] and overall functional outcome [14]. Working memory, verbal memory and processing speed, and attentional and perceptual processing account for 52% of all variance in the 9-month return to school or work following recent first-episode psychosis (FEP) [15].

Understanding the neural correlates of SZ has been a focus of extensive research. The insular cortex undergoes significant alterations in structure and connectivity over the course of the disease. With participation across broad cognitive domains, damage to the insular cortex may be involved in many observed disease deficits. We will herein discuss the structural and functional alterations undergone by the insula over the course of SZ.

27.2 Morphometric Changes of the Insula

SZ induces regional structural alterations within the brain, driving reductions in gray matter (GM), white matter (WM), cortical surface area, and corresponding ventricular enlargement [16–19]. The insula is one such region, exhibiting significant reductions in GM, WM, and cortical

C. Schmidt
Seattle Science Foundation, Seattle, WA, USA
e-mail: d.nguyen@umontreal.ca

surface area over the course of the disease [20–23].

The literature reveals variability regarding the lateralization of insular structural alterations. Reductions in insular GM have been reported to be lateralized to both the left [19, 21, 24–26] and right [27–29] insula. Reductions in insular cortical surface area have also been observed in the left [25, 30] and right insula [31]. However, the preponderance of literature reveals bilateral insular reductions [18, 23, 32–39].

It is important to consider the localization of insular reductions, given the unique functional connectivity of insular subregions. Functional connectivity studies have parcellated the insula into three distinct subregions: the posterior insula (PI), dorsal anterior insula (dAI), and ventral anterior insula (vAI) [40]. The PI is most significantly connected to the pregenual anterior cingulate cortex (ACC), supplementary motor area (SMA), and somatosensory cortex and is involved in the processing of pain, sensorimotor stimuli, and language [40–42]. The dAI is connected to the dorsal ACC (dACC) and dorsolateral prefrontal cortex (DLPFC), demonstrating involvement in executive control and tasks of higher cognition [40–42]. The vAI is connected to the amygdala, ventral tegmental area (VTA), superior temporal sulcus, and the posterolateral orbitofrontal cortex and is associated with emotion, chemosensation, and autonomic function [40].

The subregional localization of disease-related structural alterations is unclear in the literature. While GM reductions are commonly found to span the entire insular cortex [23, 34, 43], greater relative reduction has been reported in both the anterior [22–24, 44] and posterior [27, 43] aspects of the insula. Using a region-of-interest (ROI) based approach, Saze et al. showed global insular reductions in SZ patients [43]. However, a separate subdivisional analysis revealed the right PI as the only subregion with significant volume difference. Recent work has further parcellated the insular cortex, suggesting the structure to contain six functionally unique subregions [42].

Still other studies have failed to find reductions of insular GM or cortical surface size in SZ [16, 45]. Of 14 papers in a voxel-based morphometry (VBM) meta-analysis, only seven found reductions in insular GM and WM [16]. Of the seven studies, reductions were bilateral in four, localized to the left insula in two, and localized to the right insula in one [16]. In 2000, Crespo-Facorro et al. conducted a ROI analysis on 25 drug-naïve first-episode schizophrenia (FES) patients, finding significant reductions in cortical surface size and GM volume in the left insular cortex [25]. However, attempts to replicate these findings in a new study population 10 years later were unsuccessful [46]. Upon controlling for sociodemographic and clinical characteristics in a subsequent ROI analysis of drug-naïve FES patients, Crespo-Facorro and colleagues found no significant difference in insular volume between patients and healthy controls. In fact, patients exhibited non-significantly larger insular volumes.

27.2.1 Gender Effects

It has been suggested that SZ exhibits a gender-specific pathology. Duggal et al. conducted an ROI study on an equal number of male and female FES patients [47]. Significant GM reductions were found solely in the right insula of female patients [47]. Elsewhere, SZ-related morphometric changes were confined to the left insula of male patients [25, 48]. Other studies have failed to find any gender-morphometry interaction [46] and a recent meta-analysis of 15 ROI studies found no evidence for a gender effect [22].

27.2.2 Treatment Effects

Evidence suggests that antipsychotic treatment may influence structural alterations in SZ [45, 49, 50]. However, heterogeneity in study populations, imaging modalities, and analysis techniques has challenged the development of a proper understanding of this potential confound.

One such point of heterogeneity lies in the pharmacological differences between typical (first generation) and atypical (second generation) antipsychotics. In a longitudinal ROI study,

Lieberman et al. demonstrated significant differences in the progression of global GM changes between patients treated with haloperidol (typical) and olanzapine (atypical) [51]. Haloperidol-treated patients exhibited a loss of global GM, appearing to peak within the first 12 weeks of treatment, while the olanzapine-treated patients did not experience significant global GM loss. In contrast, other studies assert no difference exists between typical and atypical antipsychotics regarding their effects on global GM [52, 53]. Despite these conflicting results, converging evidence from recent meta-analyses reveals a regionally-specific effect of antipsychotic type on GM and WM loss, such that less progressive loss occurs in SZ patients treated exclusively with atypical (versus typical) antipsychotics [17, 54].

As with whole-brain analyses, studies investigating the treatment effects of antipsychotics upon the insula have been mixed. Pressler et al. found no significant difference in insular volume between SZ patients and matched controls, but patients were found to have a positive correlation between typical neuroleptic exposure and insular volume [45]. Of interest, compared to healthy controls, patients exhibited non-significantly larger insular volumes, but smaller cortical surface area. Elsewhere, authors have failed to find any correlation between daily dose or antipsychotic medication type and insular GM volume [23, 43, 44].

Antipsychotic treatment has been correlated with insular activation [55]. During an overt verbal fluency task, functional magnetic resonance imaging (fMRI) of FEP patients exposed to early atypical antipsychotic treatment demonstrated a negative correlation between medication exposure and activation of the left insula [55]

Though differences have been attributed to medication effects, genetics, or, the course of the disease, meta-analyses support the presence of bilateral insular reductions, independent of confounds [22, 33]. Fusar-Poli et al. conducted a voxel-wise analyses of 14 VBM studies of antipsychotic-naïve FEP patients [33]. Compared to healthy controls, FEP patients showed consistent GM reductions in the bilateral insula. While genetics may play a role in the heterogeneity of insular defects, Borgwardt et al.'s imaging of monozygotic twins discordant for SZ demonstrated significant bilateral GM reductions in the affected twins [56].

27.2.3 Progression of Structural Changes

SZ patients experience a regionally-specific temporal pattern of neural tissue loss [16, 17, 57, 58]. Whole-brain tissue loss is greatest during the initial phases of the disease (i.e., early years following the onset of psychosis), slowing down over subsequent time periods as patients settle into the chronic phase of the disease [17, 44]. The trajectory of this tissue loss is temporally- and regionally-specific [17, 57]. Of 14 VBM studies, 11 examined chronic patients, while three examined first-episode patients [16]. Reduced volume in the left medial frontal gyrus was reported in seven (64%) chronic patient studies, while not found in any of the three first-episode studies. Conversely, reductions in the volume of the right ACC were reported in all three first-episode patient studies, while in only 27% of the chronic patient studies.

The insula undergoes a similarly progressive deterioration of GM over the course of SZ. Compared to controls, FES patients have significantly greater reductions in bilateral insular GM over time [44]. Along with several other select regions, the insula experiences the greatest relative GM loss during the initial phases of psychosis [44]. The rate of this GM loss abates in later stages of SZ. Once in the chronic stage of the disease, greater reductions in right insular GM, relative to FES patients, are observed [28].

The insula may serve as a neuroanatomical correlate underlying the transition to psychosis [33, 48]. Fusar-Poli et al. examined structural differences between high-risk (HR) and antipsychotic-naïve FEP individuals [33]. Their voxel-wise meta-analysis revealed the onset of psychosis was characterized by significant reductions in, among other regions, insular GM. FEP subjects exhibited significant GM reductions in the bilateral insula compared with healthy controls and significant GM reductions in the left insula compared to HR subjects. These findings suggest insular reductions may reach a critical point that marks the onset of SZ symptomology.

Insular morphology also serves as an indicator of disease risk, with HR individuals exhibiting reductions in insular GM prior to disease onset [28, 33, 59]. Takahashi et al. found significantly greater volume reductions in the bilateral insula of HR patients who later developed psychosis, compared to those who did not [23]. Additional studies have demonstrated an association between HR subjects who later develop SZ and reduced GM in the right insula [59, 60]. Between ultra-high-risk patients who later developed psychosis (UHR-P) and those who did not (UHR-NP), UHR-P patients had significantly reduced bilateral insular GM compared to UHR-NP patients, and exhibited significantly reduced right insular GM compared to controls [61]. While individuals at an enhanced clinical risk for psychosis exhibit bilateral insular GM reductions, it appears individuals at an enhanced genetic risk display greater left-sided insular GM reductions [60].

27.2.4 Clinical Correlations

The insula is a broadly connected neural region. It is therefore not surprising that reductions in insular GM and cortical surface size have been associated with various symptom dimensions. Several studies report a negative correlation between the severity of positive symptoms in SZ and insular morphometry (GM volume and cortical surface area) [24, 25, 33, 45]. This correlation has been elsewhere disputed [43, 46].

Crespo-Facorro et al. found insular cortical surface area and GM volume to be negatively correlated with positive symptoms in drug-naive FES patients [25]. No such correlation existed with negative symptomology. Other research has demonstrated an association between right insular cortical surface area and hallucination severity [45]. In this study, left insular cortical surface area and bilateral GM volumes showed similar but non-significant trends. No correlations were found between delusions and insular morphology. UHR patients who later developed psychosis have been found to exhibit significant GM reductions in the right PI that correlated with negative symptom severity at baseline [61]

It is important to remember the limitations of all imaging modalities and analysis techniques, as well as the heterogeneity of study populations, as these methodological differences likely account for many observed discrepancies.

27.3 Functional Role of the Insula

The insula is involved in a broad range of functions including autonomic control, executive control, social cognition, interoception, and emotional and sensorimotor processing [62–64]. Interoception, emotional processing, and sensory processing represent three domains that underlie many cognitive and behavioral deficits in SZ. See Uddin et al. [42] for a review of insular function.

Successful social interaction hinges on social cognition, in turn served by the processes of interoception and emotional processing [64]. Dysfunction in these domains, a hallmark of SZ, can explain much of the impaired social functioning in SZ [65, 66].

27.3.1 Interoception

Interoception describes one's sensitivity to internal body states [67]. A state of interoceptive awareness is achieved upon the successful integration of current internal stimuli to form a present sense of self. Interoceptive processing requires the allocation of attention to particular internal body states and valuation of the attended-to stimulus. Craig proposes a central role for the anterior insula (AI) in these processes [64].

In line with this assertion, broad evidence supports a role for the AI as an integration hub of internal sensory information and higher-order cognitive predictions to create this state of awareness within an individual [68, 69]. The insula plays a critical role in attention to and awareness of visceral internal responses [70, 71]. Activation of the right AI predicts interoceptive accuracy in patients asked to judge the timing of their own heartbeats [70]. The same study demonstrated right AI GM volume to be directly associated with interoceptive accuracy and patient's subjective reports of interoceptive awareness [70]. Interoceptive accuracy holds significant influence over decision-making processes [72] and autonomic regulation in social situations [73]. The insula is also involved in the affective processing of anticipatory stimuli [74]. It has been suggested this anticipatory processing serves to allow the AI to assign a valuation to the anticipated stimulus and regulate the sensitivity of the PIs subsequent processing of said stimulus [74].

Interoceptive accuracy is compromised in SZ, reflected in the loss of a patient's ability to detect internal stimuli and accurately identify said stimuli as internally-generated [75]. Poor insight, characterized by impaired awareness and attribution of the origin of mental events, is common in SZ [76]. Right PI cortical surface size and WM volume have been correlated with lack of insight [76], while left insular WM abnormalities have been associated with symptom unawareness [77]. Further highlighting the effects of abnormal insular functioning on interoception, SZ patients demonstrate reduced insular response to aversive stimuli (electric shock), with insular response correlating to positive symptom severity [78].

27.3.2 Emotional Processing

Interoception is intricately intertwined with emotional processing [79]. The James-Lange theory of emotion states that emotional feelings result from internal body sensations evoked by emotional stimuli [80]. Craig similarly posits a central importance of interoception in facilitating subjective feeling, emotion, and self-awareness [67]. Accordingly, interoceptive accuracy correlates with negative emotional experience, and both are associated with activation of the right AI [70]. Right AI activation has also been associated with numerous positive and negative emotions [67].

Facial emotion recognition (FER), particularly recognition of anger and disgust, is closely associated with insular activation [81]. SZ patients have a reduced ability to process affective expressions in others [82]. A recent neuroimaging meta-analysis by Jani and Kasparek demonstrated FER impairments in SZ patients are associated with reduced activation of the right insula [83].

A complex social process such as empathy requires the recognition and internal representation of the emotional and cognitive states of others, demanding a requisite level of theory of mind (ToM). Impairments in ToM are common in SZ and correlate with emotional processing deficits, including deficits in FER and empathetic processing [84–87]. These deficits are intricately related to insular function. Focal lesions to the AI have been shown to result in impaired processing of other's pain [88]. Similarly, SZ patients demonstrate abnormal activation of the AI and dACC when asked to imagine others in pain [62]. However, compared to healthy controls, no abnormal insular activity is observed in SZ patients when observing others experience pain [62].

27.3.3 Auditory Processing

In SZ, auditory processing deficits manifest as a reduced ability to interpret and generate emotional auditory cues, and in the generation of auditory hallucinations.

In speech, nonlinguistic information transfer is facilitated by a variety of acoustic features, including duration, pitch, and intensity [89]. These nonlexical cues of speech are collectively referred to as prosody. The interpretation of emotional prosody, referred to as auditory emotion recognition (AER, or emotional prosody comprehension), allows listeners to infer significant swathes of emotional information from the speaker and adjust their behavior accordingly. The insula plays an important role in AER. The

AI is associated with the higher-level processing of vocal affect [90] and insular activation is associated with emotional versus neutral prosody [91].

AER impairments are associated with social deficits and broader psychosocial outcomes [92] and are common in SZ [93–95]. Patients exhibit lessened sensitivity to the discrimination of acoustic intensity [96] and demonstrate a tendency to overestimate the emotional intensity of weak auditory emotional prosody [97]. Larger deficits appears in the discrimination of pitch-based features of speech [95, 97, 98], and deficits in tonal discrimination have been correlated with the negative symptoms of SZ [93].

However, exactly when and where these deficits occur is unclear. Such impairments may arise from deficits in low-level acoustic processing that compromise future information processing. Alternatively, these deficits may instead be the product of impaired assignment of meaning to emotional prosody during higher-level cognitive processing [92]. A mismatch negativity (MMN) is an event-related brain potential (ERP) component that reflects pre-attentive auditory deviance detection and has been used to elucidate the neural mechanisms underlying frequency modulation (FM) tone processing [99]. Kantrowitz et al. demonstrated impaired MMN generation in SZ patient's responses to FM tones that correlated with AER deficits [95]. Localizing the auditory cortex and AI as primary MMN sources, resting-state MRI (rsMRI) revealed a significant reduction in the functional connectivity (FC) of the AI and bilateral auditory cortex in SZ patients that correlated with AER deficits. When entered into a simultaneous regression, both the variables of "MMN" and "functional connectivity" contributed significantly to AER deficits. Of relevance, only the right AI exhibited significant deficits in connectivity to the bilateral auditory cortex, suggesting the right AI holds an important role in SZ.

Auditory hallucinations represent another central feature of SZ, potentially resulting from the misattribution of stimulus origin [1]. A voxel-based meta-analysis of SZ patients identified activation in the left insula and right superior temporal gyrus to be associated with auditory hallucinations [100].

27.4 Salience Network

Rather than mapping complex cognitive functions onto localized brain regions, a network perspective understands cognition as being supported by a series of large-scale distributed networks [101]. Using functional connectivity analyses, numerous intrinsic connectivity networks (ICNs) have been described. Such analyses have provided a new window into the neurobiological underpinnings of SZ.

The default mode network (DMN), central executive network (CEN), and salience network (SN) represent three core neurocognitive networks [102]. The DMN includes the posterior cingulate cortex (PCC), medial temporal lobes (MTL), bilateral inferior parietal cortex, and ventromedial prefrontal cortex (VMPFC) and plays a role in internally-focused and self-referential processes [103–105]. The CEN includes the dorsolateral prefrontal cortex (DLPFC) and posterior parietal cortex (PPC) and is involved in goal-directed and externally-oriented tasks [106].

The salience network (SN), with key nodes at the AI, dACC, frontal operculum, and anterior prefrontal cortex (PFC), is responsible for the detection and allocation of attention to salient internal and external stimuli [107–109]. Salience refers to the effectiveness of a stimulus to stand out from its neighbors and is determined by myriad features including, visual, motivational, or emotional associations [110].

The SN exerts a causal influence over CEN and DMN activity [111]. It is believed the SN serves to regulate switches between the two networks [111]; however, others understand the insula to have a broader role in coordinating the dynamic activation and repression of functional regions of the brain as required by attentional priorities [112]. Determining the SN has a regulatory role over the CEN and DMN, Sridharan et al. performed a granger causality analysis (GCA) and identified the right AI as underlying the SN's control over the CEN and DMN [111]. GCA further revealed the right AI as a key outflow hub [111]. This view of the right AI as a key information outflow hub has been supported by additional studies [106, 113].

By Menon and Uddin's view, the right AI dynamically engages and suppresses various neural processes across a variety of domains, according to present attentional priorities. From this, Menon and Uddin propose the AI facilitates task-related information processing through its dynamic control over regions of the brain involved in attentional, working memory, and higher-order cognitive processes [112]. In this view, the SN serves as a hub of integration for bottom-up attentional priorities and top-down sensory control and biasing.

The SN, CEN, and DMN exhibit altered intra- and inter-functional connectivity (FC) in SZ [113, 119–123]. Interestingly, machine learning algorithms can identify SZ patients based on examination of SN resting-state functional connectivity (rsFC) with a 71.4% accuracy [124]. As with structural deficits, abnormal SN connectivity is present prior to disease onset in high-risk populations [82]. Given evidence of the insula's role in the detection and analysis of internal and external stimuli, and its regulation of self-referential and goal-directed processing through the CEN and DMN, dysregulation of the insula can reasonably explain several symptom dimensions in SZ [107, 114].

The aberrant salience hypothesis posits that symptoms in SZ arise from the inappropriate assignment of salience and motivational significance to otherwise irrelevant stimuli [114]. Such aberrant assignment leads to prolonged attention to select sensory inputs, creating a violation between incoming sensory information and top-down priors (expectation). Mismatches between expectation and internal/external stimuli (top-down priors and bottom-up sensory inputs) trigger a prediction error [108]. Under a hierarchical Bayesian framework, prediction errors are passed through higher levels in the hierarchy, probed by updated priors at each level, with the aim of reducing the error sufficiently enough to arrive at an appropriate inference. Importantly, a Bayesian framework assumes higher levels of the hierarchy are ever-updating lower levels to improve performance. Fletcher and Firth propose that SZ patients suffer deficits in the integration of new information [115]. Coupled with the dysregulation of both top-down priors and bottom-up sensory attentional priorities, dysfunctional prediction errors are created and propagated through higher levels of analysis. As a dysfunctional prediction error is processed (driven by a dysfunction in the ability to update inferences and beliefs about the world), the persistent uncertainty of the error demands excessive attention, gaining inappropriately significant influence over inference processes, in theory resulting in the inability to separate experience and belief that characterizes positive symptoms in SZ [115].

As a hub for the integration of internal/external information and top-down behavioral priorities, and a hub for the mediation of dynamic network interactions, the insula is uniquely positioned to sit at the center of this framework. Insular activation is associated with prediction error coding [69, 116], and patients with insular lesions have an impaired ability to update prediction frameworks [117]. Furthermore, psychosis patients demonstrate an absence of insular activation during reward prediction error [118]. It should also be noted that the role of the AI in anticipatory affective stimuli processing, which may in turn modulate the sensitivity of the PI's subsequent sensory component analysis, is of interest within a Bayesian framework [74, 108].

Important to remember, within a network perspective, the consequences of a focal neural insult may yield downstream consequences far exceeding the primary deficit. Abnormalities in the SN illustrate this concept. The normal modulatory influence of the SN over the CEN and DMN is altered in SZ [106, 121]. Disruption of the SN's intra-FC is associated with increased inter-FC between the CEN and DMN and predicts positive symptom severity [106]. In keeping with previous findings, abnormal connectivity within the right AI is specifically associated with increased inter-FC between the CEN and DMN and predicts hallucination severity [106, 113]. Abnormalities in the FC of the SN and DMN have been associated with persistent auditory hallucinations in SZ patients [125] and further study has demonstrated that SZ patients have reduced activation of the right auditory cortex, PFC, and SN [126]. The reduced activation of these regions was associated with impaired deactivation of the visual system and dorsal attention network. Driven by alterations in the FC of the right AI, impaired SN functioning and the resultant discoordination of broad associated networks may underlie ensuing neurological deficits.

SZ patients in remission demonstrate an interesting asymmetry from those in acute psychosis. Manoliu and colleagues revealed that remission patients exhibit decreased intra-FC between the CEN and DMN, associated with an increased SN and CEN inter-FC [113]. In these patients, abnormal activity within the left AI was correlated with the aberrant FC of the SN and CEN and predicted negative symptom severity. Manoliu and colleagues suggest these findings accord with the idea of an asymmetric representation of body-related interoceptive information in the AI [113]. The left and right AI hold greater association with parasympathetic and sympathetic systems, respectively [67]. The left AI is involved in the processing of positive and pleasant emotion, while the right AI is involved in processing of more taxing, biologically arousing stimuli [64]. Manoliu and colleagues propose the association of the left AI with negative symptomology may arise from impaired responses to pleasant stimuli [113].

Evidence for network dysregulation involving the AI in emotional cognition arrives from Ruiz et al. who utilized real-time fMRI to facilitate self-regulation of the bilateral AI in SZ patients [127]. Following a 2-week training interval, imaging revealed significantly greater FC of the bilateral AI that correlated with improved FER. In a similar study, patients demonstrated volitional upregulation of the left AI. This AI upregulation was associated with higher baseline connectivity between the DLPFC and dorsomedial PFC (DMPFC), as well as improved empathetic pain processing [128].

Conclusion

The insula is a target of significant structural and functional alteration over the course of schizophrenia. As a key node in the salience network, the insula serves to dynamically shift between functional brain states, activating and suppressing regional activity per contextually-relevant requirements. Alterations in insular morphometry and functional connectivity correlate with numerous functional deficits in schizophrenia patients, suggesting insular impairments may underlie these deficits. Future research will continue to parse apart the many neural underpinnings of schizophrenia in order to further elucidate the nature of these interactions.

References

1. Insel TR. Rethinking schizophrenia. Nature. 2010;468:187–93.
2. Kay SR, Fiszbein A, Opler LA. The positive and negative syndrome scale (PANSS) for schizophrenia. Schizophr Bull. 1987;13:261–76.
3. Foussias G, Agid O, Fervaha G, Remington G. Negative symptoms of schizophrenia: clinical features, relevance to real world functioning and specificity versus other CNS disorders. Eur Neuropsychopharmacol. 2014;24:693–709.
4. Bowie CR, Harvey PD. Cognitive deficits and functional outcome in schizophrenia. Neuropsychiatr Dis Treat. 2006;2:531–6.
5. Jepsen JR, Fagerlund B, Pagsberg AK, Christensen AM, Nordentoft M, Mortensen EL. Profile of cognitive deficits and associations with depressive symptoms and intelligence in chronic early-onset schizophrenia patients. Scand J Psychol. 2013;54:363–70.
6. Puig O, Baeza I, de la Serna E, Cabrera B, Mezquida G, Bioque M, Lobo A, Gonzalez-Pinto A, Parellada M, Corripio I, Vieta E, Bobes J, Usall J, Contreras F, Cuesta MJ, Bernardo M, Castro-Fornieles J, Group TP. Persistent negative symptoms in first-episode psychosis: early cognitive and social functioning correlates and differences between early and adult onset. J Clin Psychiatry. 2017;78:1414–22.
7. Mesholam-Gately RI, Giuliano AJ, Goff KP, Faraone SV, Seidman LJ. Neurocognition in first-episode schizophrenia: a meta-analytic review. Neuropsychology. 2009;23:315–36.
8. Niemi LT, Suvisaari JM, Tuulio-Henriksson A, Lonnqvist JK. Childhood developmental abnormalities in schizophrenia: evidence from high-risk studies. Schizophr Res. 2003;60:239–58.

9. Erlenmeyer-Kimling L, Rock D, Roberts SA, Janal M, Kestenbaum C, Cornblatt B, Adamo UH, Gottesman II. Attention, memory, and motor skills as childhood predictors of schizophrenia-related psychoses: the New York High-Risk Project. Am J Psychiatry. 2000;157:1416–22.
10. Matheson SL, Vijayan H, Dickson H, Shepherd AM, Carr VJ, Laurens KR. Systematic meta-analysis of childhood social withdrawal in schizophrenia, and comparison with data from at-risk children aged 9–14 years. J Psychiatr Res. 2013;47:1061–8.
11. Schiffman J, Walker E, Ekstrom M, Schulsinger F, Sorensen H, Mednick S. Childhood videotaped social and neuromotor precursors of schizophrenia: a prospective investigation. Am J Psychiatry. 2004;161:2021–7.
12. Tsuji T, Kline E, Sorensen HJ, Mortensen EL, Michelsen NM, Ekstrom M, Mednick S, Schiffman J. Premorbid teacher-rated social functioning predicts adult schizophrenia-spectrum disorder: a high-risk prospective investigation. Schizophr Res. 2013;151:270–3.
13. Bliksted V, Videbech P, Fagerlund B, Frith C. The effect of positive symptoms on social cognition in first-episode schizophrenia is modified by the presence of negative symptoms. Neuropsychology. 2017;31:209–19.
14. Green MF. Cognitive impairment and functional outcome in schizophrenia and bipolar disorder. J Clin Psychiatry. 2006;67(Suppl 9):3–8. discussion 36–42
15. Nuechterlein KH, Subotnik KL, Green MF, Ventura J, Asarnow RF, Gitlin MJ, Yee CM, Gretchen-Doorly D, Mintz J. Neurocognitive predictors of work outcome in recent-onset schizophrenia. Schizophr Bull. 2011;37(Suppl 2):S33–40.
16. Honea R, Crow TJ, Passingham D, Mackay CE. Regional deficits in brain volume in schizophrenia: a meta-analysis of voxel-based morphometry studies. Am J Psychiatry. 2005;162:2233–45.
17. Vita A, De Peri L, Deste G, Sacchetti E. Progressive loss of cortical gray matter in schizophrenia: a meta-analysis and meta-regression of longitudinal MRI studies. Transl Psychiatry. 2012;2:e190.
18. Wright IC, Ellison ZR, Sharma T, Friston KJ, Murray RM, McGuire PK. Mapping of grey matter changes in schizophrenia. Schizophr Res. 1999;35:1–14.
19. Sigmundsson T, Suckling J, Maier M, Williams S, Bullmore E, Greenwood K, Fukuda R, Ron M, Toone B. Structural abnormalities in frontal, temporal, and limbic regions and interconnecting white matter tracts in schizophrenic patients with prominent negative symptoms. Am J Psychiatry. 2001;158:234–43.
20. Bora E, Fornito A, Radua J, Walterfang M, Seal M, Wood SJ, Yucel M, Velakoulis D, Pantelis C. Neuroanatomical abnormalities in schizophrenia: a multimodal voxelwise meta-analysis and meta-regression analysis. Schizophr Res. 2011;127:46–57.
21. Fusar-Poli P, Smieskova R, Serafini G, Politi P, Borgwardt S. Neuroanatomical markers of genetic liability to psychosis and first episode psychosis: a voxelwise meta-analytical comparison. World J Biol Psychiatry. 2014;15:219–28.
22. Shepherd AM, Laurens KR, Matheson SL, Carr VJ, Green MJ. Systematic meta-review and quality assessment of the structural brain alterations in schizophrenia. Neurosci Biobehav Rev. 2012;36:1342–56.
23. Takahashi T, Wood SJ, Soulsby B, Tanino R, Wong MT, McGorry PD, Suzuki M, Velakoulis D, Pantelis C. Diagnostic specificity of the insular cortex abnormalities in first-episode psychotic disorders. Prog Neuro-Psychopharmacol Biol Psychiatry. 2009;33:651–7.
24. Makris N, Goldstein JM, Kennedy D, Hodge SM, Caviness VS, Faraone SV, Tsuang MT, Seidman LJ. Decreased volume of left and total anterior insular lobule in schizophrenia. Schizophr Res. 2006;83:155–71.
25. Crespo-Facorro B, Kim J, Andreasen NC, O'Leary DS, Bockholt HJ, Magnotta V. Insular cortex abnormalities in schizophrenia: a structural magnetic resonance imaging study of first-episode patients. Schizophr Res. 2000;46:35–43.
26. Paillere-Martinot M, Caclin A, Artiges E, Poline JB, Joliot M, Mallet L, Recasens C, Attar-Levy D, Martinot JL. Cerebral gray and white matter reductions and clinical correlates in patients with early onset schizophrenia. Schizophr Res. 2001;50:19–26.
27. Yamada M, Hirao K, Namiki C, Hanakawa T, Fukuyama H, Hayashi T, Murai T. Social cognition and frontal lobe pathology in schizophrenia: a voxel-based morphometric study. NeuroImage. 2007;35:292–8.
28. Chan RC, Di X, McAlonan GM, Gong QY. Brain anatomical abnormalities in high-risk individuals, first-episode, and chronic schizophrenia: an activation likelihood estimation meta-analysis of illness progression. Schizophr Bull. 2011;37:177–88.
29. Marcelis M, Suckling J, Woodruff P, Hofman P, Bullmore E, van Os J. Searching for a structural endophenotype in psychosis using computational morphometry. Psychiatry Res. 2003;122:153–67.
30. Kuperberg GR, Broome MR, McGuire PK, David AS, Eddy M, Ozawa F, Goff D, West WC, Williams SC, van der Kouwe AJ, Salat DH, Dale AM, Fischl B. Regionally localized thinning of the cerebral cortex in schizophrenia. Arch Gen Psychiatry. 2003;60:878–88.
31. Nesvag R, Lawyer G, Varnas K, Fjell AM, Walhovd KB, Frigessi A, Jonsson EG, Agartz I. Regional thinning of the cerebral cortex in schizophrenia: effects of diagnosis, age and antipsychotic medication. Schizophr Res. 2008;98:16–28.
32. Leung M, Cheung C, Yu K, Yip B, Sham P, Li Q, Chua S, McAlonan G. Gray matter in first-episode schizophrenia before and after antipsychotic drug treatment. Anatomical likelihood estimation meta-analyses with sample size weighting. Schizophr Bull. 2011;37:199–211.
33. Fusar-Poli P, Radua J, McGuire P, Borgwardt S. Neuroanatomical maps of psychosis onset: voxel-

wise meta-analysis of antipsychotic-naive VBM studies. Schizophr Bull. 2012;38:1297–307.
34. Kasai K, Shenton ME, Salisbury DF, Onitsuka T, Toner SK, Yurgelun-Todd D, Kikinis R, Jolesz FA, McCarley RW. Differences and similarities in insular and temporal pole MRI gray matter volume abnormalities in first-episode schizophrenia and affective psychosis. Arch Gen Psychiatry. 2003;60:1069–77.
35. Kubicki M, Shenton ME, Salisbury DF, Hirayasu Y, Kasai K, Kikinis R, Jolesz FA, McCarley RW. Voxel-based morphometric analysis of gray matter in first episode schizophrenia. NeuroImage. 2002;17:1711–9.
36. Job DE, Whalley HC, McConnell S, Glabus M, Johnstone EC, Lawrie SM. Structural gray matter differences between first-episode schizophrenics and normal controls using voxel-based morphometry. NeuroImage. 2002;17:880–9.
37. Hulshoff Pol HE, Schnack HG, Mandl RC, van Haren NE, Koning H, Collins DL, Evans AC, Kahn RS. Focal gray matter density changes in schizophrenia. Arch Gen Psychiatry. 2001;58:1118–25.
38. Wilke M, Kaufmann C, Grabner A, Putz B, Wetter TC, Auer DP. Gray matter-changes and correlates of disease severity in schizophrenia: a statistical parametric mapping study. NeuroImage. 2001;13:814–24.
39. Garcia-Marti G, Aguilar EJ, Lull JJ, Marti-Bonmati L, Escarti MJ, Manjon JV, Moratal D, Robles M, Sanjuan J. Schizophrenia with auditory hallucinations: a voxel-based morphometry study. Prog Neuro-Psychopharmacol Biol Psychiatry. 2008;32:72–80.
40. Chang LJ, Yarkoni T, Khaw MW, Sanfey AG. Decoding the role of the insula in human cognition: functional parcellation and large-scale reverse inference. Cereb Cortex. 2013;23:739–49.
41. Deen B, Pitskel NB, Pelphrey KA. Three systems of insular functional connectivity identified with cluster analysis. Cereb Cortex. 2011;21:1498–506.
42. Uddin LQ, Kinnison J, Pessoa L, Anderson ML. Beyond the tripartite cognition-emotion-interoception model of the human insular cortex. J Cogn Neurosci. 2014;26:16–27.
43. Saze T, Hirao K, Namiki C, Fukuyama H, Hayashi T, Murai T. Insular volume reduction in schizophrenia. Eur Arch Psychiatry Clin Neurosci. 2007;257:473–9.
44. Takahashi T, Wood SJ, Soulsby B, McGorry PD, Tanino R, Suzuki M, Velakoulis D, Pantelis C. Follow-up MRI study of the insular cortex in first-episode psychosis and chronic schizophrenia. Schizophr Res. 2009;108:49–56.
45. Pressler M, Nopoulos P, Ho BC, Andreasen NC. Insular cortex abnormalities in schizophrenia: relationship to symptoms and typical neuroleptic exposure. Biol Psychiatry. 2005;57:394–8.
46. Crespo-Facorro B, Roiz-Santianez R, Quintero C, Perez-Iglesias R, Tordesillas-Gutierrez D, Mata I, Rodriguez-Sanchez JM, Gutierrez A, Vazquez-Barquero JL. Insular cortex morphometry in first-episode schizophrenia-spectrum patients: diagnostic specificity and clinical correlations. J Psychiatr Res. 2010;44:314–20.
47. Duggal HS, Muddasani S, Keshavan MS. Insular volumes in first-episode schizophrenia: gender effect. Schizophr Res. 2005;73:113–20.
48. Roiz-Santianez R, Perez-Iglesias R, Quintero C, Tordesillas-Gutierrez D, Mata I, Ayesa R, Sanchez JM, Gutierrez A, Sanchez E, Vazquez-Barquero JL, Crespo-Facorro B. Insular cortex thinning in first episode schizophrenia patients. Psychiatry Res. 2010;182:216–22.
49. Smieskova R, Fusar-Poli P, Allen P, Bendfeldt K, Stieglitz RD, Drewe J, Radue EW, McGuire PK, Riecher-Rossler A, Borgwardt SJ. The effects of antipsychotics on the brain: what have we learnt from structural imaging of schizophrenia? A systematic review. Curr Pharm Des. 2009;15:2535–49.
50. Ho BC, Andreasen NC, Ziebell S, Pierson R, Magnotta V. Long-term antipsychotic treatment and brain volumes: a longitudinal study of first-episode schizophrenia. Arch Gen Psychiatry. 2011;68:128–37.
51. Lieberman JA, Tollefson GD, Charles C, Zipursky R, Sharma T, Kahn RS, Keefe RS, Green AI, Gur RE, McEvoy J, Perkins D, Hamer RM, Gu H, Tohen M. Antipsychotic drug effects on brain morphology in first-episode psychosis. Arch Gen Psychiatry. 2005;62:361–70.
52. Crespo-Facorro B, Roiz-Santianez R, Perez-Iglesias R, Pelayo-Teran JM, Rodriguez-Sanchez JM, Tordesillas-Gutierrez D, Ramirez M, Martinez O, Gutierrez A, de Lucas EM, Vazquez-Barquero JL. Effect of antipsychotic drugs on brain morphometry. A randomized controlled one-year follow-up study of haloperidol, risperidone and olanzapine. Prog Neuro-Psychopharmacol Biol Psychiatry. 2008;32:1936–43.
53. Roiz-Santianez R, Tordesillas-Gutierrez D, Ortiz-Garcia de la Foz V, Ayesa-Arriola R, Gutierrez A, Tabares-Seisdedos R, Vazquez-Barquero JL, Crespo-Facorro B. Effect of antipsychotic drugs on cortical thickness. A randomized controlled one-year follow-up study of haloperidol, risperidone and olanzapine. Schizophr Res. 2012;141:22–8.
54. Vita A, De Peri L, Deste G, Barlati S, Sacchetti E. The effect of antipsychotic treatment on cortical gray matter changes in schizophrenia: does the class matter? A meta-analysis and meta-regression of longitudinal magnetic resonance imaging studies. Biol Psychiatry. 2015;78:403–12.
55. Fusar-Poli P, Broome MR, Matthiasson P, Williams SC, Brammer M, McGuire PK. Effects of acute antipsychotic treatment on brain activation in first episode psychosis: an fMRI study. Eur Neuropsychopharmacol. 2007;17:492–500.
56. Borgwardt SJ, Picchioni MM, Ettinger U, Toulopoulou T, Murray R, McGuire PK. Regional gray matter volume in monozygotic twins con-

56. cordant and discordant for schizophrenia. Biol Psychiatry. 2010;67:956–64.
57. Andreasen NC, Nopoulos P, Magnotta V, Pierson R, Ziebell S, Ho BC. Progressive brain change in schizophrenia: a prospective longitudinal study of first-episode schizophrenia. Biol Psychiatry. 2011;70:672–9.
58. Olabi B, Ellison-Wright I, McIntosh AM, Wood SJ, Bullmore E, Lawrie SM. Are there progressive brain changes in schizophrenia? A meta-analysis of structural magnetic resonance imaging studies. Biol Psychiatry. 2011;70:88–96.
59. Borgwardt SJ, Riecher-Rossler A, Dazzan P, Chitnis X, Aston J, Drewe M, Gschwandtner U, Haller S, Pfluger M, Rechsteiner E, D'Souza M, Stieglitz RD, Radu EW, McGuire PK. Regional gray matter volume abnormalities in the at risk mental state. Biol Psychiatry. 2007;61:1148–56.
60. Fusar-Poli P, Borgwardt S, Crescini A, Deste G, Kempton MJ, Lawrie S, Mc Guire P, Sacchetti E. Neuroanatomy of vulnerability to psychosis: a voxel-based meta-analysis. Neurosci Biobehav Rev. 2011;35:1175–85.
61. Takahashi T, Wood SJ, Yung AR, Phillips LJ, Soulsby B, McGorry PD, Tanino R, Zhou SY, Suzuki M, Velakoulis D, Pantelis C. Insular cortex gray matter changes in individuals at ultra-high-risk of developing psychosis. Schizophr Res. 2009;111:94–102.
62. Horan WP, Jimenez AM, Lee J, Wynn JK, Eisenberger NI, Green MF. Pain empathy in schizophrenia: an fMRI study. Soc Cogn Affect Neurosci. 2016;11:783–92.
63. Cechetto DF. Cortical control of the autonomic nervous system. Exp Physiol. 2014;99:326–31.
64. Craig AD. How do you feel—now? The anterior insula and human awareness. Nat Rev Neurosci. 2009;10:59–70.
65. Benedetti F, Bernasconi A, Bosia M, Cavallaro R, Dallaspezia S, Falini A, Poletti S, Radaelli D, Riccaboni R, Scotti G, Smeraldi E. Functional and structural brain correlates of theory of mind and empathy deficits in schizophrenia. Schizophr Res. 2009;114:154–60.
66. Pinkham AE. Social cognition in schizophrenia. J Clin Psychiatry. 2014;75(Suppl 2):14–9.
67. Craig AD. Interoception: the sense of the physiological condition of the body. Curr Opin Neurobiol. 2003;13:500–5.
68. Gu X, Hof PR, Friston KJ, Fan J. Anterior insular cortex and emotional awareness. J Comp Neurol. 2013;521:3371–88.
69. Garrison J, Erdeniz B, Done J. Prediction error in reinforcement learning: a meta-analysis of neuroimaging studies. Neurosci Biobehav Rev. 2013;37:1297–310.
70. Critchley HD, Wiens S, Rotshtein P, Ohman A, Dolan RJ. Neural systems supporting interoceptive awareness. Nat Neurosci. 2004;7:189–95.
71. Palaniyappan L, Simmonite M, White TP, Liddle EB, Liddle PF. Neural primacy of the salience processing system in schizophrenia. Neuron. 2013;79:814–28.
72. Werner NS, Jung K, Duschek S, Schandry R. Enhanced cardiac perception is associated with benefits in decision-making. Psychophysiology. 2009;46:1123–9.
73. Ferri F, Ardizzi M, Ambrosecchia M, Gallese V. Closing the gap between the inside and the outside: interoceptive sensitivity and social distances. PLoS One. 2013;8:e75758.
74. Lovero KL, Simmons AN, Aron JL, Paulus MP. Anterior insular cortex anticipates impending stimulus significance. NeuroImage. 2009;45:976–83.
75. Ardizzi M, Ambrosecchia M, Buratta L, Ferri F, Peciccia M, Donnari S, Mazzeschi C, Gallese V. Interoception and positive symptoms in schizophrenia. Front Hum Neurosci. 2016;10:379.
76. Palaniyappan L, Mallikarjun P, Joseph V, Liddle PF. Appreciating symptoms and deficits in schizophrenia: right posterior insula and poor insight. Prog Neuro-Psychopharmacol Biol Psychiatry. 2011;35:523–7.
77. Antonius D, Prudent V, Rebani Y, D'Angelo D, Ardekani BA, Malaspina D, Hoptman MJ. White matter integrity and lack of insight in schizophrenia and schizoaffective disorder. Schizophr Res. 2011;128:76–82.
78. Linnman C, Coombs G, Goff DC, Holt DJ. Lack of insula reactivity to aversive stimuli in schizophrenia. Schizophr Res. 2013;143:150–7.
79. Pollatos O, Herbert BM, Matthias E, Schandry R. Heart rate response after emotional picture presentation is modulated by interoceptive awareness. Int J Psychophysiol. 2007;63:117–24.
80. Lang PJ. The varieties of emotional experience: a meditation on James-Lange theory. Psychol Rev. 1994;101:211–21.
81. Fusar-Poli P, Placentino A, Carletti F, Landi P, Allen P, Surguladze S, Benedetti F, Abbamonte M, Gasparotti R, Barale F, Perez J, McGuire P, Politi P. Functional atlas of emotional faces processing: a voxel-based meta-analysis of 105 functional magnetic resonance imaging studies. J Psychiatry Neurosci. 2009;34:418–32.
82. Pelletier-Baldelli A, Bernard JA, Mittal VA. Intrinsic functional connectivity in salience and default mode networks and aberrant social processes in youth at ultra-high risk for psychosis. PLoS One. 2015;10:e0134936.
83. Jani M, Kasparek T. Emotion recognition and theory of mind in schizophrenia: a meta-analysis of neuroimaging studies. World J Biol Psychiatry. 2017:1–11.
84. Mothersill O, Knee-Zaska C, Donohoe G. Emotion and theory of mind in schizophrenia-investigating the role of the cerebellum. Cerebellum. 2016;15:357–68.
85. Lamm C, Singer T. The role of anterior insular cortex in social emotions. Brain Struct Funct. 2010;214:579–91.
86. Bernhardt BC, Singer T. The neural basis of empathy. Annu Rev Neurosci. 2012;35:1–23.
87. Lamm C, Decety J, Singer T. Meta-analytic evidence for common and distinct neural networks associated

with directly experienced pain and empathy for pain. NeuroImage. 2011;54:2492–502.
88. Gu X, Gao Z, Wang X, Liu X, Knight RT, Hof PR, Fan J. Anterior insular cortex is necessary for empathetic pain perception. Brain J Neurol. 2012;135:2726–35.
89. Belin P, Fecteau S, Bedard C. Thinking the voice: neural correlates of voice perception. Trends Cogn Sci. 2004;8:129–35.
90. Bestelmeyer PEG, Maurage P, Rouger J, Latinus M, Belin P. Adaptation to vocal expressions reveals multistep perception of auditory emotion. J Neurosci. 2014;34:8098–105.
91. Bach DR, Grandjean D, Sander D, Herdener M, Strik WK, Seifritz E. The effect of appraisal level on processing of emotional prosody in meaningless speech. NeuroImage. 2008;42:919–27.
92. Hoekert M, Kahn RS, Pijnenborg M, Aleman A. Impaired recognition and expression of emotional prosody in schizophrenia: review and meta-analysis. Schizophr Res. 2007;96:135–45.
93. Kantrowitz JT, Leitman DI, Lehrfeld JM, Laukka P, Juslin PN, Butler PD, Silipo G, Javitt DC. Reduction in tonal discriminations predicts receptive emotion processing deficits in schizophrenia and schizoaffective disorder. Schizophr Bull. 2013;39:86–93.
94. Pijnenborg GH, Withaar FK, Bosch RJ, Brouwer WH. Impaired perception of negative emotional prosody in schizophrenia. Clin Neuropsychol. 2007;21:762–75.
95. Kantrowitz JT, Hoptman MJ, Leitman DI, Moreno-Ortega M, Lehrfeld JM, Dias E, Sehatpour P, Laukka P, Silipo G, Javitt DC. Neural substrates of auditory emotion recognition deficits in schizophrenia. J Neurosci. 2015;35:14909–21.
96. Bach DR, Buxtorf K, Strik WK, Neuhoff JG, Seifritz E. Evidence for impaired sound intensity processing in schizophrenia. Schizophr Bull. 2011;37:426–31.
97. Leitman DI, Laukka P, Juslin PN, Saccente E, Butler P, Javitt DC. Getting the cue: sensory contributions to auditory emotion recognition impairments in schizophrenia. Schizophr Bull. 2010;36:545–56.
98. Gold R, Butler P, Revheim N, Leitman DI, Hansen JA, Gur RC, Kantrowitz JT, Laukka P, Juslin PN, Silipo GS, Javitt DC. Auditory emotion recognition impairments in schizophrenia: relationship to acoustic features and cognition. Am J Psychiatry. 2012;169:424–32.
99. Leitman DI, Sehatpour P, Garidis C, Gomez-Ramirez M, Javitt DC. Preliminary evidence of pre-attentive distinctions of frequency-modulated tones that convey affect. Front Hum Neurosci. 2011;5:96.
100. Jardri R, Pouchet A, Pins D, Thomas P. Cortical activations during auditory verbal hallucinations in schizophrenia: a coordinate-based meta-analysis. Am J Psychiatry. 2011;168:73–81.
101. Bressler SL, Menon V. Large-scale brain networks in cognition: emerging methods and principles. Trends Cogn Sci. 2010;14:277–90.
102. Uddin LQ, Supekar KS, Ryali S, Menon V. Dynamic reconfiguration of structural and functional connectivity across core neurocognitive brain networks with development. J Neurosci. 2011;31:18578–89.
103. Greicius MD, Krasnow B, Reiss AL, Menon V. Functional connectivity in the resting brain: a network analysis of the default mode hypothesis. Proc Natl Acad Sci U S A. 2003;100:253–8.
104. Dodell-Feder D, DeLisi LE, Hooker CI. The relationship between default mode network connectivity and social functioning in individuals at familial high-risk for schizophrenia. Schizophr Res. 2014;156:87–95.
105. Andrews-Hanna JR, Smallwood J, Spreng RN. The default network and self-generated thought: component processes, dynamic control, and clinical relevance. Ann N Y Acad Sci. 2014;1316:29–52.
106. Manoliu A, Riedl V, Doll A, Bäuml JG, Mühlau M, Schwerthöffer D, Scherr M, Zimmer C, Förstl H, Bäuml J, Wohlschläger AM, Koch K, Sorg C. Insular dysfunction reflects altered between-network connectivity and severity of negative symptoms in schizophrenia during psychotic remission. Front Hum Neurosci. 2013;7:216.
107. Seeley WW, Menon V, Schatzberg AF, Keller J, Glover GH, Kenna H, Reiss AL, Greicius MD. Dissociable intrinsic connectivity networks for salience processing and executive control. J Neurosci. 2007;27:2349–56.
108. Palaniyappan L, Liddle PF. Does the salience network play a cardinal role in psychosis? An emerging hypothesis of insular dysfunction. J Psychiatry Neurosci. 2012;37:17–27.
109. Dosenbach NU, Fair DA, Miezin FM, Cohen AL, Wenger KK, Dosenbach RA, Fox MD, Snyder AZ, Vincent JL, Raichle ME, Schlaggar BL, Petersen SE. Distinct brain networks for adaptive and stable task control in humans. Proc Natl Acad Sci U S A. 2007;104:11073–8.
110. Walter A, Suenderhauf C, Smieskova R, Lenz C, Harrisberger F, Schmidt A, Vogel T, Lang UE, Riecher-Rossler A, Eckert A, Borgwardt S. Altered insular function during aberrant salience processing in relation to the severity of psychotic symptoms. Front Psych. 2016;7:189.
111. Sridharan D, Levitin DJ, Menon V. A critical role for the right fronto-insular cortex in switching between central-executive and default-mode networks. Proc Natl Acad Sci U S A. 2008;105:12569–74.
112. Menon V, Uddin LQ. Saliency, switching, attention and control: a network model of insula function. Brain Struct Funct. 2010;214:655–67.
113. Manoliu A, Riedl V, Zherdin A, Mühlau M, Schwerthöffer D, Scherr M, Peters H, Zimmer C, Förstl H, Bäuml J, Wohlschläger AM, Sorg C. Aberrant dependence of default mode/central executive network interactions on anterior insular salience network activity in schizophrenia. Schizophr Bull. 2014;40:428–37.
114. Kapur S. Psychosis as a state of aberrant salience: a framework linking biology, phenomenology, and

pharmacology in schizophrenia. Am J Psychiatry. 2003;160:13–23.
115. Fletcher PC, Frith CD. Perceiving is believing: a Bayesian approach to explaining the positive symptoms of schizophrenia. Nat Rev Neurosci. 2009;10:48–58.
116. Preuschoff K, Quartz SR, Bossaerts P. Human insula activation reflects risk prediction errors as well as risk. J Neurosci. 2008;28:2745–52.
117. Clark L, Bechara A, Damasio H, Aitken MR, Sahakian BJ, Robbins TW. Differential effects of insular and ventromedial prefrontal cortex lesions on risky decision-making. Brain J Neurol. 2008;131:1311–22.
118. Murray GK, Corlett PR, Clark L, Pessiglione M, Blackwell AD, Honey G, Jones PB, Bullmore ET, Robbins TW, Fletcher PC. Substantia nigra/ventral tegmental reward prediction error disruption in psychosis. Mol Psychiatry. 2008;13(239):267–76.
119. Mingoia G, Wagner G, Langbein K, Maitra R, Smesny S, Dietzek M, Burmeister HP, Reichenbach JR, Schlosser RG, Gaser C, Sauer H, Nenadic I. Default mode network activity in schizophrenia studied at resting state using probabilistic ICA. Schizophr Res. 2012;138:143–9.
120. Guo W, Liu F, Chen J, Wu R, Li L, Zhang Z, Chen H, Zhao J. Hyperactivity of the default-mode network in first-episode, drug-naive schizophrenia at rest revealed by family-based case–control and traditional case–control designs. Medicine. 2017;96:e6223.
121. Littow H, Huossa V, Karjalainen S, Jääskeläinen E, Haapea M, Miettunen J, Tervonen O, Isohanni M, Nikkinen J, Veijola J, Murray G, Kiviniemi VJ. Aberrant functional connectivity in the default mode and central executive networks in subjects with schizophrenia—a whole-brain resting-state ICA study. Front Psych. 2015;6:26.
122. Moran LV. Disruption of anterior insula modulation of large-scale brain networks in schizophrenia. Biol Psychiatry. 2013;74:467–74.
123. White TP, Joseph V, Francis ST, Liddle PF. Aberrant salience network (bilateral insula and anterior cingulate cortex) connectivity during information processing in schizophrenia. Schizophr Res. 2010;123:105–15.
124. Mikolas P, Melicher T, Skoch A, Matejka M, Slovakova A, Bakstein E, Hajek T, Spaniel F. Connectivity of the anterior insula differentiates participants with first-episode schizophrenia spectrum disorders from controls: a machine-learning study. Psychol Med. 2016;46:2695–704.
125. Alonso-Solis A, Vives-Gilabert Y, Grasa E, Portella MJ, Rabella M, Sauras RB, Roldan A, Nunez-Marin F, Gomez-Anson B, Perez V, Alvarez E, Corripio I. Resting-state functional connectivity alterations in the default network of schizophrenia patients with persistent auditory verbal hallucinations. Schizophr Res. 2015;161:261–8.
126. Gaebler AJ, Mathiak K, Koten JW Jr, Konig AA, Koush Y, Weyer D, Depner C, Matentzoglu S, Edgar JC, Willmes K, Zvyagintsev M. Auditory mismatch impairments are characterized by core neural dysfunctions in schizophrenia. Brain J Neurol. 2015;138:1410–23.
127. Ruiz S, Lee S, Soekadar SR, Caria A, Veit R, Kircher T, Birbaumer N, Sitaram R. Acquired self-control of insula cortex modulates emotion recognition and brain network connectivity in schizophrenia. Hum Brain Mapp. 2013;34:200–12.
128. Yao S, Becker B, Geng Y, Zhao Z, Xu X, Zhao W, Ren P, Kendrick KM. Voluntary control of anterior insula and its functional connections is feedback-independent and increases pain empathy. NeuroImage. 2016;130:230–40.

Part IV

Surgery of the Insular Cortex

Surgery of Insular Diffuse Low-Grade Gliomas

Karine Michaud and Hugues Duffau

28.1 Introduction

Insular gliomas represent a distinct and challenging entity among diffuse low-grade gliomas (DLGG). Their location buried under the opercula, the complexity of the functional anatomy of the insular lobe, its rich afferent and efferent networks, and its intimate relationships with the Sylvian vessels and the lenticulostriate branches were factors that limited their surgical resection: indeed, neurosurgeons considered for a long time those insular tumors inoperable. Remarkably, the better understanding of the natural history of DLGG, the increased knowledge of insular function and connectivity, and the improvement of surgical techniques like direct electrical stimulation for awake brain mapping led to a paradigmatic shift in the management of these gliomas. The conservative attitude is now abandoned and replaced by an early maximal surgical approach [1].

Here, we will review the recent literature on insular/paralimbic DLGG, by detailing their specific clinical and oncologic features; we will focus on their surgical management, in particular by reporting our personal experience; and finally, we will evaluate the oncological and functional outcomes of this new therapeutic attitude.

28.2 Diffuse Low-Grade Gliomas

28.2.1 Definition and Diagnosis

Gliomas are neuroepithelial tumors originating from the supporting glial cell network of the central nervous system. According to the 2016 WHO classification, DLGG are now classified according to not only neuropathological appearance but also molecular profile, and they are divided in diffuse astrocytomas and oligodendrogliomas [2]. Diffuse low-grade gliomas, i.e., WHO grade II gliomas (that are a completely distinct entity from WHO grade I gliomas, which are well demarcated and surgically curable if complete resection has been achieved), cannot be considered as "benign" anymore: they represent an initially slow progressive cancerous disease that nonetheless will inevitably become malignant [1].

K. Michaud
Department of Neurosurgery, Enfant-Jésus Hospital, CHU de Québec, Laval University,
Québec, QC, Canada

H. Duffau, M.D., Ph.D. (✉)
Department of Neurosurgery, Gui de Chauliac Hospital, Montpellier University Medical Center, Montpellier, France

Team "Plasticity of Central Nervous System, Stem Cells and Glial Tumors," National Institute for Health and Medical Research (INSERM), U1051 Laboratory, Institute for Neurosciences of Montpellier, Montpellier University Medical Center, Montpellier, France
e-mail: h-duffau@chu-montpellier.fr

28.2.2 Clinical Presentation

DLGG typically presents in high-functioning young adults in their second or third decade. Their clinical course tends to be prolonged and slowly progressive [1]. The most common symptom at presentation is epilepsy, in about 80% of cases [3]. A minority of patients may present with focal neurological symptoms from mass effect. When an appropriate neuropsychological assessment is performed, most patients will exhibit with some degree of neurocognitive deficits. Those subtle disorders are related to the tumor infiltration of white matter tracts [4]. The vast connectivity network of the insular region may explain those findings. The young age of the patients and their relative high-functioning level should always be kept in mind when treatment decisions are made for this disease, not only to improve survival but also to maintain an optimal quality of life by avoiding complication and side effects [5].

28.2.3 Radiological Presentation

DLGG have a distinctive radiological presentation on MR imaging, which is the gold standard imaging modality. They typically present as T1-hypointense and T2-hyperintense expansive lesions. Despite possible apparent radiographic margins, DLGG tend to infiltrate surrounding brain parenchyma beyond signal abnormality [6]. Imaging can show calcifications in around 20% of cases which is usually suggesting an oligodendroglial nature [7]. Some degree of contrast enhancement can be seen in up to 60% of DLGG and can be indicative of an oligodendroglioma or malignant transformation. Growth rate computed by comparing volumes calculated on two successive MRIs before any treatment is about 4 mm of mean diameter per year [8]. Newer imaging techniques such as perfusion-weighted and spectroscopy MRI might improve radiographic diagnostic potential [7], but these techniques remain currently too unreliable for most practical purposes [1].

28.2.4 Prognosis of Insular DLGG

As mentioned, DLGG represent a slow-growing tumoral disease which will progress to a higher grade of malignancy that will ultimately lead to the death. Prognostic factors include patient characteristics such as age, clinical presentation, and functional status, tumor characteristics such as volume and growth kinetics, histology and molecular profile, as well as therapeutic factors such as extent of surgical resection [5, 9, 10]. The prognosis of insular gliomas compares unfavorably to other locations. Indeed, they have a distinct histomolecular profile with a relative higher frequency of astrocytomas, as revealed by series that found a lesser rate of 1p19q codeletion gliomas [11]. It was also shown that paralimbic DLGG exhibit less IDH1/IDH2 mutation as well as a higher rate of triple-negative tumors (no 1p19q deletion, no IDH mutation, no p53 mutation) compared to pure insular gliomas [12] that could account for the worse prognosis of the paralimbic DLGG, because those tumors are indeed to be considered as wild-type diffuse astrocytomas according to the 2016 WHO classification [2], i.e., tumors that typically demonstrate a shorter survival. In addition, insular gliomas tend to present with a higher median volume, which is correlated with a higher rate of malignant transformation [1].

28.2.5 Insular Predilection of DLGG

DLGG can be found throughout the cerebrum but show a preference for insular and paralimbic structures. Indeed, although the insular lobe only accounts for 2% of the total cortical surface of the human brain [13], insular gliomas represent up to 30% of DLGG [14]. Different explanations can be considered for this phenomenon. The close proximity of the subventricular and subgranular zones of the hippocampus has led to the cancer stem cell hypothesis where the accumulation of mutations, translocations, deletions, and fusions is believed to give rise to neoplasms within a long-lived population of cells [15]. Moreover, glial tumor cells are migrating along the rich network of afferent and efferent white matter tracts connecting the insula to adjacent or remote regions [16]. Finally, DLGG seem to involve "eloquent" regions preferentially compared to malignant gliomas. Thus, although the exact mechanisms for this predilection are not proven, cytochemoarchitectonics, structural/functional anatomy, and particular glia/neuron interactions are some of the plausible hypotheses explaining the insular region involvement [14].

28.3 Classification of Insular Gliomas

Insular gliomas are classified by different authors according to their location, and this can be used to compare the lesions and their surgical approaches. Yasargil was the first to introduce a classification that includes all limbic and paralimbic gliomas. Type 3A gliomas are restricted to the insula, type 3B extend to adjacent opercula, type 5A extend to fronto-orbital and/or temporo-polar structures, and finally, type 5B extend also to mesio-temporal structures [17]. Sanai et al. described a quadrant-style classification dividing the insula into four zones in regard to a perpendicular bisection plane through the foramen of Monro and a horizontal plane through the Sylvian fissure [10]. Tumors are assigned to zones 1–4 (anterosuperior, posterosuperior, infero-posterior, and anteroinferior, respectively) depending on where the majority of the lesion lies. If the lesion involves all zones, then it is described as giant. This classification system has been validated in another series and is predictive of the extent of resection (EOR) and morbidity [18]. Mandonnet et al. proposed a different classification based on tumor extension along the subcortical fasciculi, such as uncinate fasciculus, inferior fronto-occipital fasciculus (IFOF), and arcuate fasciculus, to integrate the functional role of these pathways into therapeutic strategies—particularly with regard to the limits of surgical resection [16].

28.4 Operative Treatment

28.4.1 Role of Surgery

Early maximal safe surgery is an essential part of the treatment algorithm for diffuse gliomas of all grades, but this is especially critical in insular DLGG. Extent of resection objectively evaluated by volumetric assessment on postoperative imaging is an independent prognostic factor in DLGG, since maximizing it can delay malignant transformation and improve overall survival (OS) [1, 5, 10]. The oncological benefit of the surgery must always be weighed against the risk of functional morbidity related to the surgical procedure with the goal to preserve an optimal quality of life (QoL) [19]. In an appropriate setting with expert care in insular surgery, morbidity is kept to a minimum, and QoL is maintained and even improved following radical resection performed according to functional boundaries [20].

28.4.2 Preoperative Investigations

First, every patient should undergo a thorough neuropsychological investigation to identify possible subtle neurocognitive deficits as well as to allow a proper postoperative cognitive rehabilitation and follow-up [21]. Extensive preparation is done with the patient and speech therapist/neuropsychologist, and intraoperative tasks are chosen according to patient's educational, social, and professional background, preoperative clinical and neuropsychological evaluations, and expected pathways to be encountered depending on tumor location [22].

As with all brain tumors, gadolinium injection T1-weighted as well as T2/FLAIR-weighted MR imaging must be obtained preoperatively. Functional MRI (fMRI) and diffusion tensor imaging (DTI) have become more readily available in many centers. fMRI may help to study functional reorganization made possible thanks to mechanisms of brain plasticity induced by the slow progression of DLGG. However, fMRI is not reliable enough for surgical selection and planning, especially regarding language, and it is not capable to differentiate critical versus compensable eloquent areas [1, 23]. In the same spirit, tractography has been compared to intraoperative subcortical electrical mapping: only 81% of positive stimulations were concordant with DTI. Thus, negative tractography cannot rule out the persistence of a fiber tract, especially when invaded by tumor [24]. This is the reason why awake surgery with intraoperative real-time cognitive monitoring combined with electrical mapping is crucial to optimize the onco-functional balance [19].

28.4.3 Awake Craniotomy

In our experience, all patients are positioned in the lateral position before anesthesia with the contralateral arm positioned for easy motor monitoring. The asleep-awake-asleep anesthetic method is used [25]. A large fronto-temporo-central craniotomy is achieved to allow a good exposure of the entire Sylvian fissure and opercula. Tumor delineation is performed by ultrasonography. After dura-mater opening, the patient is awake, and the cortical mapping by means of direct electrostimulation is achieved before tumor resection [23].

28.4.4 Trans-opercular Approach

In our experience, the trans-opercular approach is favored in all insular/paralimbic tumors regardless of the hemisphere because it allows for the preservation of all blood vessels, arteries, and veins at the cortical surface and within the Sylvian fissure [20]. The transsylvian approach inevitably leads to bridging veins sacrifice and opercular retraction for adequate exposure: thus, it increases the risks of postoperative deficit [26]. Also, the subarachnoid manipulation involved in the transsylvian approach puts the arteries within the Sylvian fissure at risk of injury and spasm [17]. The trans-opercular approach also has the advantage of providing a better exposure of the insula, in particular facilitating the resection of posterosuperior tumors (notably in zone 2 of the Sanai-Berger classification system [10])—in which the dissection of the fissure is complicated by the superficial Sylvian vein bifurcation [20]. Intraoperative direct electrical stimulation (DES) of the opercula allows for identification and preservation of functional cortex and results in an individualized tailored resection of appropriate nonessential cortex—even when not infiltrated by the tumor [5, 10, 27]. This approach is safe as shown by the rate of permanent deficits that has significantly been reduced compared to the transsylvian approach [17, 28].

The exact location of the tumor within the insula and the involvement of the adjacent cortex will dictate the surgical corridor [20]. With "cortical windowing" [10], one can achieve tumor resection through different simultaneous cortical windows. Despite the existence of three "typical" cortical corridors, they will always be individually adapted according to the result of functional cortical mapping by DES [20]. The first corridor is through the inferior frontal gyrus and gives access to anterior tumors (zone 1) [27]. This is true even in the left hemisphere (Fig. 28.1). Indeed, it was clearly demonstrated by awake mapping that the main speech output is not the pars triangularis and opercularis but the ventral premotor cortex [29]. The second corridor is the superior temporal gyrus giving access to inferior tumors (zones 3 and 4). The third corridor is through the inferior parietal lobule and gives access to posterior tumors (zone 2). This corridor is particularly challenging in the left hemisphere because of its eloquence. Resection of the lateral part of the postcentral gyrus and/or the anteroinferior part of the supramarginal gyrus can be achieved safely when individually tailored by DES [5, 20, 30].

28.4.5 Subpial Resection and Axonal Mapping

Once the cortical entrance is made, tumor resection is performed by subpial aspiration to avoid coagulation and then to preserve all vessels [20, 23]. Simple aspiration by suction is favored at all

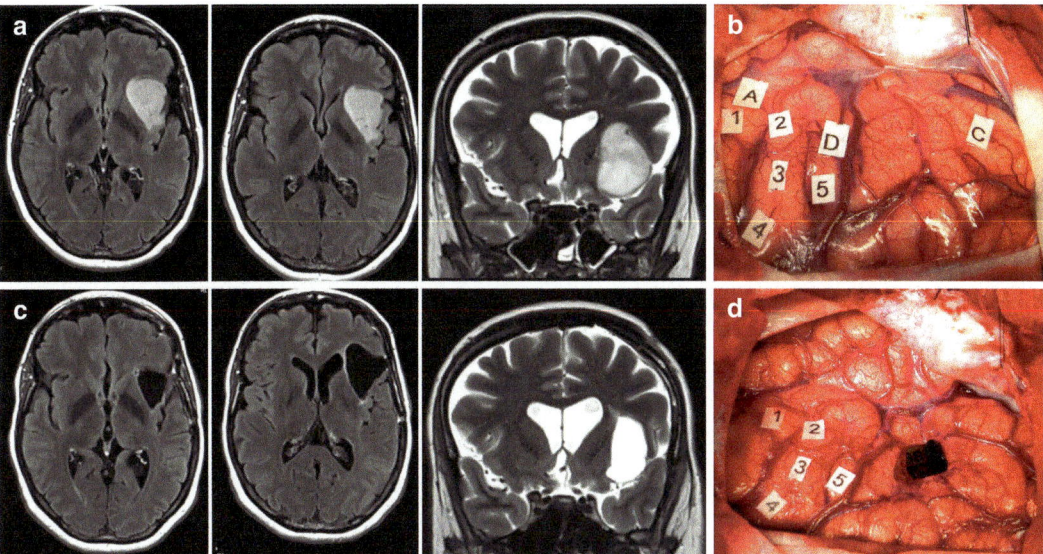

Fig. 28.1 (a) Preoperative axial FLAIR-weighted MRI and coronal T2-weighted MRI showing a left insular diffuse low-grade glioma in a 35-year-old right-handed female who experienced seizures. The neurological examination was normal. Nonetheless, the preoperative neuropsychological assessment revealed a deficit of verbal working memory as well as some missing words. (b) Intraoperative view before resection in awake patient. The anterior part of the left hemisphere is on the right, and its posterior part is on the left. Letter tags correspond to the projection on the cortical surface of the tumor limits identified using ultrasonography. Number tags show zones of positive direct electrostimulation mapping as follows: 1 and 2, ventral premotor cortex (inducing anarthria); 3 and 5, negative motor areas (eliciting both speech arrest and interruption of the movement of the left upper limb during dual task); and 4, primary motor cortex of the face (generating articulatory disorders); (c) Postoperative axial FLAIR-weighted MRI and coronal T2-weighted MRI, demonstrating a complete resection. The patient resumed a normal familial, social, and professional life within 3 months after surgery, with an improvement of the neuropsychological evaluation following a postsurgical cognitive rehabilitation. (d) Intraoperative view after resection, achieved up to eloquent structures, both at cortical and subcortical levels. Indeed, after a transcortical surgical approach through the left inferior frontal gyrus (the so-called Broca's area), direct electrostimulation of white matter enabled the identification of the left inferior fronto-occipital fasciculus that caused semantic paraphasias (tag 38): this pathway represented the deep limit of the resection

times unless contraindicated by tumor consistency. It has been shown that ultrasonic aspiration can not only cause vascular injury [31] but also a transient inhibition of axonal conduction and therefore a decrease of the sensitivity of brain mapping [32]. Indeed, the next crucial step is the axonal mapping. In fact, subcortical DES mapping is essential to identify the deeper limits of the resection while detecting and avoiding injury to subcortical fasciculi that could generate permanent deficits if damaged [33]. The posterior limb of the internal capsule is at risk, especially in zone 2 tumors [26]. It has been previously reported that the pyramidal tract might be monitored using evoked potentials under general anesthesia [34]. However, awake mapping is definitely more sensitive and will allow for the identification of the more lateral somatosensory thalamocortical tracts as well as other eloquent fasciculi that also should be identified and preserved for complication avoidance—in particular the network involved in motor control and complex movements [35]. Stimulation of the dorsal stream represented by the arcuate fasciculus/superior longitudinal fasciculus complex will lead to speech apraxia (even in the right hemisphere), phonological paraphasias, and repetition disorders during DES [27, 33]. Stimulation of the IFOF in the temporal stem (ventral stream) will lead to verbal semantic paraphasia in the left hemisphere and nonverbal semantic disorders in the right hemisphere when stimulated [36, 37]. The anterior perforated substance in which run the lenticulostriate arteries is a deep landmark for pure insular gliomas, but it can be invaded by tumor and mislead the surgeon in paralimbic DLGG. In these cases, the resections have to be limited in the depth when coming in contact with the temporal stem identified by encountering the IFOF, which runs laterally to the anterior perforated substance [20]. Leaving a small residue at this level significantly reduces the risk of permanent deficits [5]. More posteriorly, the lateral part of the lenticular nucleus is the mesial limit of the resection, and its stimulation will elicit articulatory disturbances [27].

28.4.6 Reoperation(s)

Repeated surgeries represent a safe and effective strategy in recurrent insular DLGG years after the initial resection [38–40]. The slow tumor growth and the first surgery itself can induce neuroplasticity mechanisms and then may increase the opercular cortical window for future surgeries [41]. A multistage approach is therefore appropriate to maximizing the EOR while preserving the QoL of patients with an insular/paralimbic DLGG [41].

28.5 Outcome

28.5.1 Oncological Results

EOR has been proven to be oncologically beneficial for patients with insular/paralimbic DLGG considering both OS and QoL [20]. We herein only review series with objective assessment of EOR on postoperative MRI. A series of 115 surgeries including 18.3% of repeated surgeries demonstrated a median EOR of 82% for insular DLGG: interestingly, an increased EOR had a positive predictive value of OS [10]. Another publication from the same group cumulated a total of 244 procedures with 54.3% being DLGG. The median EOR was 85%, and it was correlated to the volume of the tumor (more complete resection with smaller gliomas) and location in zone 1. No survival data was added to this second series [18]. In another experience, 51 patients were reviewed including 20% of reoperations and the median EOR was at 77%. Remarkably, 82% of the patients were still alive at 4 years, and, in 50% of second surgery, the EOR has been improved in all cases [5]. Another series looking at 74 patients, both DLGG (33.8%) and HGG, reported a median EOR of 91%. For DLGG, EOR was predictive of malignant progression-free survival and OS [42]. Three other series were reviewed [18] that did not include survival data. In a series of 33, 22, and 53 patients, they achieved 83.4% [43], 86% [26], and 82.98% [40] of median EOR, respectively. The last series was limited to reoperations and showed that the main predictive factor for tumor recurrence was EOR at the first surgery [40].

28.5.2 Functional Results

As the oncological benefit of maximal tumor resection is becoming increasingly clear, it is imperative to preserve neurological function and QoL in insular DLGG, knowing the high-functioning level and relative long-life expectancy of the population affected. An improved knowledge of insular/paralimbic functionality and connectivity in both hemispheres, combined with a rigorous application of cortical/axonal DES methodology and transcortical approach in awake patients, has led to a dramatic decrease of surgical morbidity compared to historical series—especially by allowing identification and preservation of functional fasciculi and by minimizing vascular injury [20]. In modern series reviewed, although the improved EOR was associated with a relatively high rate of transient immediate postoperative worsening (14.4% [10], 30.19% [40], 33.4% [18], 38% [43], 59% [5]), the rate of permanent deficits remains low (1.9% at 4 to 6 months [10], 1.89% at 6 months [40], 3.2% [18], 6% at 3 months [43], 3.9% [5], 2.7% [42]). In addition, in these recent series, the mortality rate was 0% [5, 10, 18, 42, 43]. Resection surgery of insular/

paralimbic DLGG was also shown to be safe for neurocognitive functions [43].

Functional improvement can even be seen in patients with preoperative deficits by relief of mass effect and postoperative rehabilitation. One series demonstrated such improvement in up to 83% [5]. Surgical resection of insular/paralimbic DLGG can also improve epilepsy control as shown in two series of about 50 patients, which reported 78% and 87.88% of seizure relief, respectively [5, 44]. Increased EOR is correlated to a better seizure control [44], and epilepsy existing for less than 1 year is more responsive to surgery [5]. Removal of mesio-temporal structures, even when not invaded by tumor, is also improving seizure control in patients with intractable epilepsy due to paralimbic DLGG [45].

To sum up, favorable functional results and preservation of QoL can thus be obtained with maximal surgical resection achieved up to eloquent structures in awake patients. Because of the high-functioning level of DLGG patients, outcomes have to be evaluated by postoperative neuropsychological assessment, and expectations should be a return to a normal active lifestyle. To expedite this process, early intensive rehabilitation has to be considered in the immediate postoperative period, whenever needed [21].

Conclusion

The first treatment in insular/paralimbic DLGG is early maximal resection. With an expert multidisciplinary team and a rigorous application of the methodology of cortical and subcortical DES mapping as well as transcortical approach in awake patients, tumor resection in this location is safe and effective. Preservation of brain functions is currently very reliable, and it is even possible to improve QoL, particularly thanks to seizure control, while avoiding mortality. In addition, improvement of the EOR has resulted in prolonged OS due to a delay of malignant transformation. In other words, insular/paralimbic DLGG surgery is now able to optimize the onco-functional balance and has to be proposed in a more systematic manner.

References

1. Duffau H. In: Duffau H, editor. Diffuse low-grade gliomas in adults: natural history, interaction with the brain, and new individualized therapeutic strategies. 2nd ed. London: Springer; 2017.
2. Louis DN, Perry A, Reifenberger G, von Deimling A, Figarella-Branger D, Cavenee WK, et al. The 2016 World Health Organization classification of tumors of the central nervous system: a summary. Acta Neuropathol. 2016;131:803–20.
3. Forst DA, Nahed BV, Loeffler JS, Batchelor TT. Low-grade gliomas. Oncoologist. 2014;19:403–13.
4. Almairac F, Herbet G, Moritz-Gasser S, de Champfleur NM, Duffau H. The left inferior fronto-occipital fasciculus subserves language semantics: a multilevel lesion study. Brain Struct Funct. 2015;220:1983–95.
5. Duffau H. A personal consecutive series of surgically treated 51 cases of insular WHO grade II glioma: advances and limitations. J Neurosurg. 2009;110:696–708.
6. Pallud J, Varlet P, Devaux B, Geha S, Badoual M, Deroulers C, et al. Diffuse low-grade oligodendrogliomas extend beyond MRI-defined abnormalities. Neurology. 2010;74:1724–31.
7. Pirzkall A, Nelson SJ, McKnight TR, Takahashi MM, Li X, Graves EE, et al. Metabolic imaging of low-grade gliomas with three-dimensional magnetic resonance spectroscopy. Int J Radiat Oncol Biol Phys. 2002;53:1254–64.
8. Pallud J, Mandonnet E, Duffau H, Kujas M, Guillevin R, Galanaud D, et al. Prognostic value of initial magnetic resonance imaging growth rates for World Health Organization grade II gliomas. Ann Neurol. 2006;60:380–3.
9. Pouratian N, Asthagiri A, Jagannathan J, Shaffrey ME, Schiff D. Surgery insight: the role of surgery in the management of low-grade gliomas. Nat Clin Pract Neurol. 2007;3:628–39.
10. Sanai N, Polley MY, Berger MS. Insular glioma resection: assessment of patient morbidity, survival, and tumor progression. J Neurosurg. 2010;112:1–9.
11. Gozé C, Rigau V, Gibert L, Maudelonde T, Duffau H, et al. J Neuro-Oncol. 2009;91:1–5.
12. Gozé C, Mansour L, Rigau V, Duffau H. Distinct IDH1/IDH2 mutation profiles in purely insular versus paralimbic WHO Grade II gliomas. J Neurosurg. 2013;118:866–72.
13. Nieuwenhuys R. The insular cortex: a review. Prog Brain Res. 2012;195:123–63.
14. Duffau H, Capelle L. Preferential brain locations of low-grade gliomas. Cancer. 2004;100:2622–6.
15. Kalani MY, Kalani MA, Gwinn R, Keogh B, Tse VC. Embryological development of the human insula and its implications for the spread and resection of insular gliomas. Neurosurg Focus. 2009;27(2):E2.
16. Mandonnet E, Capelle L, Duffau H. Extension of paralimbic low grade glioma: toward an anatomical

16. classification based on white matter invasion pattern. J Neuro-Oncol. 2006;78:179–85.
17. Yasargil MG, von Ammon K, Cavazos E, Doczi T, Reeves JD, Roth P. Tumours of the limbic and paralimbic systems. Acta Neurochir. 1992;118:40–52.
18. Hervey-Jumper SL, Li J, Osorio JA, Lau D, Molinaro AM, Benet A, et al. Surgical assessment of the insula. Part 2: validation of the Berger-Sanai zone classification system for predicting extent of glioma resection. J Neurosurg. 2016;124:482–8.
19. Duffau H, Mandonnet E. The "onco-functional balance" in surgery for diffuse low-grade glioma: integrating the extent of resection with quality of life. Acta Neurochir. 2013;155:951–7.
20. Michaud K, Duffau H. Surgery of insular and paralimbic diffuse low grade gliomas. J Neuro-Oncol. 2016; 130:289–98.
21. Rofes A, Mandonnet E, Godden J, Baron MH, Colle H, Darlix A, et al. Survey on current cognitive practices within the European Low-Grade Glioma Network: towards a European assessment protocol. Acta Neurochir. 2017;159:1167–78.
22. Fernández Coello A, Moritz-Gasser S, Martino J, Martinoni M, Matsuda R, Duffau H. Selection of intraoperative tasks for awake mapping based on relationships between tumor location and functional networks. J Neurosurg. 2013;119:1380–94.
23. Duffau H. A new concept of diffuse (low-grade) glioma surgery. Adv Tech Stand Neurosurg. 2012;38: 3–27.
24. Leclercq D, Duffau H, Delmaire C, Capelle L, Gatignol P, Ducros M, et al. Comparison of diffusion tensor imaging tractography of language tracts and intraoperative subcortical stimulations. J Neurosurg. 2010;112:503–11.
25. Deras P, Moulinié G, Maldonado IL, Moritz-Gasser S, Duffau H, Bertram L. Intermittent general anesthesia with controlled ventilation for asleep-awake-asleep brain surgery: a prospective series of 140 gliomas in eloquent areas. Neurosurgery. 2012;71:764–71.
26. Lang FF, Olansen NE, DeMonte F, Gokaslan ZL, Holland EC, Kalhorn C, et al. Surgical resection of intrinsic insular tumors: complication avoidance. J Neurosurg. 2001;95:638–50.
27. Duffau H, Moritz-Gasser S, Gatignol P. Functional outcome after language mapping for insular World Health Organization grade II gliomas in the dominant hemisphere: experience with 24 patients. Neurosurg Focus. 2009;27:E7.
28. Vanaclocha V, Saiz-Sapena N, Garcia-Casasola C. Surgical treatment of insular gliomas. Acta Neurochir. 1997;139:1126–35.
29. Tate MC, Herbet G, Moritz-Gasser S, Tate JE, Duffau H. Probabilistic map of critical functional regions of the human cerebral cortex: Broca's area revisited. Brain. 2014;137:2773–82.
30. Duffau H, Capelle L, Lopes M, Faillot T, Sichez JP, Fohanno D. The insular lobe: physiopathological and surgical considerations. Neurosurgery. 2000;47:801–10.
31. Rey-Dios R, Cohen-Gadol AA. Technical nuances for surgery of insular gliomas: lessons learned. Neurosurg Focus. 2013;34(2):E6.
32. Carrabba G, Mandonnet E, Fava E, Capelle L, Gaini SM, Duffau H, et al. Transient inhibition of motor function induced by the cavitron ultrasonic surgical aspirator during brain mapping. Neurosurgery. 2008; 63:E178–9.
33. Duffau H. Stimulation mapping of white matter tracts to study brain functional connectivity. Nat Rev Neurol. 2015;11:255–65.
34. Neuloh G, Pechstein U, Schramm J. Motor tract monitoring during insular glioma surgery. J Neurosurg. 2007;106:582–92.
35. Schucht P, Moritz-Gasser S, Herbet G, Raabe A, Duffau H. Subcortical electrostimulation to identify network subserving motor control. Hum Brain Mapp. 2013;34:3023–30.
36. Moritz-Gasser S, Herbet G, Duffau H. Mapping the connectivity underlying multimodal (verbal and non-verbal) semantic processing: a brain electrostimulation study. Neuropsychologia. 2013;51:1814–22.
37. Herbet G, Moritz-Gasser S, Duffau H. Direct evidence for the contributive role of the right inferior fronto-occipital fasciculus in non-verbal semantic cognition. Brain Struct Funct. 2017;222:1597–610.
38. Martino J, Taillandier L, Moritz-Gasser S, Gatignol P, Duffau H. Re-operation is a safe and effective therapeutic strategy in recurrent WHO grade II gliomas within eloquent areas. Acta Neurochir. 2009; 151:427–36.
39. Schmidt MH, Berger MS, Lamborn KR, Aldape K, McDermott MW, Prados MD, et al. Repeated operations for infiltrative low-grade gliomas without intervening therapy. J Neurosurg. 2003;98:1165–9.
40. Ius T, Pauletto G, Cesselli D, Isola M, Turella L, Budai R, et al. Second surgery in insular low-grade gliomas. Biomed Res Int. 2015;2015:497610.
41. Duffau H. The huge plastic potential of adult brain and the role of connectomics: new insights provided by serial mappings in glioma surgery. Cortex. 2014;58:325–37.
42. Eseonu CI, ReFaey K, Garcia O, Raghuraman G, Quinones-Hinojosa A. Volumetric analysis of extent of resection, survival, and surgical outcomes for insular gliomas. World Neurosurg. 2017;103: 265–74.
43. Wu AS, Witgert ME, Lang FF, Xiao L, Bekele BN, Meyers CA, et al. Neurocognitive function before and after surgery for insular gliomas. J Neurosurg. 2011;115:1115–25.
44. Ius T, Pauletto G, Isola M, Gregoraci G, Budai R, Lettieri C, et al. Surgery for insular low-grade glioma: predictors of postoperative seizure outcome. J Neurosurg. 2014;120:12–23.
45. Ghareeb F, Duffau H. Intractable epilepsy in paralimbic World Health Organization Grade II gliomas: should the hippocampus be resected when not invaded by the tumor? J Neurosurg. 2012;116:1226–34.

Surgical Strategy for Insular Cavernomas

29

Mehmet Turgut, Paulo Roberto Lacerda Leal, and Evelyne Emery

29.1 Introduction

The insular lobe, one of the five cerebral lobes, is a complex anatomic structure between the neocortex and the allocortex, known as the "paralimbic system," and the insular cortex is situated deeply within each cerebral hemisphere. Among various pathologies located in the insular lobe, cavernomas (CAs), which are composed of vascular channels lined by endothelium known as "cavern" are rarely seen [1]. CAs are equally seen among males and females and are usually present between 20 and 40 years of age. These lesions are easily diagnosed by computed tomography (CT) and magnetic resonance imaging (MRI), although they are angiographically occult vascular lesions. Classically, treatment options for cerebral CAs include conservative management, microneurosurgery, and radiosurgery.

The understanding of functional roles of the insula has been historically the insufficient, and the surgical approach for CAs in this region is technically complex with high risks and morbidity resulting from surgery. Therefore, most neurosurgeons avoided performing any surgical intervention to the region of the insula, especially in cases located in the dominant cerebral hemisphere, until the last 1–2 decades [2–4]. Nowadays, the surgical indications for these challenging vascular lesions have been revised after the use of multimodal management approaches including image guidance systems with the aid of the ultrasound (US) system and intraoperative stimulation for subcortical language pathway mapping. In this chapter, we review the natural history of insular CA and the current surgical strategies using the image guidance systems for the selection of the ideal trajectory of dissection of the insular cortex and planning of the surgical approach.

29.2 Incidence

With the development of modern neuroimaging, the detection rate of CAs has increased significantly; 70–80% of all cerebral CAs are supratentorial, and 10–20% of the cerebral CAs are located in the temporal lobes [5, 6]. The size of the cerebral CAs may vary from a few millimeters to few centimeters with an average size of one cm, while the ones located in the insula range from three to five cm in size [7]. Clinically, patients with CA frequently have epileptic seizures and focal neurological deficits due to

M. Turgut, M.D., Ph.D. (✉)
Department of Neurosurgery, Adnan Menderes University School of Medicine, Aydın, Turkey

P. R. L. Leal, M.D., Ph.D.
Department of Neurosurgery, Faculty of Medicine of Sobral, Federal University of Ceará, Sobral, Brazil

E. Emery, M.D., Ph.D.
Department of Neurosurgery, University of Caen, Caen, France

Table 29.1 Authors and their technical aids for surgery

Authors	Patients/side	Technical aids for surgery
Bertalanffy et al. [8]	6 (4R/2L)	
Casazza et al. [9]	3 (3R)	
Cukiert et al. [11]	1 (1R)	Corticography
Chaskis et al. [10]	1 (1R)	
Woydt et al. [12]	1 (1R)	US
Schmitt et al. [13]	1 (1R)	
Tirakotai et al. [14]	8 (5R/3L)	Neuronavigation
Duffau and Fontaine [4]	1 (1L)	Neuronavigation, intraoperative mapping
Esposito et al. [15]	1 (1L)	Corticography with MRI
von Lehe et al. [16]	2 (2 NS)	
Leal et al. [7]	4 (2L, 2R)	Neuronavigation, intraoperative mapping, US
Chevrier et al. [17]	2 (1L/1 NS)	
Cossu et al. [18]	1 (NS)	DTI tractography, fMRI

US ultrasound, *MRI* magnetic resonance imaging, *DTI* diffusion tensor imaging, *fMRI* functional MRI, *NS* not stated

repeated hemorrhage or mass effect. To the best of our knowledge, there are a total of 32 insular CAs who were treated surgically in the literature to date (Table 29.1) [4, 7–18].

29.3 Natural History

There is a considerable variability in the natural history (epidemiology, hemorrhage rates, and risk factors for hemorrhage) of cerebral CAs, although their biology is usually accepted as benign [7, 19]. In patients with CA located within the dominant hemisphere, the natural history of these vascular lesions should be evaluated in detail against their high surgical risk and surgical morbidity [7, 19]. Clinically, epileptic seizures were the most common symptom of CAs at the supratentorial location followed by headache and various neurological deficits [20]. Insular CAs are frequently associated with an epileptic seizure, termed insular epilepsy, and the risk of drug-resistant epilepsy is high in lesions involving insula [18]. Not surprisingly, insular epilepsy which is frequently refractory to antiepileptic drugs may have similar clinical and electroencephalographic features consistent with temporal lobe epilepsy [11]. Moreover, it has been suggested that epilepsy following the intralesional hemorrhage is caused by the hemosiderin in the adjacent brain tissue [21]. Typically, electrocorticography (ECoG) of patients with insular CA reveals the existence of insular spiking and/or temporal lobe spiking originating from the insula. Based on these findings, Cukiert et al. [11] suggested that there was involvement of temporal lobe in the epileptogenic process in patients with insular epilepsy.

29.4 Imaging Studies

Radiologically, CAs may simulate various neoplastic, vascular, or inflammatory pathologies. Today, MRI is frequently the main available diagnostic study in differential diagnosis of insular CA from other pathologies located in the insula. In patients with insular CAs, MRI with T1- and T2-weighted gradient-echo and contrast-enhanced images used a 1.5-T MR imager as a diagnostic tool in the preoperative period. CT and/or MRI-based image-guided techniques are used for planning the craniotomy [7]. The presence of hemosiderin, blood degradation products, and calcification in CT and MRI may be helpful in the diagnosis of some cases of insular CA, but the correct diagnosis is not possible until there is a histopathological evaluation of the surgical specimen. Furthermore, functional MRI (fMRI) and diffusion tensor imaging (DTI) tractography play a crucial role in the evaluation of the relationships of motor tracts and language pathways with CAs of the insula. Moreover, 3-D cerebral angiograms are used to eliminate lesions such as aneurysms and other vascular malformations.

29.5 Language Examination

The speech/language pathology test is used by speech therapists using the aphasia examination [22]. In patients with insular CA in the dominant cerebral hemisphere, fMRI is combined to determine language dominance.

29.6 Surgery for Insular Cavernoma

As a general rule, surgical intervention is the treatment of choice for all CAs including insular CA. In current neurosurgery, microsurgical removal of CAs of the insula is a safe and effective method, and most authors recommend total excision of the CA of the insula with some precautions. Technically, a stereotactic craniotomy device is useful in surgeries involving deep lesions of an insular CA [23]. Following craniotomy, surgery via transsylvian-transinsular approach is usually used in these patients. The surgical dissection should be done to prevent any vascular complication related with the branches of the middle cerebral artery. Many authors proposed image guidance systems in surgery of patients with CA of the insula. When the lesion is located within the dominant cerebral hemisphere, intraoperative mapping of the subcortical language pathways using stimulations with/without the use of fMRI and/or DTI tractography is also indicated for preservation of the language functional areas. Specifically, DTI tractography reveals whether the CA has critical relationships with language and motor tracts or not. In patients presenting with epilepsy, the CA of the insula and the adjacent gliotic tissue is removed, but the temporal lobe is preserved [7, 11].

29.7 Image Guidance System

29.7.1 Intraoperative US

In patients with insular CA, intraoperative US is used to determine the anatomical landmarks of the CA before and after opening of the dura. With the aid of an intraoperative US, the Sylvian fissure is identified first, the insular cortex is dissected without any difficulty, and then an incision is made for complete excision of the CA. During the last 1–2 decades, the US study was used as a guiding device independent of patient position during neurosurgical procedures for exact localization of various lesions including CA of the brain [12]. It is well-known that US is the cheapest and the quickest imaging technique for intracranial navigation. Leal et al. [7] used a US in only one of their cases, and they reported that the US allowed them to do a minimal incision for the removal of a CA located in the right insular cortex (Fig. 29.1).

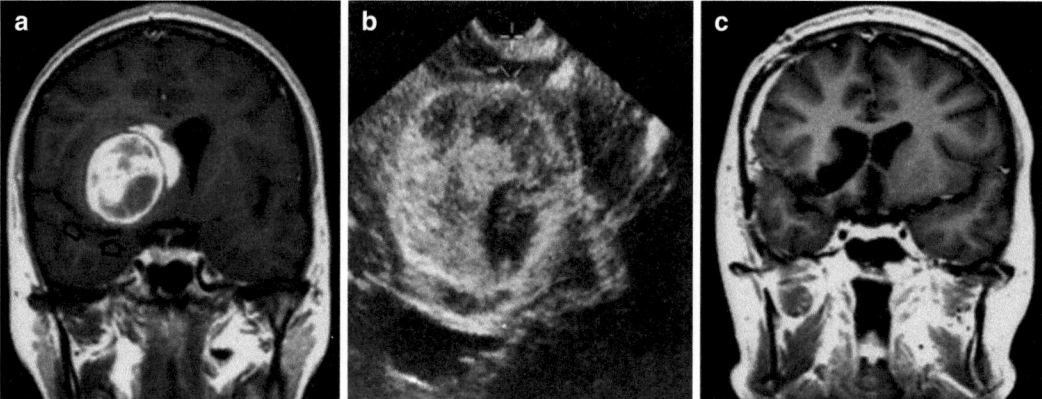

Fig. 29.1 Magnetic resonance imaging (MRI) and ultrasound (**a**) preoperative coronal T1-weighted MRI demonstrating the cavernoma (CA), (**b**) intraoperative ultrasound image confirming the CA, and (**c**) postoperative coronal T1-weighted MRI 3-month postsurgery (reproduced, with permission, from Leal et al.: Acta Neurochir 2010;152:1653–1659)

29.7.2 Neuronavigation

The neuronavigation provides an important guide for the preservation of the language functional areas during the corticotomy of the insula [7]. With the aid of neuronavigation, the insula is inspected for determination of adjacent cortical hemosiderin-stained tissue following the opening of the Sylvian fissure, and the CA is excised. From the surgical point of view, it is important to note that neuronavigation may help for the visualization of the CA next to the efferent motor tracts as well as for reduction of surgical morbidity [7]. In particular, in cases with the normal insular cortex, the shortest and safest corridor for removal of the CA is determined according to functional neuronavigation (Fig. 29.2) [7]. Many authors used the image guidance system during the removal of the insular CA [4, 12, 14, 15]. Recently, Leal et al. [7] reported three new patients with insular CA operated on by using neuronavigation. They stated that all patients had an uneventful postoperative course and there were not any surgical complication in these patients. As a useful guide for neurosurgeon during surgery, it is possible to detect incomplete resection by the intraoperative MRI in patients with insular CA [24]. Surgically, it is also important to note that the complete removal of CA and its perilesional hemosiderin rim is necessary following the corticotomy because of its potential epileptogenic effect [6, 7, 24, 25].

29.8 Intraoperative Mapping of the Subcortical Language Pathways

In addition to the visualization of the eloquent language area near the CA by DTI tractography and/or fMRI, intraoperative mapping of the subcortical language pathways using stimulations is used on patients with deep insular CA in the dominant cerebral hemisphere. In patients with insular CA located in the dominant hemisphere, the use of intraoperative mapping of cortical and subcortical language pathways is necessary to determine the eloquent cortical structures prior to insular corticotomy. At present, it is well-known that the intraoperative stimulation technique is a safe, sensitive, and reliable method of detection of the functional areas [3, 4, 26, 27]. Following awake craniotomy after local anesthesia, the patient is asked to do a picture test termed,"the DO (dénomination d'objet) 80 test" and number counting for identification of the cortical language sites, which may be inhibited by electrical stimulation using a bipolar

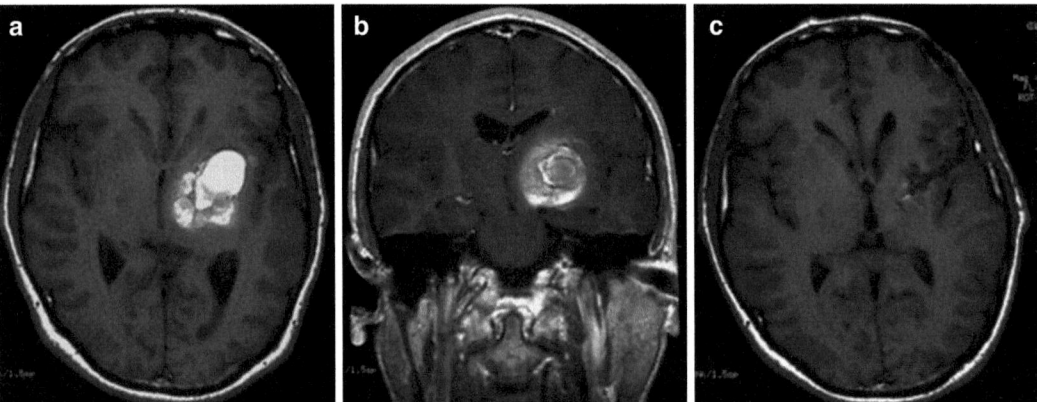

Fig. 29.2 Magnetic resonance imaging (MRI) (**a**) preoperative axial T1-weighted MRI, (**b**) preoperative coronal-T1 MRI, and (**c**) postoperative axial T1-weighted MRI 3 month postsurgery. Please note that the insular corticotomy and the removal of the cavernoma are guided by neuronavigation (reproduced, with permission, from Leal et al.: Acta Neurochir 2010;152: 1653–1659)

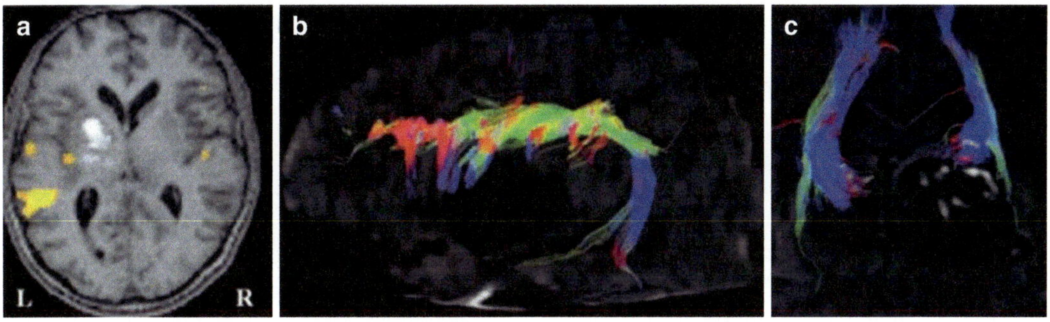

Fig. 29.3 Functional magnetic resonance imaging (fMRI) and diffusion tensor imaging (DTI) with tractography (**a**) fMRI demonstrating speech-activated regions adjacent to cavernoma (CA) in the left insula in a patient operated with an awake surgical intervention, (**b**) preoperative DTI of language tract revealing the relationships between the arcuate fasciculus and the CA, and (**c**) preoperative DTI of motor tract revealing the corticospinal pyramidal tract involved by the CA (reproduced, with permission, from Leal et al.: Acta Neurochir 2010;152:1653–1659)

stimulation electrode which is applied upon the left insular lobe [3, 4, 27]. In 2005, Duffau and Fontaine [4] first used an intraoperative mapping of the subcortical language pathways with direct stimulation of the insula; they found that the left dominant insula is important for speech planning because of its important connections with the frontal, parietal, and temporal lobes, limbic regions, and basal nuclei. It is important to know that the removal of the CA should be abandoned when there is a speech inhibition after bipolar stimulation. Some authors used intraoperative mapping of the subcortical language pathways during surgery of the CA of the insula [4, 7]. In 2010, Leal et al. [7] used subcortical stimulations for preservation of the functional language areas in one of their cases (Fig. 29.3).

29.9 Intraoperative Electrocorticography

Following the split of the Sylvian fissure, intraoperative ECoG is useful for identification of epileptic focus in patients with insular CA [28]. In patients presenting with epilepsy, ECoG and the depth electrodes will demonstrate epileptic discharges. It is important to know that complete resection of epileptic focus is necessary in the patients with insular CM causing epilepsy to achieve a better prognosis, although von Lehe et al. [16] suggested that subtotal resection also provides seizure control [6, 16, 29, 30].

29.10 Experiences from Current Surgical Strategies

In the review of the literature, there are several authors who reported their experience from current surgical strategies about insular CA. Today, the use of intraoperative neuronavigations and mapping cortical and subcortical language pathways makes possible the lesions located subcortically in eloquent areas of the insula [3]. Recent surgical techniques with the combined use of subcortical mapping, DTI tractography, and intraoperative neuronavigation provide an opportunity for protection of the tracts related with language areas and motor pathways in patients with low-grade gliomas of the insular lobe [26]. Bertalanffy et al. [8] removed the insular CA in their six patients by microsurgical intervention without intraoperative neuronavigation; transient and permanent neurological deficits developed in the third and first of their cases, respectively, due to damage to the lenticulostriate artery of the internal capsule. On the other hand, Tirokatai et al. [14] operated a total of eight patients with insular

CA using intraoperative neuronavigation; no surgical complications occurred in their series. Recently, Leal et al. [7] reported that the neuronavigation system was used in their three cases, providing the use of a smaller craniotomy and the shortest insular corticotomy in a CA located in the nondominant cerebral hemisphere. With the use of neuronavigation systems in surgical intervention for insular CA, minimal retraction and careful exploration will be sufficient for correct localization of subcortical and deep lesions, with shorter operative time, low complication rate, and shorter length of hospital stay.

Conclusion

In summary, MRI study is sensitive for diagnosis of insular CA. Complete removal is crucial for management of the insular CA using functional mapping of the subcortical language pathways and intraoperative neuronavigation to preserve the eloquent area in the insular cortex. We advocate a surgical strategy for insular CAs following a topographical analysis of the CA within the insular cortex. In patients with insular CA within the nondominant hemisphere, described by the help of functional MRI study and neuropsychological tests, the intraoperative neuronavigation is vital for the insular corticotomy and determination of the shortest approach to the vascular lesion within the insula. In the insular CA within the dominant cerebral hemisphere, the intraoperative neuronavigation is necessary, in addition to the use of an intraoperative mapping of the subcortical language pathways. With preoperative analysis of neuropsychological tests and functional MRI and DTI tractography, the complete removal of the insular CA as well as its surrounding hemosiderin rim, in particular for those located in the dominant hemisphere, is performed by microsurgical techniques. Complete removal of frequently medically intractable insular CA using a multimodal approach, even subtotal resection, provides good seizure control; thus surgical therapy for insular lesions is therefore a promising option.

References

1. Aiba T, Tanaka R, Koike T, Kameyama S, Takeda N, Komata T. Natural history of intracranial cavernous malformations. J Neurosurg. 1995;83:56–9.
2. Duffau H, Capelle L, Lopes M, Faillot T, Sichez JP, Fohanno D. The insular lobe: physiopathological and surgical considerations. Neurosurgery. 2000;47:801–11.
3. Duffau H, Capelle L, Sichez N, Denvil D, Lopes M, Sichez JP, Bitar A, Fohanno D. Intraoperative mapping of the subcortical language pathways using direct stimulations. An anatomo-functional study. Brain. 2002;125:199–214.
4. Duffau H, Fontaine D. Successful resection of a left insular cavernous angioma using neuronavigation and intraoperative language mapping. Case report. Acta Neurochir. 2005;147:205–8.
5. Batra S, Lin D, Recinos PF, Zhang J, Rigamonti D. Cavernous malformations: natural history, diagnosis and treatment. Nat Rev Neurol. 2009;5:659–570.
6. Shan YZ, Fan XT, Meng L, An Y, Xu JK, Zhao GG. Treatment and outcome of epileptogenic temporal cavernous malformations. Chin Med J. 2015;128:909–13.
7. Leal PRL, Houtteville JP, Etard O, Emery E. Surgical strategy for insular cavernomas. Acta Neurochir. 2010;152:1653–9.
8. Bertalanffy H, Gilsbach JM, Eggert HR, Seeger W. Microsurgery of deep-seated cavernomas: report of 26 cases. Acta Neurochir. 1991;108:91–9.
9. Casazza M, Broggi G, Franzini A, Avanzini G, Spreafico R, Bracchi M, Valentini MC. Supratentorial cavernomas and epileptic seizures: preoperative course and postoperative outcome. Neurosurgery. 1996;39:26–32.
10. Chaskis C, Brotchi J. The surgical management of cerebral cavernous angioma. Neurol Res. 1998;20:597–606.
11. Cukiert A, Forster C, Andrioli MS, Frayman L. Insular epilepsy. Similarities to temporal lobe epilepsy. Case report. Arq Neuropsiquiatr. 1998;56:126–8.
12. Woydt M, Horowski A, Krone A, Soerensen N, Roosen K. Localization and characterization of intracerebral cavernomas by intraoperative high-resolution colour-duplex-sonography. Acta Neurochir. 1999;141:143–52.
13. Schmitt JJ, Ebner A. The anatomic substrate of the epigastric aura-a case report. Nervenarzt. 2000;71:485–8.
14. Tirakotai W, Sure U, Benes L, Krischek B, Siegfried B, Bertalanffy H. Image-guided transsylvian, transinsular approach for insular cavernomas. Neurosurgery. 2003;53:1299–305.
15. Esposito V, Paolini S, Morace R. Resection of a left insular cavernoma aided by a simple navigational tool. Neurosurg Focus. 2006;21:1–4.
16. von Lehe M, Wellmer J, Urbach H, Schramm J, Elger CE, Clusmann H. Insular lesionectomy for

refractory epilepsy: management and outcome. Brain. 2009;132:1048–56.
17. Chevrier MC, Bard C, Guilbert F, Nguyen DK. Structural abnormalities in patients with insular/peri-insular epilepsy: spectrum, frequency, and pharmacoresistance. AJNR Am J Neuroradiol. 2013;34:2152–6.
18. Cossu M, Raneri F, Casaceli G, Gozzo F, Pelliccia V, Lo Russo G. Surgical treatment of cavernoma-related epilepsy. J Neurosurg Sci. 2015;59:237–53.
19. Gross BA, Lin N, Du R, Day AL. The natural history of intracranial cavernous malformations. Neurosurg Focus. 2011;30:E24.
20. Sawarkar DP, Janmatti S, Kumar R, Singh PK, Gurjar HK, Kale SS, Sharma BS, Mahapatra AK. Cavernous malformations of central nervous system in pediatric patients: our single-centered experience in 50 patients and review of literature. Childs Nerv Syst. 2017;33(9):1525–38. https://doi.org/10.1007/s00381-017-3429-7.
21. Raabe A, Schmitz AK, Pernhorst K, Grote A, von der Brelie C, Urbach H, Friedman A, Becker AJ, Elger CE, Niehusmann P. Cliniconeuropathologic correlations show astroglial albumin storage as a common factor in epileptogenic vascular lesions. Epilepsia. 2012;53:539–48.
22. Goodglass H, Kaplan E. The assessment of aphasia and related disorders. Philadelphia: Lea & Febiger; 1972. p. 1–80.
23. Ota Y, Araki O, Oki S, Inagawa T, Emoto K, Shibukawa M, Yamasaki H, Kano Y, Tani I. A case report of stereotactic craniotomy for the cerebral cavernous angioma (in Japanese). No Shinkei Geka. 2016;44:149–54.
24. Sommer B, Kasper BS, Coras R, Blumcke I, Hamer HM, Buchfelder M, Roessler K. Surgical management of epilepsy due to cerebral cavernomas using neuronavigation and intraoperative MR imaging. Neurol Res. 2013;35:1076–83.
25. Bauman CR, Schuknecht B, Russo GL, Cossu M, Citterio A, Andermann F, Siegel AM. Seizure outcome after resection of cavernous malformations is better when surrounding hemosiderin-stained brain also is removed. Epilepsia. 2006;47:563–6.
26. Bello L, Gambini A, Castellano A, Carrabba G, Acerbi F, Fava E, Giussani C, Cadioli M, Blasi V, Casarotti A, Papagno C, Gupta AK, Gaini S, Scotti G, Falini A. Motor and language DTI fiber tracking with intraoperative subcortical mapping for surgical removal gliomas. NeuroImage. 2008;39:369–82.
27. Ojemann G, Ojemann JG, Lettich E, Berger M. Cortical language localization in left, dominant hemisphere. An electrical stimulation mapping investigation in 117 patients. J Neurosurg. 1989;71:316–26.
28. Sugano H, Shimizu H, Sunaga S. Efficacy of intraoperative electrocorticography for assessing seizure outcomes in intractable epilepsy patients with temporal-lobe-mass lesions. Seizure. 2007;16:120–7.
29. Van Gompel JJ, Rubio J, Cascino GD, Worrell GA, Meyer FB. Electrocorticography-guided resection of temporal cavernoma: is electrocorticography warranted and does it alter the surgical approach? J Neurosurg. 2009;110:1179–85.
30. Yeon JY, Kim JS, Choi SJ, Seo DW, Hong SB, Hong SC. Supratentorial cavernous angiomas presenting with seizures: surgical outcomes in 60 consecutive patients. Seizure. 2009;18:14–20.

Role of the Insula in Temporal Lobe Epilepsy Surgery Failure

30

Vamsi Krishna Yerramneni, Alain Bouthillier, and Dang Khoa Nguyen

Abbreviations

EEG   Electroencephalography
MEG   Magnetoencephalography
MRI   Magnetic resonance imaging
PET   Positron emission tomography
SEEG Stereoencephalography
SPECT Single-photon emission computed tomography

30.1 Introduction

30.1.1 Epilepsy Surgery

Epilepsy is a chronic condition characterized by recurrent seizures. Between these "ictal" events, asymptomatic brief discharges (called spikes or interictal epileptiform discharges) can also be observed on EEG. Common causes for epilepsy include acquired brain lesions (stroke, head trauma, brain infection), tumors, vascular malformations, hippocampal sclerosis, malformations of cortical development, and genetic mutations. Despite appropriate therapy with antiepileptic drugs, up to 30% of patients continue to suffer from frequent seizures [1]. For these patients, epilepsy surgery must seriously be considered since multiple studies have shown that cure of seizures may be achieved by removing the epileptogenic zone [2]. In order to localize the epileptic focus, patients first undergo a noninvasive comprehensive presurgical evaluation typically comprising a good clinical history (seizure semiology can provide clues to focus localization), an MRI (looking for epileptogenic lesions), video-EEG monitoring (to observe seizure semiology and characterize epileptic activity between and during seizures), and PET (to reveal areas of abnormal glucose use) [3]. Some centers have the setup to perform ictal SPECT (to reveal areas of increased blood flow during seizures). When these complementary noninvasive studies fail to adequately localize the epileptogenic zone, an invasive EEG study is performed which consists in the implantation of intracranial electrodes under general anesthesia either with a stereotactic frame (SEEG), by frameless MRI-guided stereotaxy, or via open craniotomy [4]. Patients are then transferred to the epilepsy monitoring unit for continuous video-intracranial EEG monitoring, awaiting seizures. Once a sufficient number

V. K. Yerramneni, M.D. · A. Bouthillier, M.D., M.Sc.
Service de Neurochirurgie, Centre Hospitalier de l'Université de Montréal, Montreal, QC, Canada

D. K. Nguyen, M.D., Ph.D. (✉)
Centre de Recherche du Centre Hospitalier de l'Université de Montréal, Montreal, QC, Canada

Département de Neurosciences, Université de Montréal, Montreal, QC, Canada

Service de Neurologie, Centre Hospitalier de l'Université de Montréal, Montreal, QC, Canada
e-mail: d.nguyen@umontreal.ca

of seizures have been recorded, the patient is brought back to the operating room for the removal of electrodes and resection of the epileptic focus (if successfully identified by these recordings). In the end, identification of the epileptic focus is based on multimodal analysis of whatever clinical, structural, electrophysiological, and functional data is available. Foci identified by this relatively standard presurgical evaluation are most frequently located in the temporal (50–75%) and frontal lobes (25%) [5, 6]. Hence, most epilepsy surgeries are performed in these lobes.

30.1.2 Insular Seizures as a Cause for Epilepsy Surgery Failure

Despite the wealth of knowledge provided by decades of clinical and basis science research on temporal and later on frontal lobe epilepsies, complete seizure control remains elusive. In temporal lobe epilepsy surgeries, the probability of becoming seizure-free is ~75% in lesional (i.e., an epileptogenic lesion is identified on MRI) cases and only 51% in nonlesional cases [7]. In frontal lobe epilepsy surgeries, the probability of becoming seizure-free is only 60% in lesional and a mere 35% in nonlesional cases. The most obvious explanations for surgical failures are inaccurate localization or incomplete removal of the epileptic focus [8]. This is partly explained by limitations of current localization techniques: MRI fails to identify an epileptogenic lesion in ~25% of temporal lobe epilepsy cases and ~45% of frontal lobe epilepsy cases [9]; scalp EEG, ictal SPECT, and PET lack sufficient spatial and/or temporal resolution [10]; intracerebral electrodes overcome the sensitivity limitations of scalp electrodes because they are closer to bioelectric sources of epileptiform activity but only a limited number may be safely implanted to minimize risk of hemorrhage, edema, or infection. Inadequate coverage may produce a false electrographic picture as the first signal recorded may simply represent propagated activity if there is no electrode over the actual seizure onset zone [4]. Using novel noninvasive localization techniques and with invasive sampling of the insula with depth electrodes, work from our group and others has shown that failing to detect seizures from the insular lobe is responsible for a number of these surgical unfavorable outcomes [8, 11, 12]. Indeed, we have encountered patients with insular epilepsy misdiagnosed with temporal, frontal, or even parietal lobe epilepsy that were unfortunately operated in the wrong area. Other patients had temporo-insular epilepsy with persisting seizures after temporal lobe surgery because it was not recognized that they suffered from a more complex epileptogenic network that included not only the temporal lobe but also the neighboring insula [13].

30.2 Anatomical/Functional Considerations

The insular cortex, the fifth and smallest lobe of the brain, is a complex structure enclosed in the depth of the Sylvian fissure covered by the frontal, parietal, and temporal opercula [14]. It is divided by the central insular sulcus into an anterior portion made of three to four short gyri and a posterior portion made of two long gyri [15]. Seminal work by Mesulam in the 1980s has exposed the widespread connections of the macaque insula using tracing techniques [16], while more recently, our group [17] and a few others have done the same in humans using tractography (see Chap. 5 in this book). Briefly, the insula is connected with the frontal lobe (inferior frontal gyrus, orbitofrontal cortex, prefrontal cortex, cingulate gyrus, and supplementary motor area), the temporal lobe (temporal pole, superior temporal gyrus, amygdala, and entorhinal cortex), the parietal lobe (primary and secondary somatosensory cortices), the basal ganglia, and the thalamus. Cumulative work over years has shown that the insula has many roles [18, 19]. It is a multimodal area involved in the processing of several sensory stimuli from viscerosensory and somatosensory stimuli to special senses such as hearing, taste, and smell. It is also part of the language, pain, vestibular, pain, autonomic, and limbic networks. Additionally, it is thought to be involved in several cognitive processes such as attention, verbal memory, social cognition, and higher-order executive functions. Electrical stimulation of the insula performed by us [20] and other groups [21–27] has evoked a variety of responses: somatosensory (e.g., tingling, pain), viscerosensory (e.g., nausea, epigastric sensation), visceromotor (e.g., eructation), auditory (e.g., buzzing, echoing sounds), gustatory (e.g., metallic taste), vestibular (e.g., vertigo, dizziness), and speech (e.g., speech arrest) symptoms.

Considering the wealth of insular connections to surrounding brain areas and more distant sites and the variety of symptoms it may generate when stimulated, it should come as no surprise that clinical manifestations of insular seizures are quite various. Based mainly on case reports and a few small series (with intracerebral depth recording to confirm the insular onset) by us [20, 28, 29] and other groups [21, 30–33], we know insular seizures may feature early somatosensory symptoms (similar to parietal lobe seizures), hypermotor symptoms/complex motor behaviors (e.g., body rocking, pedaling) (resembling mesial frontal lobe seizures), and early visceral symptoms or dysphasia (suggesting temporal lobe seizures). Such mimicry most likely has fooled some clinicians into thinking that their patients had temporal, frontal, or parietal lobe seizures leading to the resection of the wrong cortical area [8, 28].

30.3 Historical Perspective

In a 1949 publication, Guillaume and Mazars reported their first seven insular resections for epilepsy, either because intraoperative electrocorticography revealed the insula to be site of the epileptic focus or the epileptogenic lesion extended to it. Authors mentioned that their experience was unique except for unpublished cases operated by Penfield [34]. In 1955, Penfield and Faulk reported their observations on the insula based on the examination of all cases in which positive results of insular stimulation were recorded during surgery for focal epilepsy at the Montreal Neurological Institute from 1945 to 1953. In the end, they concluded that "it is not surprising that the results of stimulation of the insula are confusingly varied since it is surrounded by such a remarkable variety of functional areas…it follows, too, that local epileptic discharges often produce a great variety of seizure patterns" [24]. Hence, these authors had already recognized that seizures originating from the insula may be difficult to recognize as they may resemble other types of seizures, notably temporal lobe seizures. Furthermore, speculating that persistent seizures after temporal lobe surgery might be explained by the failure to remove a residual focus within the insula, they started performing complementary insular resections if the electrocorticogram taken from the exposed cortex after temporal lobectomy showed robust insular spikes. This surgical attitude eventually declined as the procedure prolonged an already time-consuming operative procedure, increased the risk of complications, and most importantly did not clearly improve seizure outcome [35]. After comparing 58 cases on whom a total or partial insulectomy was performed and 48 patients from the Montreal Neurological Institute finally concluded that the presence of insular epileptiform abnormality in the post-excision cortical EEG did not necessarily represented an indication for removal of insular cortex. Indeed, while satisfactory control of epilepsy was relatively similar in both groups (45% with insulectomy versus 42% without), insulectomy carried a considerably higher incidence of neurological complications (20.6 versus 2.8%) [36]. These findings are not necessarily surprising since the value of intraoperative electrocorticography to tailor resections in mesial temporal epilepsy is controversial [37–40]. Intraoperative electrocorticography is short in durations and more frequently records interictal spikes rather than seizures. However, these spikes may represent only a partial representation of the epileptic network, consist of propagated spikes, or be "injury spikes" due to surgical manipulation or post-excisional activation spikes [41]. The occurrence of multiple divergent spike populations in patients despite a single epileptogenic zone as proven by seizure freedom following surgical resection has led most to rely on ictal studies rather than interictal spike analyses to guide the extent of surgical resection [42, 43]. When one closely looks at the study of Silfvenius et al. [36], it must be noted that in the entire study, electrocorticographic seizures were recorded exclusively from the insula in only four cases, three of which had an unsuccessful outcome. Furthermore, it was stated that "in patients with persisting epileptiform abnormality recorded from the insula after removal of the temporal lobe, both the spike frequency per minute and the amplitude of spikes in the postexcision cortical electrocorticogram was greater in the unsuccessful than in the successful follow-up categories." Finally, it should be pointed out that "when patients who previously had had an operation were reoperated upon and had an insulectomy, the results remained unsatisfactory in only 46.2% of the patients as compared to in 83.3% in whom no insular ablation was performed," indicating that insular ablation was beneficial for some.

Although there were some reports of epileptic patients who benefited from the removal of an epileptogenic lesion in the insula [44–46], interest in insula was mainly renewed in the epilepsy community when Isnard et al. [12] managed to record independent insular seizures using orthogonal depth electrodes sampling the insula in 2 out of 21 patients with suspected temporal lobe epilepsy (7 nonlesional/14 with temporal MR abnormality) undergoing SEEG because they had ictal symptoms or scalp EEG suggesting an early spread of seizures either to the suprasylvian or infrasylvian opercular cortex. Out of the 17 operated patients, a favorable outcome was obtained after a temporal cortectomy (sparing the insula) in the 15 patients who had temporal lobe seizures only (7 of whom had insular spikes but no seizures), whereas the 2 patients with independent temporal and insular seizures had a poor outcome (Engel IV). This led the authors to conclude that insular seizures likely explained why some temporal lobectomies fail. Since then, cases of insular seizures have been accumulating in the recent literature, suggesting that they may be more frequent than previously thought.

30.4 Failed Temporal Lobe Surgeries in the Context of Extended Temporo-Insular Epilepsy

As mentioned above, some patients suffer from a more complex epileptogenic network which not only include structures in the temporal lobe but also neighboring lobes such as the orbitofrontal cortex, the insula, the frontal and parietal operculum, and the temporo-parieto-occipital junction [47]. The term "temporal plus epilepsy" was proposed by Ryvlin and Kahane in 2005 to regroup these patients who, despite prominent ictal involvement of the temporal lobe, electroclinical features primarily suggestive of temporal lobe epilepsy, and even frequently hippocampal sclerosis on MRI, have in fact multilobar epilepsy. Depending on the structures involved, patients may be subdivided in the temporal-frontal subgroup, the temporo-parieto-occipital subgroup or the temporo-perisylvian subgroup [48]. In the temporo-perisylvian subgroup, it is possible to encounter seizures that start independently or simultaneous from mesial temporal and insular structures (temporo-insular epilepsy); however, in other instances of insular involvement, the seizure onset zone also extends to one or more overlying opercula (temporo-operculo-insular epilepsy).

A relatively recent study has clearly established that patients with temporal plus epilepsy were a major determinant of temporal lobe surgery failures [11]. In this series of 168 patients (108 had SEEG; 131 had hippocampal sclerosis), risk of temporal lobe surgery (encompassing the anatomical boundaries of a standard anterior temporal lobectomy) failure was 5.06 greater in patients with SEEG-confirmed temporal plus epilepsy ($n = 18$, 2 fronto-basal cortex, 4 suprasylvian operculum, 7 *insula*, 5 temporo-parieto-occipital junction) than in those with unilateral temporal lobe epilepsy ($n = 149$) (mean follow-up 7 years with a minimum of 2 years). Kaplan-Meier estimates of Engel class I at 10 years after surgery was 74.5% for unilateral temporal lobe epilepsy and 14.8% for temporal plus epilepsy.

Unfortunately, a priori identification of temporal plus epilepsy cases is arduous on the basis of general clinical features and noninvasive paraclinical tests. In a retrospective study comparing 58 patients with SEEG-confirmed purely temporal seizures and 22 with SEEG-confirmed temporal plus epilepsy (9 temporo-frontal, 7 temporo-sylvian, and 6 temporo-parieto-occipital junction), Barba et al. [49] concluded that the two groups were difficult to differentiate on the basis of general clinical features (sex, age of onset, febrile seizures, disease duration, seizure frequency, tonic-clonic generalizations, nocturnal seizures) or MRI data. The presence of hippocampal sclerosis did not distinguish both groups. However, patients with temporal lobe epilepsies more frequently presented the ability to warn at seizure onset, an abdominal aura, gestural automatisms, and postictal amnesia, while patients with temporal plus epilepsies more frequently had gustatory hallucinations, rotatory vertigo and auditory illusions, eye/head contraversion, piloerection, ipsilateral tonic motor signs, and postictal dysphoria. On surface EEG, temporal plus patients more frequently exhibited interictal bilateral or precentral abnormalities and ictal anterior frontal, temporo-parietal and precentral changes [49]. Looking more specifically at the temporo-perisylvian/temporo-insular subgroup, the presence of atypical ictal features such as throat constriction, limb paresthesia (including pain), and gustatory or auditory hallucinations (especially if they occur in combination) is an important clue [48]. This is exemplified by six cases reported by Aghakhani et al. [50] who did not benefit from an anterior temporal resection (three also had parietal lobe surgeries) despite well-defined focal anterior and infero-mesial temporal epileptic discharges; it is probable that most of these patients had an epileptogenic zone which included the insula as five had somatosensory auras (although this could not be proven as the insula was not sampled during invasive EEG recordings) [50].

In regard to other noninvasive tests, the value of ictal SPECT and PET remains uncertain to identify temporal plus epilepsies (including temporo-insular epilepsies). For example, studies have shown that in mesial temporal lobe epilepsies, ictal SPECT may reveal hyperperfusion of the ipsilateral insula in addition to the anterior temporal lobe without affecting seizure outcome after temporal lobe surgery, suggesting secondary activation from propagated epileptic activity [51–53]. Whether patients with temporo-insular epilepsy have a different pattern of activation in terms of distribution or intensity compared to patients with mesial temporal lobe epilepsy has not been adequately studied. As for PET, a recent large study of 97 mesial temporal lobe epilepsy cases with unilateral hippocampal sclerosis found that seizure freedom (Engel IA; mean follow-up >6 years) was associated with a focal anteromesial temporal hypometabolism whereas non-1A outcome correlated with extratemporal metabolic changes (such as in the insula) [54]. Until more data is available, the detection of temporo-insular epilepsy still heavily relies on invasive EEG recordings with adequate sampling of structures mentioned above. However, there is no clear consensus on who should undergo an invasive EEG and decisions are made on an individual basis on available multimodal data. Fortunately, for patients who are eventually diagnosed with temporo-insular epilepsy, there is some limited evidence that larger resections of temporal plus epileptogenic zones may result in good outcome [55, 56]; this needs to be further investigated with larger series however.

30.4.1 Illustrative Case 1

A 33-year-old right-handed man was referred to our center for a history of drug-resistant seizures since age 27 years. Past medical history was significant for an atypical febrile seizure with left Todd's paresis at age 6 months. Seizures were characterized by a combination of viscerosensory, gustatory and olfactory symptoms followed by impaired awareness. Presurgical evaluation revealed right hippocampal sclerosis on MRI (Fig. 30.1); hypometabolism of the medial, polar, and anterior lateral portions of the right temporal lobe; and ictal right hemispheric (maximum over the temporal region) rhythmic activity on EEG. He underwent a selective right amygdalohippocampectomy followed 6 months later by an

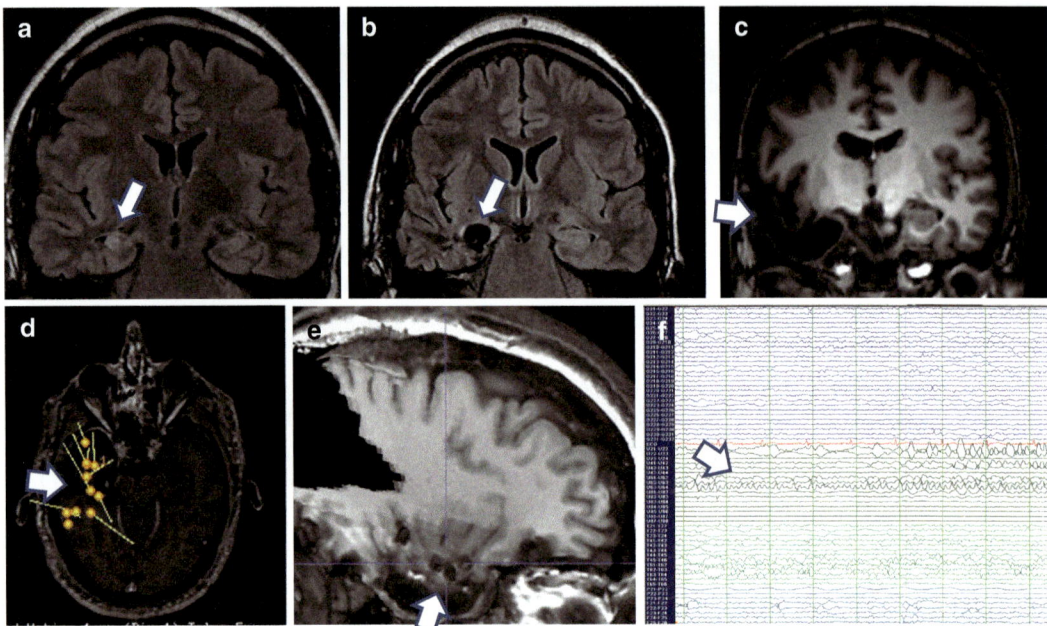

Fig. 30.1 (Case 1): Patient with R hippocampal sclerosis (**a**; *white arrow*) who continued to have seizures after selective amygdalo-hippocampectomy (**b**; *white arrow*) and anterior temporal lobectomy (**c**; *white arrow*); MEG disclosed sources in the insula (**d**; *white arrow*); invasive EEG with insular samplig (**e**; *white arrow*) confirmed the presence of insular seizures (**f**; *white arrow*)

anterior temporal lobectomy due to persisting seizures. Unfortunately, he continued to have seizures consisting in a combination of somatosensory and olfactory auras, hypersalivation, and impaired awareness with or without convulsions. A 2-week intracranial EEG study at age 41 years disclosed right insular seizures but also very active spiking (no seizures) in the orbitofrontal cortex; at this point, it was decided to only remove the right insula. Unfortunately, he experienced three seizures (starting with an olfactory aura) in the first 3 months and then was found dead one morning by his son.

30.5 Failed Temporal Lobe Surgeries in the Context of Restricted Operculo-Insular Epilepsy

Distinct from temporal plus epilepsies, some patients have isolated pure insular or operculo-insular epilepsies. Because operculo-insular seizures may not only mimic temporal but also frontal and parietal lobe seizures, is it likely that some surgeries targeting the temporal, parietal or frontal lobe have failed because the insular generator was not resected [57]. Indeed, cumulative data obtained from case reports, small case series and electrical insular stimulations have shown that insular seizures may feature a variety of symptoms depending on the subinsular area of onset and propagation pathways to connected structures [13, 30, 58–63]. Ictal semiology may include sensory (somatosensory, olfactory, auditory, gustatory, vestibular), emotional (anxiety, fear, laughing), autonomic (flushing, nausea, palpitations, respiratory changes, tachycardia or bradycardia), cognitive (dysphasia, responsiveness impairment), automatisms (pedaling, pelvic thrusting) and/or motor (dystonic) symptoms [57].

Obviously, identification of an insular epileptogenic lesion on MRI greatly helps in recognizing insular epilepsy [64]. This is well exemplified by Mortati et al. who reported an epileptic patient cured by the resection of a right insular ganglioglioma despite the fact he had left (not right) anterior temporal spikes and right then left temporal rhythmic changes during seizures [65]. Similarly, Kaido et al. [31] opted to remove the insula (rather than the temporal lobe even if invasive EEG without insular sampling revealed a temporal lobe seizure onset) because MRI had identified a slight high signal change in the posterior insula. Recognition of operculo-insular epilepsy in patients with no clear lesion is challenging. The value of noninvasive techniques such as surface EEG, SPECT, PET, MRS, and MEG has recently been reviewed by Obaid et al. [57]. In brief, surface EEG is helpful to lateralize the focus but is unable to distinguish a deep-seated insular focus from a more medial temporal or lateral neocortical focus [60], ictal SPECT and interictal FDG-PET have moderate sensitivity (~65 and 47%, respectively) [66], MEG was highly useful [67–70], and proton MR spectroscopy was not useful [71]. Similar to temporal plus epilepsies, use of intracranial EEG with adequate insular/perisylvian sampling is frequently required to confirm and better delineate the seizure onset zone in suspected cases when MRI is negative [72].

While epilepsy surgery in the operculo-insular region is technically challenging, several recent series indicate that good seizure outcome may be obtained with low rates of permanent complications [13, 58, 73, 74].

30.5.1 Illustrative Case 2

A 41-year-old man was referred to our center for drug-resistant seizures since age 35 years. Seizures were characterized by an epigastric sensation of guilt, flushing with or without impaired awareness and rarely ending in tonic-clonic seizures. MRI was normal. EEG revealed interictal bifrontal spikes with right predominance and diffuse ictal onset changes with maximal right-sided evolution. Ictal SPECT showed diffuse non-localizing changes. PET disclosed a right fronto-polar area of hypometabolism. A right fronto-polar/anterior medial frontal cortectomy guided by intracranial EEG findings (without insular sampling) failed to control his seizures. Four years later, he underwent a MEG study which identified a cluster of sources in the right fronto-operculo-insular region (Fig. 30.2).

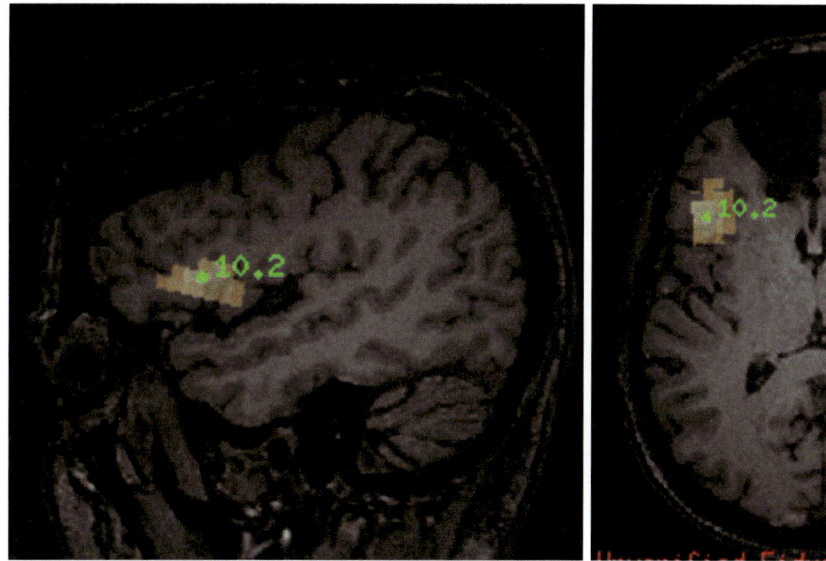
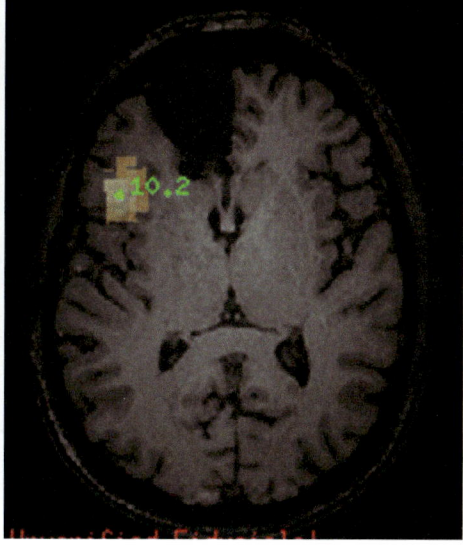

Fig. 30.2 (Case 2): MEG recording performed after first failed epilepsy surgery: spike sources were localized in the anterior insula using an electrical current dipole model (not shown); burst of focal gamma activity were localized to the same area using beamformer technique

Subsequent resection of this region led to seizure freedom (FU 4 years).

Conclusion

Although exact prevalence is yet unclear, it has become clear that a fair number of patients have an epileptogenic zone in or involving the insula. Although our knowledge of insular seizure semiology has significantly improved, the value of noninvasive tests to confirm clinical suspicion of temporo-insular or operculo-insular cases remains either limited or not yet well established. Hence, for the time being, intracranial EEG studies remains pertinent provided operculo-insular structures are adequately sampled. Further work is necessary to better establish the value of more recent noninvasive tools (such as electrical magnetic source imaging, EEG-fMRI, multiparametric quantitative magnetic resonance imaging including diffusion and relaxometry, machine learning, connectivity measures) to refine the identification of temporo-insular or operculo-insular cases, clarify indications for intracranial EEG (especially for patients with hippocampal sclerosis), and predict surgical outcome.

References

1. Kwan P, Brodie MJ. Early identification of refractory epilepsy. N Engl J Med. 2000;342(5):314–9.
2. Wiebe S, Blume WT, Girvin JP, Eliasziw M, Effectiveness, Efficiency of Surgery for Temporal Lobe Epilepsy Study Group. A randomized, controlled trial of surgery for temporal-lobe epilepsy. N Engl J Med. 2001;345(5):311–8.
3. Rosenow F, Luders H. Presurgical evaluation of epilepsy. Brain J Neurol. 2001;124(Pt 9):1683–700.
4. Spencer SS, Nguyen DK, Duckrow RB. Invasive EEG evaluation for epilepsy surgery. In: Shorvon S, Fish D, Dodson E, editors. The treatment of epilepsy. 3rd ed. Chichester: Blackwell Publishing; 2009. p. 767–98.
5. Tellez-Zenteno JF, Hernandez-Ronquillo L. A review of the epidemiology of temporal lobe epilepsy. Epilepsy Res Treat. 2012;2012:630853.
6. Beleza P, Pinho J. Frontal lobe epilepsy. J Clin Neurosci. 2011;18(5):593–600.
7. Tellez-Zenteno JF, Hernandez Ronquillo L, Moien-Afshari F, Wiebe S. Surgical outcomes in lesional and non-lesional epilepsy: a systematic review and meta-analysis. Epilepsy Res. 2010;89(2–3):310–8.
8. Harroud A, Bouthillier A, Weil AG, Nguyen DK. Temporal lobe epilepsy surgery failures: a review. Epilepsy Res Treat. 2012;2012:201651.
9. Nguyen DK, Mbacfou MT, Nguyen DB, Lassonde M. Prevalence of nonlesional focal epilepsy in an adult epilepsy clinic. Can J Neurol Sci. 2013;40(2):198–202.
10. Burch J, Hinde S, Palmer S, Beyer F, Minton J, Marson A, et al. The clinical effectiveness and cost-effectiveness of technologies used to visualise the seizure focus in people with refractory epilepsy being considered for surgery: a systematic review and decision-analytical model. Health Technol Assess. 2012;16(34):1–157. iii–iv
11. Barba C, Rheims S, Minotti L, Guenot M, Hoffmann D, Chabardes S, et al. Temporal plus epilepsy is a major determinant of temporal lobe surgery failures. Brain J Neurol. 2016;139(Pt 2):444–51.
12. Isnard J, Guenot M, Ostrowsky K, Sindou M, Mauguiere F. The role of the insular cortex in temporal lobe epilepsy. Ann Neurol. 2000;48(4):614–23.
13. Bouthillier A, Nguyen DK. Epilepsy surgeries requiring an operculoinsular cortectomy: operative technique and results. Neurosurgery. 2017;81(4):602–12.
14. Surbeck W, Bouthillier A, Nguyen DK. Refractory insular cortex epilepsy: clinical features, investigation and treatment. Future Neurol. 2010;5(4):491–9.
15. Surbeck W, Nguyen DK, Bouthillier A. Insular and peri-rolandic epilepsy surgery: techniques. In: Extratemporal lobe epilepsy surgery. London: John Libbey Eurotext; 2011. p. 371–92.
16. Mesulam MM, Mufson EJ. Insula of the old world monkey. I. Architectonics in the insulo-orbito-temporal component of the paralimbic brain. J Comp Neurol. 1982;212(1):1–22.
17. Ghaziri J, Tucholka A, Girard G, Houde JC, Boucher O, Gilbert G, et al. The corticocortical structural connectivity of the human insula. Cereb Cortex. 2017;27(2):1216–28.
18. Augustine JR. Circuitry and functional aspects of the insular lobe in primates including humans. Brain Res Brain Res Rev. 1996;22(3):229–44.
19. Boucher O, Citherlet D, Ghaziri J, Hébert-Seropian B, Von Siebenthal Z, Nguyen DK. Insula: neuropsychologie du cinquième lobe du cerveau. Rev Neuropsychol. 2017;9(3):154–61.
20. Nguyen DK, Nguyen DB, Malak R, Leroux JM, Carmant L, Saint-Hilaire JM, et al. Revisiting the role of the insula in refractory partial epilepsy. Epilepsia. 2009;50(3):510–20.
21. Isnard J, Guenot M, Sindou M, Mauguiere F. Clinical manifestations of insular lobe seizures: a stereo-electroencephalographic study. Epilepsia. 2004;45(9):1079–90.
22. Mazzola L, Isnard J, Peyron R, Mauguiere F. Stimulation of the human cortex and the experience of pain: Wilder Penfield's observations revisited. Brain J Neurol. 2012;135(Pt 2):631–40.
23. Mazzola L, Mauguiere F, Isnard J. Electrical stimulations of the human insula: their contribution to the ictal

semiology of insular seizures. J Clin Neurophysiol. 2017;34(4):307–14.
24. Penfield W, Faulk ME Jr. The insula; further observations on its function. Brain J Neurol. 1955;78(4):445–70.
25. Penfield W, Jasper H. Epilepsy and the functional anatomy of the human brain. Boston: Little, Brown; 1955.
26. Pugnaghi M, Meletti S, Castana L, Francione S, Nobili L, Mai R, et al. Features of somatosensory manifestations induced by intracranial electrical stimulations of the human insula. Clin Neurophysiol. 2011;122(10):2049–58.
27. Stephani C, Fernandez-Baca Vaca G, Maciunas R, Koubeissi M, Luders HO. Functional neuroanatomy of the insular lobe. Brain Struct Funct. 2011;216(2):137–49.
28. Nguyen DK, Nguyen DB, Malak R, Bouthillier A. Insular cortex epilepsy: an overview. Can J Neurol Sci. 2009;36(Suppl 2):S58–62.
29. Nguyen DK, Surbeck W, Weil AG, Villemure JG, Bouthillier A. Insular epilepsy: the Montreal experience. Rev Neurol. 2009;165(10):750–4.
30. Dobesberger J, Ortler M, Unterberger I, Walser G, Falkenstetter T, Bodner T, et al. Successful surgical treatment of insular epilepsy with nocturnal hypermotor seizures. Epilepsia. 2008;49(1):159–62.
31. Kaido T, Otsuki T, Nakama H, Kaneko Y, Kubota Y, Sugai K, et al. Complex behavioral automatism arising from insular cortex. Epilepsy Behav. 2006;8(1):315–9.
32. Proserpio P, Cossu M, Francione S, Tassi L, Mai R, Didato G, et al. Insular-opercular seizures manifesting with sleep-related paroxysmal motor behaviors: a stereo-EEG study. Epilepsia. 2011;52(10):1781–91.
33. Ryvlin P. Avoid falling into the depths of the insular trap. Epileptic Disord. 2006;8(Suppl 2):S37–56.
34. Guillaume J, Mazars G. Technique de résection de l'insula dans les épilepsies insulaires. Rev Neurol. 1949;81:900–3.
35. Ajmone-Marsan C, Baldwin M. Electrocorticography. In: Baldwin M, Bailey P, editors. Temporal lobe epilepsy. Springfield, IL: Ch. C Thomas; 1958.
36. Silfvenius H, Gloor P, Rasmussen T. Evaluation of insular ablation in surgical treatment of temporal lobe epilepsy. Epilepsia. 1964;5:307–20.
37. Kanazawa O, Blume WT, Girvin JP. Significance of spikes at temporal lobe electrocorticography. Epilepsia. 1996;37(1):50–5.
38. Schwartz TH, Bazil CW, Walczak TS, Chan S, Pedley TA, Goodman RR. The predictive value of intraoperative electrocorticography in resections for limbic epilepsy associated with mesial temporal sclerosis. Neurosurgery. 1997;40(2):302–9. discussion 9–11
39. Tran TA, Spencer SS, Javidan M, Pacia S, Marks D, Spencer DD. Significance of spikes recorded on intraoperative electrocorticography in patients with brain tumor and epilepsy. Epilepsia. 1997;38(10):1132–9.
40. Tran TA, Spencer SS, Marks D, Javidan M, Pacia S, Spencer DD. Significance of spikes recorded on electrocorticography in nonlesional medial temporal lobe epilepsy. Ann Neurol. 1995;38(5):763–70.
41. Chatrian GE, Quesney LF. Intraoperative electrocorticography. In: Engel J Timothy A Pedley . Epilepsy- A comprehensive Textbook. 2. Philadelphia Lippincott-Raven; 1998. p. 1749-1765.
42. Engel J Jr, Rausch R, Lieb JP, Kuhl DE, Crandall PH. Correlation of criteria used for localizing epileptic foci in patients considered for surgical therapy of epilepsy. Ann Neurol. 1981;9(3):215–24.
43. Ojemann LM, Ojemann GA, Baughbookman C. What is the optimal extent of the medial resection in anterior temporal lobe epilepsy? Epilepsia. 1986;27(5):636.
44. Cukiert A, Forster C, Andrioli MS, Frayman L. Insular epilepsy. Similarities to temporal lobe epilepsy. Case report. Arq Neuropsiquiatr. 1998;56(1):126–8.
45. Hatashita S, Sakakibara T, Ishii S. Lipoma of the insula. Case report. J Neurosurg. 1983;58(2):300–2.
46. Roper SN, Levesque MF, Sutherling WW, Engel J Jr. Surgical treatment of partial epilepsy arising from the insular cortex. Report of two cases. J Neurosurg. 1993;79(2):266–9.
47. Ryvlin P, Kahane P. The hidden causes of surgery-resistant temporal lobe epilepsy: extratemporal or temporal plus? Curr Opin Neurol. 2005;18(2):125–7.
48. Barba C, Minotti L, Job AS, Kahane P. The insula in temporal plus epilepsy. J Clin Neurophysiol. 2017;34(4):324–7.
49. Barba C, Barbati G, Minotti L, Hoffmann D, Kahane P. Ictal clinical and scalp-EEG findings differentiating temporal lobe epilepsies from temporal 'plus' epilepsies. Brain J Neurol. 2007;130(Pt 7):1957–67.
50. Aghakhani Y, Rosati A, Dubeau F, Olivier A, Andermann F. Patients with temporoparietal ictal symptoms and inferomesial EEG do not benefit from anterior temporal resection. Epilepsia. 2004;45(3):230–6.
51. Bouilleret V, Valenti MP, Hirsch E, Semah F, Namer IJ. Correlation between PET and SISCOM in temporal lobe epilepsy. J Nucl Med. 2002;43(8):991–8.
52. Hur JA, Kang JW, Kang HC, Kim HD, Kim JT, Lee JS. The significance of insular hypometabolism in temporal lobe epilepsy in children. J Epilepsy Res. 2013;3(2):54–62.
53. Tae WS, Joo EY, Kim JH, Han SJ, Suh YL, Kim BT, et al. Cerebral perfusion changes in mesial temporal lobe epilepsy: SPM analysis of ictal and interictal SPECT. NeuroImage. 2005;24(1):101–10.
54. Chassoux F, Artiges E, Semah F, Laurent A, Landre E, Turak B, et al. 18F-FDG-PET patterns of surgical success and failure in mesial temporal lobe epilepsy. Neurology. 2017;88(11):1045–53.
55. Kahane P, Huot JC, Hoffmann D, Lo Russo G, Benabid AL, Munari C. Perisylvian cortex involvement in seizures affecting the temporal lobe. In: Avanzini G, Beaumanoir A, Mira L, editors. Limbic seizures in children. Eastleigh: John Libbey & Company Ltd; 2001. p. 115–27.
56. Ribaric I, Sekulovic N. Experience with orbital electrodes in the patients operated on for epilepsy--results

of temporofrontal resections. Acta Neurochir Suppl (Wien). 1989;46:21–4.
57. Obaid S, Zerouali Y, Nguyen DK. Insular epilepsy: semiology and noninvasive investigations. J Clin Neurophysiol. 2017;34(4):315–23.
58. Freri E, Matricardi S, Gozzo F, Cossu M, Granata T, Tassi L. Perisylvian, including insular, childhood epilepsy: presurgical workup and surgical outcome. Epilepsia. 2017;58(8):1360–9.
59. Hagiwara K, Jung J, Bouet R, Abdallah C, Guenot M, Garcia-Larrea L, et al. How can we explain the frontal presentation of insular lobe epilepsy? The impact of non-linear analysis of insular seizures. Clin Neurophysiol. 2017;128(5):780–91.
60. Levy A, Tran Phuoc Y, Boucher O, Bouthillier A, Nguyen DK. Operculo-insular epilepsy: scalp and intracranial electroencephalographic findings. J Clin Neurophysiol. 2017;34(5):438–47.
61. Ryvlin P, Minotti L, Demarquay G, Hirsch E, Arzimanoglou A, Hoffman D, et al. Nocturnal hypermotor seizures, suggesting frontal lobe epilepsy, can originate in the insula. Epilepsia. 2006;47(4):755–65.
62. Tayah T, Savard M, Desbiens R, Nguyen DK. Ictal bradycardia and asystole in an adult with a focal left insular lesion. Clin Neurol Neurosurg. 2013;115(9):1885–7.
63. Tran TP, Truong VT, Wilk M, Tayah T, Bouthillier A, Mohamed I, et al. Different localizations underlying cortical gelastic epilepsy: case series and review of literature. Epilepsy Behav. 2014;35:34–41.
64. Chevrier MC, Bard C, Guilbert F, Nguyen DK. Structural abnormalities in patients with insular/peri-insular epilepsy: spectrum, frequency, and pharmacoresistance. AJNR Am J Neuroradiol. 2013;34(11):2152–6.
65. Mortati KA, Arnedo V, Post N, Jimenez E, Grant AC. Sutton's law in epilepsy: because that is where the lesion is. Epilepsy Behav. 2012;24(2):279–82.
66. Fei P, Soucy J-P, Obaid S, Boucher O, Bouthillier A, Nguyen DK. The value of rCBF SPECT and FDG PET in operculo-insular epilepsy submitted to Clinical Nuclear Imaging. 2017.
67. Heers M, Rampp S, Stefan H, Urbach H, Elger CE, von Lehe M, et al. MEG-based identification of the epileptogenic zone in occult peri-insular epilepsy. Seizure. 2012;21(2):128–33.
68. Mohamed IS, Gibbs SA, Robert M, Bouthillier A, Leroux JM, Khoa Nguyen D. The utility of magnetoencephalography in the presurgical evaluation of refractory insular epilepsy. Epilepsia. 2013;54(11):1950–9.
69. Park HM, Nakasato N, Tominaga T. Localization of abnormal discharges causing insular epilepsy by magnetoencephalography. Tohoku J Exp Med. 2012;226(3):207–11.
70. Zerouali Y, Pouliot P, Robert M, Mohamed I, Bouthillier A, Lesage F, et al. Magnetoencephalographic signatures of insular epileptic spikes based on functional connectivity. Hum Brain Mapp. 2016;37(9):3250–61.
71. Aitouche Y, Gibbs SA, Gilbert G, Boucher O, Bouthillier A, Nguyen DK. Proton MR spectroscopy in patients with nonlesional insular cortex epilepsy confirmed by invasive EEG recordings. J Neuroimaging. 2017;27(5):517–23.
72. Ryvlin P, Picard F. Invasive investigation of insular cortex epilepsy. J Clin Neurophysiol. 2017;34(4):328–32.
73. Gras-Combe G, Minotti L, Hoffmann D, Krainik A, Kahane P, Chabardes S. Surgery for nontumoral insular epilepsy explored by stereoelectroencephalography. Neurosurgery. 2016;79(4):578–88.
74. Weil AG, Le NM, Jayakar P, Resnick T, Miller I, Fallah A, et al. Medically resistant pediatric insular-opercular/perisylvian epilepsy. Part 2: outcome following resective surgery. J Neurosurg Pediatr. 2016;18(5):523–35.

31

Neuropsychology in Insular Lesions Prior-During and After Brain Surgery

Barbara Tomasino, Dario Marin, Tamara Ius, and Miran Skrap

31.1 Neuropsychology of the Insula Cortex in Neurosurgical Patients

For its connectivity, the insula has been defined as an integrative multimodal area [1–3]. Tracing studies in nonhuman primates [4, 5] evidenced that the middorsal sector receives input from the thalamic taste area [6], the anterior-basal sector, connected with limbic areas, is part of a frontal-entorhinal, piriform, and olfactory cortex network, and the mid-posterior sector is connected with somatosensory areas [4, 5]. This indicates that a neuropsychological examination of patients with lesions involving the insula should assess the emotion-related processing and sensorimotor sympthoms or olfactory/gustatory related changes, in addition to assessing their cognitive processing. In a quantitative meta-analysis of functional activations found in the insula [3], it has been found that emotional-social tasks activate the anterior-ventral insula and the right central region, gustatory-olfactory stimuli activate the right central region, and the anterior-dorsal insula is activated by cognitive tasks. In the same study [3], a conjunction analysis showed that the anterior-dorsal insula was activated by social-emotional, gustatory-olfactory, and cognitive tasks.

However, the neuropsychological symptoms observed in patients with lesions to the insular cortex are often difficult to interpret. First, selective lesions affecting the insula alone are extremely rare. Thus, the neuropsychological deficit observed could be due to the lesion affecting surrounding areas or to the subcortical connections passing through the insula. Data on stroke patients indicated that only 4 out of 4800 examined lesions [7] or, in another study, 0 out of 72 examined lesions [8] selectively involved the insula without involvement of surrounding areas. Neurosurgical lesions are usually more selective than those due to stroke, the interpretation of cognitive deficits observed in patients with glioma affecting the insular cortex might consider the connectivity between the insula and other areas.

The importance of neuropsychological evaluation in patients undergoing the resection of a tumor involving the insula is related to both the functional integration nature of this area and to neurosurgical aspects of performing a resection in this area.

B. Tomasino (✉) · D. Marin
IRCCS "E. Medea",
San Vito al Tagliamento, Pordenone, Italy
e-mail: btomasino@ud.lnf.it

T. Ius · M. Skrap
Unità Operativa di Neurochirurgia,
Azienda Sanitaria Universitaria Integrata
S. Maria della Misericordia, Udine, Italy

31.1.1 Insula of Reil Surgical Considerations

Despite the recent advantages in microsurgical and brain mapping techniques, the insula still remains to be a challenging area for neurosurgical resection due to complex anatomy. Insular gliomas have a clear propensity to spread along the intricate network of the afferent and efferent connections within the insula and with the cortical structure [9], involving the subcortical pathways and numerous critical vascular structures. These represent the major limitations in obtaining total tumoral resection [10, 11–13]. Notwithstanding these technical difficulties, there is growing evidence in literature that the extent of tumoral resection tends to provide increased overall survival and progression-free survival, in addition to improving seizure control [10, 11, 12, 14].

There are several basic surgical approaches to insular tumors: (1) transsylvian, (2) transcortical (transfrontal or transtemporal), and (3) combined (transcortical + transsylvian). The transsylvian approach was used in the innovative work of M. Yaşargil et al. [15]. However, a review of studies suggest that, there is no general consensus about the best approach in terms of safety and maximal resection.

It is well known that the transsylvian approach is associated with the possibility of injury to veins and arteries of the Sylvian fissure, which leads to ischemia and, as a consequence, to postoperative deterioration of neurologic functions. Traction of the opercular area during this approach can also lead to postoperative deterioration [16].

Thus, some authors used the transsylvian approach only for isolated insular tumors [13, 17]. In cases where tumors spread to the frontal and temporal areas, they began resection with transcortical approach and only then used transsylvian approach. Other authors preferred to use only the transsylvian approach even for fronto-insulo-temporal tumors [15].

At the beginning of this series, we used a transsylvian approach, which followed the methodological technique described by Yasargil et al. [15].

Afterward, on the bases of data provided by cortical mapping, we have preferred to use a transcortical transfrontal and/or transtemporal approach on the basis of the "windowing technique" described by Berger et al. [12].

This technique allowed us to obtain more operative space, which made it easier to dissect the *medial cerebral artery* and its branches subpially. We preferred to do this at the end of the procedure. By increasing the space around the vessels, we decreased the risk of evoking pain in manipulating the MCA in an early phase of surgery. Finally, to reduce the risk of ischemia, we now prefer to leave a minimal part of the tumor around the lenticulostriate arteries (*LSAs*) in case they are encased.

Regarding the preoperative surgical planning, in all cases, we overlapped the fMRI and DTI data and analyzed the 3D relationship between the tumor and the cortical funcional areas and subcortical pathways. The images were loaded in the Neuronavigation System (Stealth Station, Medtronic, USA; or BrainLab Curve 1.0 Cranial 3.x).

Regarding the selection of the anesthesiological protocol, it was based on the preoperative evaluation of hemispheric dominance. Awake craniotomy was performed in all dominant locations, following the methodology previously described by Skrap and colleagues [10]. Awake craniotomy was performed also in the right hemisphere, when patients were highly motivated and collaborated. The procedure is interesting for the neuropsychological aspect in light of the very complex network of this area.

In the present chapter, we considered a patient series for which all surgical procedures were conducted under intraoperative cortical and subcortical electrical stimulation (IES), following the intraoperative methodology previously described by Berger and colleagues [11]. Before the resection was started, with the help of the navigator and the bipolar IES, we put some methylene blue-colored small markers in the depth along the medial border of the temporal and frontal extension of the lesion to have a better anatomical orientation during the most advanced phases of surgery, when brain shifting occurs.

During the cortical and subcortical brain mapping, we rarely needed more than 4 mA of our current intensity for the cortex, while we usually started with 6 or 8 mA for subcortical simulation. Proceeding to the depth, we alternated resection of thin tumoral layers with subcortical stimulation (to detect the pyramidal tract) particularly in the posterior part of the lesion and subcortical pathways of language (AF and IFOF) for tumors harboring on the dominant hemisphere.

Moreover, in all cases, we routinely used the neurophysiological monitoring of motor-evoked potentials (MEPs) and somatosensory-evoked potentials (SEPs), which constitutes a tool in guiding the extent of resection and preventing or minimizing direct injury to the posterior limb of the internal capsule and the superior limit toward the corona radiate (64-channel Eclipse Neurovascular Workstation, Axon Systems, Inc.; 32-channel video polygraphic station, Brain Quick SystemPlus, MicroMed).

Changes in MEP amplitude represent a warning sign, while sudden loss or a significant reduction of MEP amplitude probably indicates that an injury to perforating vessels has already occurred. In our experience, MEP is useful to predict a direct trauma of the fibers, but they do not guarantee prevention of vascular damage. For ECoG, silicon strips with four or eight electrodes and an intercontact distance of 10 mm were placed on the exposed lesional tissue and its surroundings, after opening of the dura mater. ECoG was recorded during a pre-resection phase (for at least 10 min), a resection phase, and at the end of resection (for at least 10 min). The low-frequency filter was set at 1 Hz, the high-frequency filter was set at 80 Hz, and the gain was between 200 and 400 mV, depending on the amplitude of the background and discharges. We think that the use of intraoperative ECoG is essential to record after discharge phenomena (ADs), electrical and clinical seizures, particularly if they are short-lasting focal seizures that may interfere with responses induced by mapping with the loss of the patient's collaboration.

In the present chapter, our described experience is based on a patient population operated on within the last 10 years.

31.2 Intra-surgery Neuropsychology

Studies reporting neuropsychological testing during surgery exclusively regard the description of the effects of direct cortical stimulation (DES) mappings of the insula and surrounding areas. These studies mainly report visceral/somesthetic/emotion-related symptoms evoked by stimulation. For example, gustatory and olfactory sensations were evoked by intraoperative direct stimulation of the anterior insula [18], viscerosensitive and visceromotor responses were evoked by stimulation of the anterior insula [19], while somesthetic sensations were evoked by posterior insula stimulation [19].

On the cognitive side, DES studies have suggested that the insula forms part of the language network [20]. For example, dysarthria and speech arrest [11, 21] and hypophonia [22] have been observed during routine DES suggesting an insular role in spontaneous speech [19, 23].

None of the studies addressed cognitive status while resection occurs. We recently developed a new neuropsychological method called real-time neuropsychological testing (RTNT) [24] that allows the immediate detection of neuropsychological dysfunction while resection occurs. In the RTNT, as previously described [24], we selected a series of cognitive tasks according to the functional role of the insula, fMRI and DTI results, and to patients' preoperative neuropsychological profile. The task sequence followed a fixed order. The sequence of tasks was repeated (presenting a different stimuli list for each sequence) until the end of the resection. Each task had a duration of about 30 s/1 min. In this way, task assessment and task switching were quick and dynamic for immediate dysfunction detection. As soon as the patient exhibited a decrement, the neurosurgeon was immediately informed. If the RTNT showed a decrease in a test, the preordered task sequence was interrupted, and the neuropsychologist followed the principles of differential diagnosis. For example, if a patient manifests a decrease in nonword repetition, word repetition and speech comprehension are assessed to understand the nature of the decrease in performance.

31.2.1 Real-Time Neuropsychological Testing

The list of tasks we use for patients with a right temporo-insular glioma involves the following task (Table 31.1).

The list of tasks we use for patients with a left temporo-insular glioma is shown in Table 31.2.

31.2.2 Example of RTNT During a Right Anterior Insular Resection

Here, we report as an example a single case of a patient undergoing a right anterior insular resection who performed the RTNT. SL is a 52-year-old, right-handed, administrative employee, with 13 years of schooling. Recently she complained of sissiness and nausea. She later felt a sense described as "confusion" and "panic" that after some minutes disappeared. After some months of persistence of these symptoms, SL underwent a neurological visit and MRI investigation. Magnetic resonance revealed the presence of a right temporo-insular lesion (see Fig. 31.1). After the discovery of the lesion, the patient started the pharmacological treatment with antiepileptic drugs that stopped the partial seizures (with neurovegetative symptoms) reported above. This case is an interesting example of a voluminous LGG involving temporo-insular cortex that evidenced good neuropsychological performances during all RTNT (see Table 31.3). However, during the RTNT the patient showed some "non-cognitive" symptoms that resolved. Among them, the tendency to fall asleep and a strange taste in her mouth.

Interestingly, the patient's neuropsychological evaluations before and after (at one week and after four months) surgery confirmed a normal cognitive profile (see Table 31.4).

Table 31.1 List of RTNT tasks for a right insular resection

Task n.	Task	N. Items	Instructions
1	Milner landmark test [25]	14	The patient sees bisected lines and must decide which part of the line is longer or shorter. The answer is verbal
2	*Emotion processing* [26]	10	The patient sees color photographs representing positive (such as a happy child), neutral (an object), or negative (a car destroyed after an accident) scenes. Two Likert scales ranging from 1 to 9 are presented below each picture. The patient has to say how much the picture he sees makes her/him happy-sad (valence) and how much the same image makes her/him agitated-calm (arousal)
3	Picture reality decision [27]	10	The pictures are presented one at a time, asking the subject about the existence of the exemplar
4	Visual recognition and identification of celebrities [28]	10	Ten photographs (five of famous characters, five of unknown people) are presented. The patient is asked to say whether the person is famous or not: if he/she recognizes his/her as famous, he/she must say the name
5	Visual scanning: reading words	8 rows	The patient is asked to read the words that appear on the screen from the first to the last row starting from left to right. For each stimulus array, 21 words are displayed
6	Visual scanning: objects	1 picture	The patient is asked to describe all the visual elements he sees in a full screen scene
7	Visual scanning: reading sentences	8 rows	The patient is asked to read some sentences that are displayed in different positions of the screen
8	Theory of mind (emotion recognition) [29]	10	The patient reads short stories with a protagonist. In the end, he must say the emotion felt by the character of the story in the situation
9	Clock test [30]	10	The patient is told 1 h. He is asked to indicate whether the hour and minute hands are located on the right, left, or on both clock faces

Table 31.2 List of RTNT tasks for a left insular resection

Task no.	Task	N. items	Instructions
1	Naming nouns [31]	10	Say the correct name of the picture representing an animal/plant/object he/she sees
2	Reading words [32]	10	Read the words he/she sees
3	Repetition of words [32]	10	The neuropsychologist pronounces a word and asks to the patient to repeat it correctly
4	Auditory grammar comprehension [33]	10	The patient sees two scenes. The neuropsychologist reads a sentence (e.g., the dog follows the ball) and the patient has to tell which of the two scenes corresponds to the one read
5	Reading pseudowords [32]	10	Read the pseudowords he/she sees on the screen
6	Repetition of pseudowords [32]	10	The neuropsychologist pronounces a pseudoword and asks the patient to repeat it correctly
7	Phonological discrimination, adapted from [34]	16	On the screen there are four figures that have a similar phonological sound (e.g., bat cat rat fat). The neuropsychologist reads one of the four words and the patient must indicate the correct one
8	Digit span Digits forward and digit span backward [35]	9	The examiner reads pairs of numbers in a pre-fixated sequence; when the sequence is repeated by the subject correctly, the examiner reads the next one, which is longer than a number from the previous one, and continues until the subject fails a sequence of pairs
9	Visual lexical decision	10	Words and pseudowords are presented randomly, one at a time. The patient has to decide if the word he/she reads is a real word
10	Naming verbs [31]	10	Say the correct name of the action that he/she sees on the screen
11	Adapted token test	5	The patient sees on the screen five squares and five circles of five different colors (red, green, yellow, and black). Each stimulus corresponds to a number from 1 to 10. The neuropsychologist asks which number matches the mentioned stimuli (e.g., red circle and black square)
12	Production of narratives	3	The patient sees on the screen four ordered pictures that represent a story. The patient has to tell a story that has logic and follows the images

31.2.3 Example of RTNT During a Left Anterior Insular Resection

We report, as an example, a single case of a patient undergoing a left anterior insular resection who performed the RTNT (see Table 31.5). RP is a 35-year-old, right-handed, housewife, with 13 years of schooling. During sleep, she suffered a generalized seizure with loss of consciousness. She was hospitalized for a left low-grade glioma involving the temporo-insular cortex (see Fig. 31.2). This is an interesting example of dissociation between language functions that remain stable for all the RTNT and the emergence of pain due to partial seizures arising from the temporo-insular area, as evinced by the electrocorticography (ECoG).

When the patient was reevaluated at one week after surgery, she could not remember the sensation of pain described during the partial seizures reported above. The neuropsychological profile at the immediate postsurgery (at one week) and at follow-up (fourmonths after surgery) remained as at one week before surgery (see Table 31.6).

31.3 Emotional Processing Evaluation

The emotional sphere is one additional function that we address while testing patients with insular lesions.

It is well known that emotion regulation is normally regulated in the human brain by a complex circuit that includes several regions of

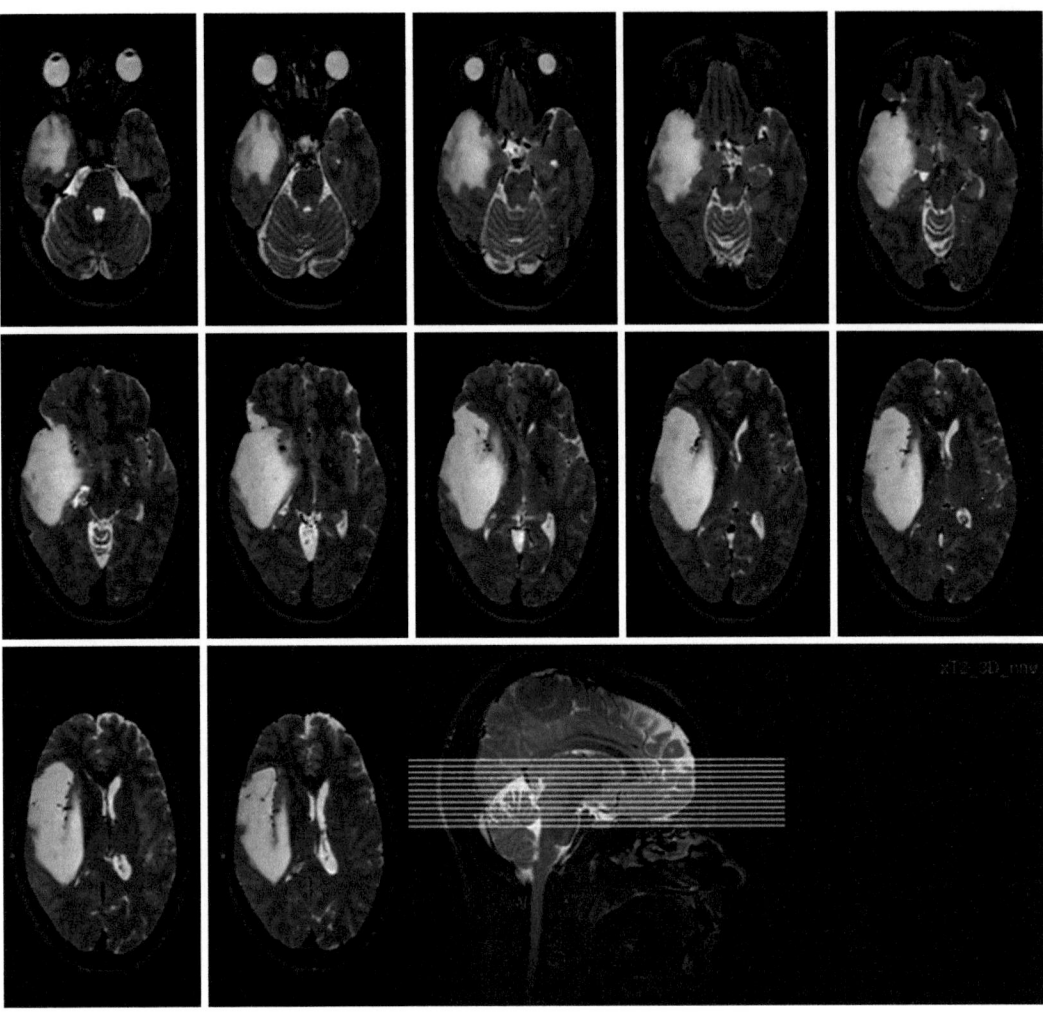

Fig. 31.1 Patient's T2-weighted MRI image showing a right temporo-insular lesion

Table 31.3 RTNT performance of patient SL with a lesion to the right insular cortex

RTNT	1 run of RTNT	2 run of RTNT	3 run of RTNT	4 run of RTNT
The Milner Landmark test	10/14 sleepiness	12/14	13/14	12/141
Emotion processing (valence) (arousal)	(9/10) (9/10)	(8/10) (9/10)	(9/10) (10/10)	(9/10) (8/10) sleepiness
Picture reality decision	9/10	10/10	8/10	10/10
Visual recognition and identification of celebrities	6/10	7/10	7/10	6/10
Visual scanning (Reading words)	21/21	21/21	21/21	21/21
Visual scanning (scenes)	10/10	10/10	10/10 pain	10/10
Visual scanning (reading sentences)	3/3	3/3	3/3 strange taste sensation	end of resection
Theory of mind	8/10	8/10	8/10	
Clock test (imaginative section)	8/10	10/10	8/10	

Table 31.4 Pre, immediate post, and follow-up performance of patient SL with a lesion to the right insular cortex

Neuropsychological test	Before surgery (raw data)	±	1 week after surgery	±	Follow-up (4 months)	±
Raven colored matrices	34/36	+	–		34/36	+
Spatial orientation	5/5	+	5/5	+	5/5	+
Temporal orientation	5/5	+	5/5	+	5/5	+
Corsi span forward	4/9	+	5/9	+	5/9	+
Corsi span backward	4/9	+	4/9	+	4/9	+
Constructional apraxia	14/14	+	14/14	+	14/14	+
Little man (mental rotation)	32/32	+	31/32	+	31/32	+
Apple cancelation test	50/50	+	50/50	+	50/50	+
Balloon test (A)	22/22	+	22/22	+	22/22	+
Balloon test (B)	21/22	+	20/22	+	21/22	+
Line bisection (BIT)	9/9	+	8/9	+	9/9	+
Reading neglect (BIT)	28/28	+	28/28	+	28/28	+
The Milner landmark test (PB)	54.8	+	40.5	–	53.2	+
The Milner landmark test (RB)	52.4	+	50.0	+	50.4	+
Clock test	10/10	+	10/10	+	10/10	+
Clock test (imaginative section) (BIP)	30/30	+	30/30	+	30/30	+
Clock test (perceptive section) (BIP)	30/30	+	30/30	+	30/30	+
TMT-A	25″; 0e	+	26″; 0e	+	28″; 0e	+
TMT-B	64″; 0e	+	45″; 0e	+	53″; 0e	+
Digit symbol substitution test	43 (90″)	+	42 (90″)	+	44 (90″)	+
Rey complex figure test (copy)	36/36	+	36/36	+	36/36	+
Rey complex figure test (recall)	10/36	+	22/36	+	20/36	+
BORB object decision	30/32	+	28/32	+	29/32	+
Visual recognition and identification of celebrities	61/78	+	72.5/78	+	72.5/78	+

Table 31.5 RTNT performance of patient RP with a lesion to the left insular cortex

RTNT	1 run of RTNT	2 run of RTNT	3 run of RTNT	4 run of RTNT	5 run of RTNT
Naming nouns	10/10	10/10	9/10	10/10	
Reading words	10/10	10/10 Sleepiness	10/10	10/10	
Repetition words	10/10	10/10	10/10	End of resection	
Reading pseudowords	10/10	10/10	10/10		
Repetition pseudowords	10/10	10/10	10/10		
Phonological discrimination	16/16	16/16	15/16		
Digit span forward	4/9	4/9	4/9 Pain—seizure		
Visual lexical decision	9/10	9/10	10/10		
Naming verbs	10/10	10/10	10/10		
Token (revised)	9/10 Pain—seizure	10/10	9/10		

the prefrontal cortex, the amygdala, hippocampus, hypothalamus, anterior cingulate cortex, insular cortex, ventral striatum, and other interconnected structures. Many patients during clinical interviews prior to surgery complain of a series of physical/emotional symptoms that may be related to a lesion involving insular areas. For example, patients describe a series of symptoms related to neurovegetative reactions such as increased heart rate, excessive sweat-

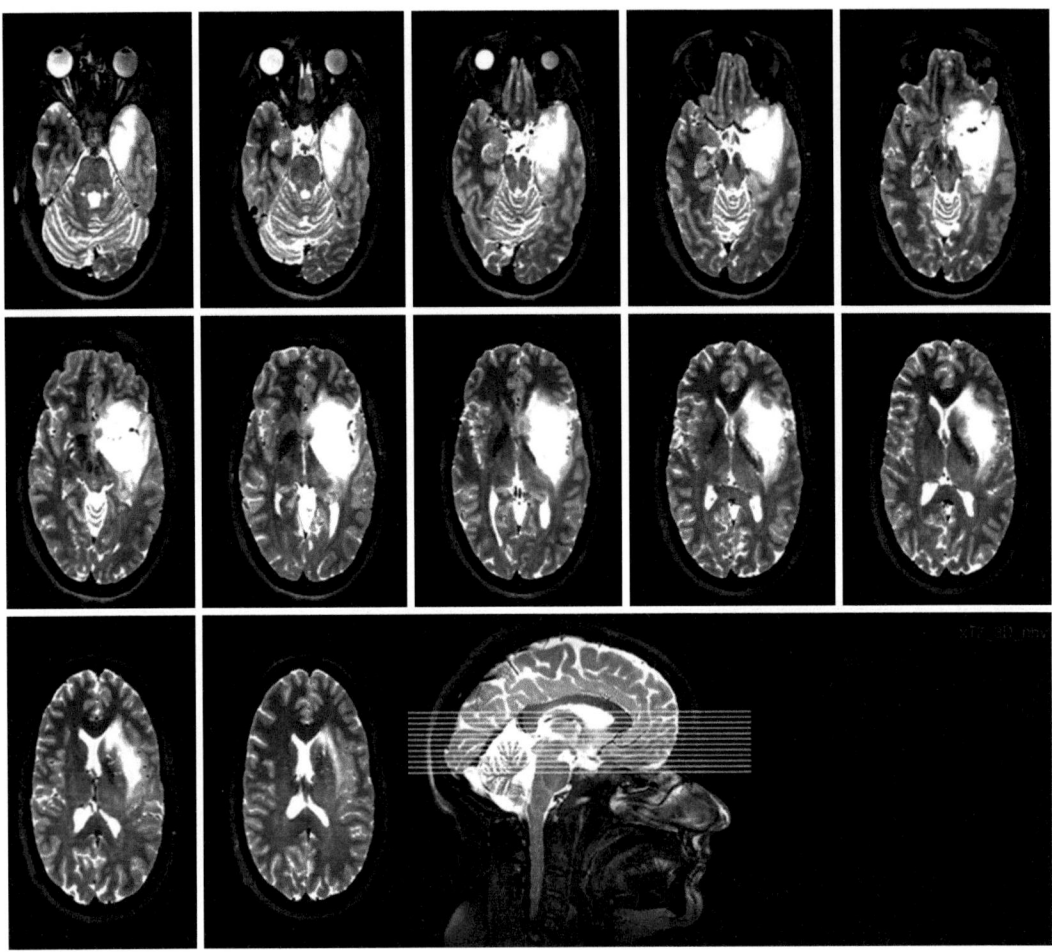

Fig. 31.2 Patient's T2-weighted MRI showing a left temporo-insular lesion. Images are shown in radiological convention

ing, dizziness, stomach discomfort, and confusion which may be a result from partial seizures involving the insula and/or from anxiety reactions linked to a stressful period. It is important to understand whether the symptoms are of organic origin or due to a psychological reaction due to the discovery of the disease. We used the Symptom Checklist-90-Revised (SCL-90-R). This scale is easy to administer and has rapid filling times (about 10 min). The SCL-90-R is a 90-item self-report symptom inventory developed by L. R. Derogatis to measure psychological symptoms and psychological distress. It is planned to be suitable for use with individuals from the community, as well as individuals with either medical or psychiatric conditions. The SCL-90-R assesses psychological distress in terms of ten primary symptom dimensions. The principal symptom dimensions are categorized as somatization (SOM), obsessive-compulsive (O-C), interpersonal sensitivity (INT), depression (DEP), anxiety (ANX), hostility (HOS), phobic anxiety (PHOB), paranoid ideation (PAR), psychoticism (PSY), and sleep difficulties (SLEEP).

We administered the SCL-90-R to patients with left and right insular lesions. In Table 31.7, we report as an example the SCL-90-R performance of a representative group of patients with right and left lesion. Table 31.7 highlights the

Table 31.6 Pre, immediate post, and follow-up performance of patient RP with a lesion to the left insular cortex

Neuropsychological test	Before surgery (raw data)	±	1 week after surgery	±	Follow-up (4 months)	±
Raven colored matrices	30/36	+	–		31/36	+
Digit span forward	5/9	+	5/9	+	5/9	+
Digit span backward	4/8	+	4/8	+	4/8	+
Orofacial apraxia	20/20	+	20/20	+	20/20	+
Ideomotor apraxia	72/72	+	72/72	+	72/72	+
Letter fluency (3 min)	28	+	28	+	27	+
Category fluency (3 min)	33	+	26	+	31	+
Naming nouns	29/30	+	28/30	+	29/30	+
Naming verbs	26/28	+	26/28	+	27/28	+
Reading words	80/80	+	80/80	+	80/80	+
Reading pseudowords	20/20	+	20/20	+	20/20	+
Reading sentences	6/6	+	6/6	+	6/6	+
Repetition words	80/80	+	80/80	+	80/80	+
Repetition pseudowords	20/20	+	20/20	+	20/20	+
Repetition sentences	20/20	+	20/20	+	20/20	+
Writing words	157/158	+	155/158	+	157/158	+
Writing pseudowords	25/25	+	21/25	–	25/25	+
Auditory lexical decision	117/127	–	126/127	+	127/127	+
Visually lexical decision	122/144	–	128/144	–	144/144	+
Token test	35/36	+	34/36	+	35/36	+
Nouns comprehension	40/40	+	40/40	+	40/40	+
Verbs comprehension	20/20	+	20/20	+	20/20	+
Palm and pyramid test	50/52	+	50/52		51/52	+
Rey complex figure test (copy)	36/36	+	36/36	+	–	
Rey complex figure test (recall)	25/36	+	21/36	+	–	
Rey auditory verbal learning test (immediate)	45/75	+	38/75	+	50/75	+
Rey auditory verbal learning test (recall)	9/15	+	6/15	–	9/15	+

main symptom: sleep disturbances. In the remaining scales, patients with right lesions exhibit higher scores in obsessive-compulsive disorders and somatization, while patients with left brain lesions exhibit more internalizing symptoms (anxiety and depression symptomatology).

Our experience allowed highlighting how often the emotional symptomatology reported by patients corresponded to partial epileptic seizures at the beginning is misunderstood. Usually, the symptomatology, following the onset of pharmacological therapy (antiepileptic drugs), doesn't recur. In fact, the scores on the SCL90R scale (which is given a week before surgery, when patients are already in antiepileptic therapy) show tendencies of low scores in all measured dimensions.

31.4 Discussion and Conclusions

Symptoms recorded prior to surgery of insular glioma often involve manifestations such as sensations of dizziness, a sense of vacuum followed by indisposition and nausea, and "confusion" and "panic" that after some minutes disappear. This was also the case of the patients described above, and it is consistent with the description reported in a previous study [36] investigating the effects of anterior insula lesion of the left hemisphere (LH) in 22 patients and of the right hemisphere (RH) in 18 patients. In that study, it was found that presurgery the symptomatology involved anxiety/fear/panic attacks (38.88 of the RH and 18.18% of the LH), taste sensations (27.77% of

Table 31.7 SCL90R scale performance of left and right insular patients (LEFT and RIGHT, respectively) at the different subscales: somatization (SOM), obsessive-compulsive (O-C), interpersonal sensitivity (INT), depression (DEP), anxiety (ANX), hostility (HOS), phobic anxiety (PHOB), paranoid ideation (PAR), psychoticism (PSY), and sleep difficulties (SLEEP)

GROUP	(SOM)	(O-C)	(INT)	(DEP)	(ANX)	(HOS)	(PHOB)	(PAR)	(PSY)	(SLEEP)
LEFT1	0.00	0.10	0.00	0.08	0.30	0.00	0.14	0.00	0.10	0.33
LEFT2	1.50	1.40	0.33	1.46	1.20	0.33	0.00	0.67	0.40	1.00
LEFT3	1.42	1.80	0.78	1.23	1.70	1.00	1.29	0.67	1.80	1.00
LEFT4	0.42	0.80	0.67	1.15	1.20	1.17	0.29	0.83	1.10	2.67
LEFT5	2.00	1.30	2.56	2.38	2.70	2.00	2.14	2.17	1.50	2.33
LEFT6	0.92	1.50	0.44	0.62	1.60	0.50	0.71	0.17	0.50	3.00
LEFT7	0.58	0.60	0.56	0.77	0.60	0.50	0.00	0.67	0.30	1.00
LEFT8	0.50	0.40	1.00	0.00	0.30	0.33	0.00	1.00	0.10	0.33
LEFT9	0.50	0.20	0.11	0.62	0.50	0.00	0.00	0.17	0.50	0.67
LEFT10	0.00	0.50	0.44	0.77	0.40	0.33	0.00	0.17	0.10	1.00
M	0.78	0.86	0.69	0.91	1.05	0.62	0.46	0.65	0.64	1.33
SD	0.66	0.57	0.72	0.70	0.78	0.61	0.73	0.63	0.61	0.97
RIGHT1	0.75	1.40	0.11	0.38	0.40	0.17	0.00	0.50	0.20	0.67
RIGHT2	0.50	0.30	0.67	0.31	0.30	0.50	0.14	0.50	0.30	1.00
RIGHT3	0.42	0.10	0.00	0.15	0.00	0.00	0.00	0.00	0.00	0.00
RIGHT4	0.50	0.10	0.44	0.23	0.20	0.33	0.14	0.00	0.20	0.33
RIGHT5	2.25	1.60	0.78	1.46	1.30	0.50	0.57	0.83	0.70	4.00
RIGHT6	0.50	0.60	0.44	1.15	0.90	0.83	0.14	0.33	0.60	1.00
RIGHT7	1.17	1.30	0.00	1.00	0.70	0.00	0.43	0.33	0.60	1.33
RIGHT8	0.67	1.40	1.00	1.54	1.20	1.67	1.00	1.33	1.20	1.00
RIGHT9	0.17	0.70	0.22	0.23	0.40	0.17	0.00	0.33	0.60	0.67
RIGHT10	1.08	1.20	2.44	1.23	1.40	0.67	1.86	0.17	1.30	1.67
M	*0.80*	*0.87*	*0.61*	*0.77*	*0.68*	*0.48*	*0.43*	*0.43*	*0.57*	*1.17*
SD	*0.59*	*0.58*	*0.73*	*0.56*	*0.50*	*0.50*	*0.59*	*0.40*	*0.42*	*1.10*

the RH and 4.5% of the LH), olfactory sensations (11.11% of the RH), gastric sensations (22.22% of the RH and 9% of the LH), auditory sensations (11.11% of the RH), paresthesia (22.22% of the RH), and motor symptoms (27.27% of the LH).

On the cognitive side, our patients did not show impairments. This is not consistent with previous data [36] showing that presurgery 31.81% of the LH patients had a noun naming deficit, 36.26% had a verb naming deficit, and 27.27% had an impaired performance at phonological fluency. RH patients, with the exception of one case showing an impaired performance at the visuospatial planning test, were within the normal range in performing all the tests. One possibility is that there is preoperative reshaping due to plasticity [37] so that patients are proficient at the preoperative examination. Low-grade glioma grows slowly, and this allowed for a potential reshaping.

Our preliminary experience indicates that the RTNT evidences a proficient cognitive performance during resection in insular cortex. The changes we observe often involve sleepiness and pain complaining. Similarly, in their presurgical profile (as well as postsurgery), the two indicative cases reported above showed a proficient cognitive functioning. This result is not fully comparable with data from literature since no intra-resection neuropsychological testing has been reported yet. The type of data one can record during DES represents the effect of stimulation on the insula are different from the type of data one can record while resection in the insula occurs [24].

Postsurgery often it is possible to observe neuropsychological deficits. In one study [36] at follow-up, the examined patients ($N = 9$ LH and $N = 8$ RH) were still pathological. To date, the impaired patients had lesions involving other regions in addition to the anterior; thus, a direct causal relation between the deficit and the presence of a lesion involving the anterior insula cannot be proved. In another study [10] of a series of 66 neurosurgical patients (19 right insula and 47 left insula) with insular nonenhancing gliomas, language disorders were reported in the immediate postoperative phase in 11 patients (16.7 articulatory disorders in 1 patient, phonemic paraphasia without comprehension deficit in 8 patients, speech disorders with comprehension deficits in 2 patients). At a three-month follow-up examination, only 3% of the patients had speech disorders [10]. Similarly, in another study, it has been reported that 10/42 neurosurgical patients with low-grade insular glioma had articulatory disorders immediately following resection, which resolved at a three month postsurgery examination [38]. In the two cases we reported in the present chapter, there were no postsurgical neuropsychological deficits.

However, on the emotion processing side, we reported, an example, the SCL-90-R performance of a representative group of patients with right and left lesion. It was found that patients with right lesions exhibit higher scores in obsessive-compulsive disorders and somatization, while patients with left brain lesions exhibit more internalizing symptoms (anxiety and depression symptomatology).

One interesting approach to understanding these differences in the literature showing pathological neuropsychological performance, or spared cognitive functioning, or emotional processing changes, would be considering the functional parcelization of the insula. It has been found [3] that different sectors of the insula are involved in different functions: emotional-social tasks activate the anterior-ventral insula and the right central region, gustatory-olfactory stimuli activate the right central region, and the anterior-dorsal insula is activated by cognitive tasks. In the same study [3], a conjunction analysis showed that the anterior-dorsal insula was activated by social-emotional, gustatory-olfactory, and cognitive tasks. Therefore, these data deserve further investigation by analyzing the anatomical portion involved in the pathological tissue.

References

1. Chen LM. Imaging of pain. Int Anesthesiol Clin. 2007;45:39–57.
2. Craig AD. How do you feel? Interoception: the sense of the physiological condition of the body. Nat Rev Neurosci. 2002;3:655–66.
3. Kurth F, Zilles K, Fox PT, Laird AR, Eickhoff SB. A link between the systems: functional differentiation and integration within the human insula revealed by meta-analysis. Brain Struct Funct. 2010;214:519–34.
4. Mesulam MM, Mufson EJ. Insula of the old world monkey. III: Efferent cortical output and comments on function. J Comp Neurol. 1982;212:38–52.
5. Mufson EJ, Mesulam MM. Insula of the old world monkey. II: Afferent cortical input and comments on the claustrum. J Comp Neurol. 1982;212:23–37.
6. Pritchard TC, Hamilton RB, Morse JR, Norgren R. Projections of thalamic gustatory and lingual areas in the monkey, Macaca fascicularis. J Comp Neurol. 1986;244:213–28.
7. Cereda C, Ghika J, Maeder P, Bogousslavsky J. Strokes restricted to the insular cortex. Neurology. 2002;59:1950–5.
8. Fink JN, et al. Insular cortex infarction in acute middle cerebral artery territory stroke: predictor of stroke severity and vascular lesion. Arch Neurol. 2005;62:1081–5.
9. Kalani MY, Kalani MA, Gwinn R, Keogh B, Tse VC. Embryological development of the human insula and its implications for the spread and resection of insular gliomas. Neurosurg Focus. 2009;27:E2.
10. Skrap M, et al. Surgery of insular non-enhancing gliomas: volumetric analysis of tumoral resection, clinical outcome and survival in a consecutive series of 66 cases. Neurosurgery. 2011;70:1081–93.
11. Duffau H. A personal consecutive series of surgically treated 51 cases of insular WHO Grade II glioma: advances and limitations. J Neurosurg. 2009;110:696–708.
12. Sanai N, Polley MY, Berger MS. Insular glioma resection: assessment of patient morbidity, survival, and tumor progression. J Neurosurg. 2010;112:1–9.
13. Lang FF, Olansen NE, DeMonte F, et al. Surgical resection of intrinsic insular tumors: complication avoidance. J Neurosurg. 2001;95:638–50.
14. Ius T, et al. Surgery for insular low-grade glioma: predictors of postoperative seizure outcome. J Neurosurg. 2014;120:12–23.
15. Yasargil MG, et al. Tumours of the limbic and paralimbic systems. Acta Neurochir (Wien). 1992;118(1–2):40–52.
16. Schätz C, et al. Interstitial 125-iodine radiosurgery of low-grade gliomas of the insula of Reil. Acta Neurochir. 1994;130:80–9.
17. Vanaclocha V, Saiz-Sapena N, Garcia-Casasola C. Surgical treatment of insular gliomas. Acta Neurochir. 1997;139:1126–34.

18. Penfield W, Faulk ME. The insula; further observations on its function. Brain. 1955;78:445–70.
19. Ostrowsky K, et al. Functional mapping of the insula cortex: clinical implication in temporal lobe epilepsy. Epilepsia. 2000;41:681–6.
20. Ojemann GA, Whitaker HA. Language localization and variability. Brain Lang. 1978;6:239–60.
21. Duffau H, et al. The insular lobe: physiopathological and surgical considerations. Neurosurgery. 2000;47:801–10.
22. Afif A, Minotti L, Kahane P, Hoffmann D. Middle short gyrus of the insula implicated in speech production: intracerebral electric stimulation of patients with epilepsy. Epilepsia. 2010;51:206–13.
23. Isnard J, Guenot M, Sindou M, Mauguiere F. Clinical manifestations of insular lobe seizures: a stereo-electroencephalographic study. Epilepsia. 2004;45:1079–90.
24. Skrap M, Marin D, Ius T, Fabbro F, Tomasino B. Brain mapping: A novel intraoperative neuropsychological approach. J Neurosurg. 2016;5:1–11.
25. Capitani E, Neppi-Modona M, Bisiach E. Verbal-response and manual-response versions of the Milner Landmark task: normative data. Cortex. 2000;36:593–600.
26. Lang PJ, Bradley MM, Cuthbert BN. International affective picture system (IAPS): affective ratings of pictures and instruction manual. Gainesville, FL: University of Florida; 2008.
27. Barbarotto R, Laiacona M, Macchi V, Capitani E. Picture reality decision, semantic categories and gender. A new set of pictures, with norms and an experimental study. Neuropsychologia. 2002;40:1637–53.
28. Bizzozero I, Ferrari F, Pozzoli S, Saetti MC, Spinnler H. Who is who: Italian norms for visual recognition and identification of celebrities. Neurol Sci. 2005;26:296.
29. Prior M, Marchi S, Sartori G. Cognizione sociale e comportamento-Volume I. Domeneghini Editore UPSEL; 2003.
30. Antonietti A, Bartolomeo P, Colombi A, Incorpora C, Olivieri S. BIP – Batteria Immaginazione e Percezione (Test) per la valutazione della cognizione visuo-spaziale. Milano; 2017.
31. Crepaldi D, et al. Noun-verb dissociation in aphasia: the role of imageability and functional locus of the lesion. Neuropsychologia. 2006;44:73–89.
32. Ripamonti E, et al. The anatomical foundations of acquired reading disorders: A neuropsychological verification of the dual-route model of reading. Brain Lang. 2014;134C:44–67.
33. Miceli G, Laudanna A, Burani C, Capasso R. Batteria per l'analisi dei deficit afasici. B. A. D. A. [B. A. D. A.: A Battery for the assessment of aphasic disorders.]. Roma: CEPSAG; 1994.
34. Paradis M. Bilingual aphasia test (Italian version). Bologna: EMS; 1986.
35. Orsini A, et al. Verbal and spatial immediate memory span: normative data from 1355 adults and 1112 children. Ital J Neurol Sci. 1987;8:539–48.
36. Tomasino B, et al. Neuropsychological patterns following lesions of the anterior insula in a series of forty neurosurgical patients. AIMS Neurosci. 2014;1:225–44.
37. Duffau H, Bauchet L, Lehericy S, Cappelle L. Functional compensation of the left dominant insula for language. Neuroreport. 2001;12:2159–63.
38. Duffau H, Taillandier L, Gatignol P, Capelle L. The insular lobe and brain plasticity: lessons from tumor surgery. Clin Neurol Neurosurg. 2006;108:543–8.

Suggested Reading

39. Jones CL, Ward J, Critchley HD. The neuropsychological impact of insular cortex lesions. J Neurol Neurosurg Psychiatry. 2010;81:611–8.

Index

A
Accessory and transverse gyri (Ag and Tg), 51
Activation likelihood estimation (ALE) technique, 126
Agranular insula, 8
Alexithymia, 163
Alzheimer's disease (AD), 171, 172
    affects, 169
    autonomic instability, 171
    case study, 169, 170
    clinical diagnosis, 170
    histology, 170
    imaging, 170
    insular cortex, 169
    insular pathology, 171
        clinical symptoms, 171, 172
        historical characteristics, 172
        pathological symptoms, 172
    neuropsychiatric symptoms, 171
    special sensory dysfunction, 171
Amusia, 226
Angular artery (Ang), 50
Angular gyrus (AG), 72
Anterior and posterior insular points (AIP and PIP), 52
Anterior cingulate cortex (ACC), 8
Anterior insular vein (AI), 60
Anterior limiting sulcus (ALs), 51, 66
Anterior long gyrus (ALg), 66
Anterior parietal artery (APr), 50
Anterior short gyrus (ASg), 51, 66
Anterior temporal artery (ATm), 50
Anterior vs. posterior insular cortex, 8
Anxiety disorders, 113
Aphasia, 123, 124
Apraxia of speech, 124, 125
Astrocytoma, 97, 98
Attention, 214
Auditory perception
    auditory pattern, 152
    auditory perception pathology, 153
    music, 153
    sound localization, 152
    speech, 153
Autism spectrum disorder (ASD), 118
    advent of fMRI and experimental study, 158, 159
    anterior insula, 157, 158
    neurobiological research, 157
    notion of connectivity, 157
Awake craniotomy, 282

B
Bipolar disorder (BD), 104
Blood oxygen level-dependent (BOLD), 176
Brain language area, 123
Brain mapping, 283
Brain prediction, 179
Broca's aphasia, 124, 125

C
Capsulo-insular vein (CIV), 56
Cavern, 263
Cavernomas (CAs), 265, 266
    electrocorticography, 267
    imaging, 264
        intraoperative US, 265
        neuronavigation, 266
    incidence, 263
    intraoperative mapping, 266
    language examination, 265
    natural history, 264
    surgery, 265
    surgical strategy, 267
Cells movement, neuroepithelium, 6
Central executive network (CEN), 244
Central insular sulcus (Cns), 66
Central insular sulcus and anterior long gyrus (Cns and ALg), 51
Central insular vein (CnI), 63
Central sulcal artery (Cn), 50
Cerebral artery
    ICA, 40
    MCA
        accessory, 41, 45
        cortical branches, 42
        early branches, 45
        stem artery, 42
        termination and branches, 42
    parenchymal and perforating artery, 45, 46

Cerebral cortex, 18
Cerebral veins
  classification, 56, 58
  deep and superficial parenchymal veins, 56, 57
  DMC, 58
  SMC, 57
Circle of Willis, 40
Circulus arteriosus, 40
Clinical neuroscience, 147
Cognition, 147
Conduction aphasia, 124
Connectivities, 148
Contemporary neuroimaging technique, 125, 126
Cortical folding, 5
Cortical (M4) segment, 42

D
Deep middle cerebral (DMC) vein, 58
Deep parenchymal veins, 56
Deep veins, 59, 60
Deep venous system, 56
Default mode network (DMN), 244
Diffuse low-grade gliomas (DLGG)
  clinical presentation, 256
  definition and diagnosis, 255
  insular predilection, 256
  operative treatment, 260
    awake craniotomy, 258
    outcome
    functional results, 260
    oncological results, 260
    preoperative investigations, 257
    reoperation, 260
    role of surgery, 257
    subpial resection and axonal mapping, 258
    trans-opercular approach, 258
  prognosis, 256
  radiological presentation, 256
Diffusion tractography, 78
Direct cortical stimulation (DES)
  cognitive side, 283
  mappings, 283
Dopamine receptors, 87, 88

E
ECoG, 283
Electrical mapping, 258
Electrical stimulation, 273
Embryology, insula, 4, 5
Emotion, 147, 182
Emotion regulation, 287
Emotional awareness
  AIC and ACC, 163
  brain mechanisms, 162
  clinical considerations, 163, 164
  conscious vs. unconscious, 161
  empathy, 162, 163
  hypothetical model, 164
  interception, 161, 162
  top-down and bottom-up integration, 163, 164
Emotional processing, 129
Empathy, 162, 163
Epilepsy surgery
  complementary noninvasive studies, 271
  epileptic focus, 272
Epilepsy surgery failure
  cause for, 272
  historical perspective, 273, 274
  temporo-insular epilepsy, 274–278
ERK mitogen-activated protein kinase (MAPK), 193
Executive functioning, 170
Explicit motivation, 148

F
Flavour system, 133
Free energy-efficient system, 181, 182
Frontal lobe epilepsy surgery, 272
Frontoinsular cortex (FIC), 8
Frontotemporal dementia (FTD), 171
Functional MRI (fMRI), 148
  meta-analytic studies, 126
  theoretical analysis, 126
Functional specificity, 148
Fusiform face area (FFA), 154, 155

G
GABA receptors, 89
Granular insular cortex, 6

H
Herpes encephalitis, 93, 96
Heschl's gyrus, 77
Histology, insula, 6, 8, 9
Hub, 148
Hyperacusis, 226

I
IBASPM (Individual Brain Atlas Using Statistical Parametric Mapping), 105, 106
Inferior fronto-occipital fasciculus (IFOF), 73, 74
Inferior limiting sulcus (ILs), 52, 66
Inferior trunk (IT), 50
Insula, 243, 244
  functional role, 242
    auditory processing, 243, 244
    emotional processing, 243
  neuroimaging studies, 223
  salience network, 244–246
Insula cortex
  autonomic functions, 129
  bilateral activation, 130
  lateralization of function, 129

Index

modulation, 131
Insula modularity and connectivity, 186, 187
Insular (M2) segment, 42
Insular apex (A), 52
Insular aphasia, 123
Insular artery, 47, 48
   classification, 46
   cortical and central branches, 46
   hemisphere operculum, 43, 44
   insuloopercular artery, 48
   LPr artery, 46
      cortical and subcortical course of, 48
      location and penetration side of, 47
   supplier arterial pattern of, 50, 51
   supply of region, 51, 52
   terminal artery, 48
Insular cortex
   activity, 214
   anatomical connections, 4, 10
   anatomy, 15, 16, 19
      anterior aspect, 17
      anterior limiting sulcus, 16
      cerebral central core, 15
      circular/peri-insular sulcus, 16
      deep and superficial anatomical relationships, 20
      deep anterior limiting sulcus, 16
      form of, 16
      function, 272, 273
      fronto-orbital portion of the superior operculum, 20
      frontoparietal operculum, 21
      gyrus, 17, 18
      inferior limiting sulcus, 16, 18
      lateral aspect, 17
      medial orbital gyrus, 18
      posterior lobule, 18
      preinsular topography, 15
      retroinsular segment, 15
      subopercular gyrus, 21
      subtriangular gyrus, 21
      sulcus, 17
      superior limiting sulcus, 16, 18
      transverse gyrus, 18
   attention, 214, 215
   consciousness and emotions, 3
   definition, 3
   development, 5
   functional aspects, 22
   functions, 3
   gustation, 133, 134
      function, 134, 135
      taste hedonics, 137, 138
      taste intensity, 136, 137
      taste quality, 135, 136
   homeostasis of body, 3
   IBASPM, 105, 106
   lateralisation, 138, 139
   mediates, 116
   meditation, 216–218
   morphological concepts, 11
   MriCloud, 106–108
   multifaceted function, 213, 214
   multimodal process, 139
      cross-modal anticipatory gustatory responses, 141
      multimodal flavour processing, 139, 140
   neuroimaging and neurosurgery, 22
   nomenclature, 22
   operative techniques, 22
   radiological interpretations, 22
   salience, 215, 216
   self-awareness, 215
   volume of, 103–105
   volumetric methods, 105
Insular cortex, 22
Insular cortex epilepsy (ICE), 199
   evaluation, 199, 200
   intracranial EEG investigation, 199
   invasive monitoring, 197
   noninvasive imaging study, 200
   refractoriness to antiepileptic drug treatment, 200
   seizures, 197
   semiological characterization, 198
   surgical anatomy, 198
   surgical safety, 200
Insular damage
   cognition
      attention and executive functions, 231
      language, 229, 230
      memory, 230, 231
   emotion and empathy
      decition-making, 234
      emotional experience, 231–233
      social cognition, 233, 234
   sensory processing
      audition, 225, 226
      chemosensation, 226, 227
      interoceptive function, body representation, and self-awareness, 228, 229
      neclet, 229
      somatosensory deficits, 224, 225
      vestibular function, 227, 228
Insular electrode implantation, 199
Insular epilepsy, 272
Insular gliomas, 257, 282
Insular vascularizsation, 10
Insular veins, 56, 58, 65–67
   anastomoses, 67, 68
   blood drainage of, blood, 65
   confluence of, 65
   deep veins, 59, 60
   drainageof, 66, 67
   number and size, 65
   relationships, 64
   superficial veins, 60–64
   three-dimensional configuration, 57
Insuloopercular artery (IO), 48
Intercellular stimulus transmission, 85, 86
Intermedius primus, 72
Internal carotid arteries (ICA), 40

Interoception, 117, 242
   awareness, 118
   brain regions, 115
   definition, 180
   dimensions, 180
   emotional awareness, 161, 162
   insular cortex, 115
   mediation analysis, 115
   neural substrates, 114
   neural substrates of, 115
   neuroimaging studies, 114
   sensibility modulates, 115
   subcategories, 117
Interoceptive information, 148
Intraoperative electrocorticography, 273
Ischemic stroke
   auditory deficits, 208
   cardiovascular effects, 208, 209
   clinical presentation, 205
   dysphagia, 207
   etiology, 204
   isolation, 204
   motor deficits, 207
   neclect, 207
   somatosensory deficits, 206, 207
   speech and language deficits, 205, 206
   subinsular region, 205
   taste and olfaction deficits, 207
   vestibular deficits, 207

J
James-Lange theory, 243

K
Klingler's technique, 71, 73

L
Language
   insula in, 123
   motivation and affect, 126, 127
   processing, 126
   production, 130
   understanding, 126
Lateral orbitofrontal artery (LOF), 50
Limbic cortex, 147
Limbic system, 171
Lobar corticography, 199
Long perforating arteries (LPr), 46
   cortical and subcortical course of, 48
   location and penetration side of, 47
   subcortical course of, 49
Loss of insular ribbon sign, 93, 95

M
Major depressive disorder (MDD), 104, 171
Medial temporal lobe epilepsy (MTLE), 197

Meditation, 216–218
Meditation, scientific study of, 213
Memory, 169, 231
Middle cerebral artery (MCA), 40, 203, 223
   accessory, 41, 45
   cortical branches, 42
   early branches, 45
   stem artery, 42
   termination and branches, 42
Middle longitudinal fasciculus (MdLF)
   diffusion imaging techniques, 71
   IFOF, 73, 74
   overview, 71
   radioisotopic tracing studies, 74
   temporal lobectomy, 74
   temporo-parieto-occipital white matter dissection, 71–73
   topography, 74
   tractography, 74
   trajectory, 74
Middle perforating arteries (MPr), 46
Middle short gyrus (MSg), 51, 66
Middle temporal artery (MTm), 50
Mindfulness, 213
Motor-evoked potentials (MEPs), 283
MriCloud, 106–108
Multimodal function, 151
   audiovisual illusions perception, 152
   speech perception, 151
Muscarinic receptors, 86

N
Neurofibrillary tangle (NFT), 170
Neuroimaging techniques
   angiography, 92
   CT, 92–94
   fMRI, 91, 93
   hemorrhages, 96, 97
   herpes encephalitis, 96
   MCA acute stroke, 93, 95
   MRI, 92–94
   neoplasies, 97, 98
   radiography, 92
Neuroimaging techniques, 98
Neuropsychology
   direct cortical stimulation, 283
   emotional processing evaluation, 285–289
   insula cortex
      quantitative meta-analysis, 281
      reil surgical approaches, 282, 283
      selective lesions, 281
   RTNT, 283
      left anterior insular resection, 285–288
      left insular resection, 285
      right anterior insular resection, 284, 285
      right insular resection, 284
Neurovascular unit, 46
NMDA receptors, 86, 87

Nonfluent progressive aphasia (NFPA), 172
Non-motor symptoms (NMS), 191, 192

O

Obsessive-compulsive disorder (OCD), 104
Oligodendroglioma, 97
Opercular (M3) segment, 42
Operculo-insular epilepsy, 276–278
Opioid receptors, 88

P

Paralimbic cortex, 147
Paralimbic system, 263
Parenchymal artery, 45, 46
Parkinson's disease (PD), 191–194
    insula
        function, 191
        treatments, 193, 194
    non-motor symptoms, 191–193
Perforans artery, 45, 46
Perforating artery, 45, 46
Polar temporal artery, 51
Postcentral insular sulcus (PstCns), 52, 63
Posterior insular point, 18
Posterior insular vein (PI), 64
Posterior long gyrus (Plg), 52
Posterior long gyrus (PLg), 66
Posterior parietal artery (PPr), 50
Posterior short gyrus (PSg), 51, 66
Posterior temporal artery (PTm), 50
Postprandial hypotension (PPH), 191
Precentral artery (PreCn), 50
Precentral insular sulcus (PreCns), 51
Precentral insular vein (PreCnI), 60
Prediction, 181, 182
    cognition, 182–184
    emotion and motivation, 182
    error, 185, 186
    subjective experience, 184, 185
Prefrontal artery (PreF), 50
Preinsular area, 21
Proliferative precursor cells, 6

Q

Quality of life (QoL), 257

R

Radial glial fiber system, 6
Radial migration pathway, 6
Radiography, 92
Real-time neuropsychological testing (RTNT), 283
    left anterior insular resection, 285
    left insular resection, 285
    right anterior insular resection, 284, 285
    right insular resection, 284
Regional cerebral blood flow (rCBF), 103
Retroinsular area, 21
RTNT, *see* real-time neuropsychological testing (RTNT)

S

Salience, 215, 216
Salience network (SN), 244
Salience networks, 157
Schizophrenia (SZ), 240–242
    cognitive deficits, 239
    cognitive functioning, 239
    morphometric changes, 239, 240
        clinical correlations, 242
        gender effects, 240
        structural changes, 241, 242
        treatment effects, 240, 241
    symptoms, 239
Self-awareness, 215
Sequential embryonic development, 10
Short insular sulcus (Ss), 51
Short perforating arteries (SPr), 46
Social anxiety disorder (SAD)
    interoception
        fMRI study, 115–117
        self-referential thought and, 113–115
        subcategories, 117, 118
    modified functional connectivity, 118
Social phobia, *see* Social anxiety disorder (SAD)
Speech apraxia, 175
Speech articulation, 230
Speech production, 175, 176
    apraxia, 175
    insula, 175
    insular speech and language function
        clinical study, 175, 176
        functional imaging study, 176
    speech motor control, 177
Sphenoidal (M1) segment, 42
Structural connectivity of insula
    anatomy, 77, 78
    diffusion tractography, 78–80
    frontal lobe, 78
    in non-human primates, 78
    occipital lobe, 81
    parietal lobe, 81
    subcortical regions, 81
    temporal lobe, 79, 81
Superficial insular veins, 58
Superficial middle cerebral vein (SMC), 57
Superficial parenchymal veins, 56
Superficial sylvian vein, 57
Superficial veins, 60–64
Superficial venous system, 56

Superior trunk (ST), 50
Supramarginal gyrus (GSM), 72
Susceptibility-weighted phase-sensitive magnetic resonance imaging (SW-MRI), 65
Sylvian fossa, 40
Symptom Checklist-90-Revised (SCL-90-R), 288

T
Tastants, 135
Temporal lobe surgery
    operculo-insular epilepsy, 276–278
    temporo-insular epilepsy, 274–276
Temporal plus epilepsy, 274
Temporo-insular epilepsy, 274–276
Temporooccipital artery (TmO), 50
Terminal insular arteries (TrI), 48
Thalamic nuclei, 85
Tinnitus distress, 153
Transcerebral vein, 57
Transsylvian approach, 282

V
Ventromedial prefrontal cortex (VMPFC), 115
Virchow-Robin space, 46
Visual perception
    anterior insular cortex, 154
    decision-making, 154
    facial recognition, 154
    frontoparietal network, 153
von Economo neurons (VENs), 6, 8, 9, 86
Voxel-based morphometry (VBM), 105

W
Wernicke's aphasia, 124
Western scientific community, 213
Windowing technique, 282